EXTENSION MANAGEMENT IN THE INFORMATION AGE

(Initiatives and Impacts)

NIPA® GENX ELECTRONIC RESOURCES & SOLUTIONS P. LTD.
New Delhi-110 034

EXTENSION MANAGEMENT IN THE INFORMATION AGE

(Initiatives and Impacts)

Edited by

H. Philip
T. Rathakrishnan
Ravi Kumar Theodre

NIPA® GENX ELECTRONIC RESOURCES & SOLUTIONS P. LTD.
New Delhi-110 034

NIPA® GENX ELECTRONIC RESOURCES & SOLUTIONS P. LTD.

101,103, Vikas Surya Plaza, CU Block
L.S.C. Market, Pitam Pura, New Delhi-110 034
Ph : +91 11 27341616, 27341717, 27341718
E-mail: newindiapublishingagency@gmail.com
www: www.nipabooks.com

For customer assistance, please contact
Phone: + 91-11-27 34 17 17
Fax: + 91-11- 27 34 16 16
E-Mail: feedbacks@nipabooks.com

ISBN: 978-81-19103-88-1

Composed and Designed by NIPA®.

Contents

Section IV : Gender Issues

Section V : Policy Issues

Preface

A series of interventions initiated in mid — 1960's that led to the green revolution in cereal production transformed the country from a situation of food deficiency to food sufficiency. During the last two decaes, Indian Agriculture has largely been influenced by globalization, trade liberalization and increasing role of private sector. Agriculture is becoming more knowledge intensive and market driven and no more closed and protected, but globalised and open. Agricultural extension needs to assume new challenges and reforms interms of content, approach, success and processes. Adequate focuss have to be given on effective technology selection, optimization, application and management. The approaches and models evolved during 60-80's and followed hither to are insufficient to deal with the current concerns emerging out of globalization, sustainability and other dimensions of agricultural development envisaged to meet the current challenges. The different section dealt in the book is addressing all those dimensions.

Section I : Covers on the lead papers focussed towards the major themes of strengthening Technology Application and Delivery systems, ICT, BLESS, NRM and women empowerment, Section II : Dealt with the extension strategies in development departments, institutional mechanism including NGOs, new media tools and impact assessment, followed, by management of Natural Resources including ITKS in Section III : Gender mainstreaming, women empowerment, international experiences and policy issues have been synthesized under Section IV and V respectively. This book would enable in exposing the extension professionals on a cross cultural approach in various development spheres of extension reforms in India. The project experiences of the extension managers have been presented in the appropriate areas. The out come of the research endeavours incorporated in the text would serve as a strong data base and provide a viable platform to initiate network of research projects and schemes on consortium mode among extension scientists from State Agicultural Universities and National institutions.

The editors believe sincerely that the content of the book would be much helpful to incorporate appropriate interventions in extension efforts in frontiers of developmental scenario.

Editors

Section I

Introduction

1

Framework for Strengthening Technology Application and Delivery System in Agricultural Extension

V. Venkatasubramanian
Assistant Director General (Extn.), ICAR, New Delhi

The Context

A series of interventions initiated in mid-1960s that led to the green revolution in cereal production transformed the country from a situation of food def iciency to self sufficiency. The green revolution was however, restricted to productivity improvements in cereals especially wheat and rice in the initial decades, primarily grown in irrigated regions. In subsequent decades, productivity increased in other crops, namely oilseeds, sugarcane, cotton, fruits and vegetables. The Green Revolution generally by passed India's vast rain fed tracts, especially arid zones, hill and mountain ecosystems and coastal regions, thus exacerbating agro-ecological and economic disparities. Although self sufficiency was attained in food grain production, potential of different crop varieties are yet to be attained. The main features of the green revolution were:

- exclusive focus on food grains
- intensive use of water, fertilizer and pesticides to maximize yields from High Yielding Varieties (HYVs)
- practically no focus on sustainability issues
- benefited the resource-rich and large farmers and
- dependent on few technological options focusing on hybrids, changes in plant architecture.

The equity, efficiency and sustainability of these approaches became a matter of serious concern since the 80's. Over a period of time, especially during the last two decades, the course of Indian agriculture has largely been influenced by factors such as globalization, trade liberalization and increasing role of private sector. Apart from these the agricultural sector has been challenged by the following factors:

- Shrinking resource base
- Changes in demand and consumption pattern
- Changes in farming systems including increasing diversification to high-value crops
- Declining public investments in agriculture
- International developments- WTO (subsidies in agriculture and trade liberalization)
- Climate change (global warming, seasonal variations, changes in rainfall pattern and increasing occurrence of natural disasters)

Owing to globalization and liberalization, agriculture in India need to change and change for better than the best. Diversification of production is fast happening along with widespread dietary evolution. Commodity based production is giving way to system based production and there is a paradigm shift using farming system to production to consortium system of operation. Private sector participation is increasing. Agriculture is becoming more and more knowledge-intensive and market-driven. Hence for more innovative research, efficient policies and effective delivery of services, supplies and markets are imperative. Agriculture is no more closed and protected, but globalized and open.

Agricultural Development : An Overview

With fifteen discrete agro climatic regions, country's agriculture is very diverse and so the distribution of livestock population. The impact of research and development efforts is reflected in remarkable increase in an all round productivity w.r.t. crop, horticulture, animal and fishery sectors. Compared to 1950, the productivity in 2005 became 3.3 times in case of food grains, 2.1 for vegetables, 1.6 for fruits, 5.7 for fish, 4.8 for eggs and 1.8 for milk. In spite of these remarkable achievements, country faces great challenges in further increasing the productivity of all these sectors to match the population demand. The livestock and fisheries sector play an important role in generating income and employment for marginal farmers and landless labourers in meeting nutritional and livelihood security.

Against the anticipated annual growth rate of 4% plus, the agricultural sector grew only at 2% p.a. during the X Plan. The demand for food is expected to double by 2050, and this will only be met if small holders contribute to increasing production. There are over 600 million small holders farmers in India alone, and therefore, extension is a strategically important link for ensuring that small holder demands are at the centre of technology development and delivery system. National Commission on Farmers (NCF) in its 3rd Report suggested attention to soil health care, water harvesting and management, credit and insurance, quality and safety, technology and inputs and farmer-friendly marketing in order to help our hard working farm families to help the country to achieve 4% plus growth rate in agriculture.

Suggesting to measure progress in agriculture by the growth rate in the net income against the current measure of physical production of food grains and other farm commodities, the NCF recommended review and reform of the service, support, research, extension and input supply and market system. It is therefore, essential to promote a broader perspective to support farmers in addressing their changing needs and challenges characterized by uncertainty, unpredictability and uncontrollability. Thus, to achieve a growth rate of 4% p.a. plus in agriculture as well as to serve farmers and save farming from increasing distress, a new model for technology development and delivery system in agriculture is necessary.

India has only 2.4% of world's land and 4.2% of world's renewable water supply. With this, Indian Agriculture needs to sustain 16.8% of world's population and 11% of world's livestock. Out of 142 million ha of cultivated land only 55 million ha is irrigated. About 65% of the population is engaged in agriculture contributing to 21% of gross domestic product (GDP) and 11.2% of the total export earnings. Agriculture is supporting a large number of agro-industry, generating 55 billion US dollars in the food sector only.

The plan out lay for the agriculture and allied sector has increased about 156% substantially from Rs.7,431 crore in 2006-07 to Rs .19,070 crore in 2010-11. In recent years there has been an increase in the gross capital formation in agriculture as a proportion of agricultural GDP which has gone up from 14.1 percent in 2004-05 to 21.3 in 2008-09; a substantial jump from the production of about 198 million tons of food grains in 2004-05. Presently, India is the second largest producer of fruits and vegetables. Due to interventions under National Horticultural Mission and Technology Mission for NE states the production of fruits, vegetables and spices has been increased by 27%,22% and 12% respectively during 2009-10 over 2005-06.

The National Food Security Mission (NFSM) was launched in 2007-08 to enhance the production of rice, wheat and pulses by 10 million tons, 8 million tons and 2 million tons respectively by the end of 11th plan. The mision has helped to widen the fod basket of country with significant contributions coming from NFSM districts. During 2008-09 nearly 50 % of the NFSM rice and pulses districts, and 33% of NFSM wheat districts have recorded 10-20% increase in productivity compared to 2006-07.

The livestock sector has maintained a respectable growth of 4-5% per annum in spite of several constraints, contributing about 112 million tonnes of milk, 59.8 billion eggs, 43.2 million kg of wool and 3.8 million tonnes of meat (Economic Survey 2009-10). The per capita milk availability increased from 112 gm/ day in 1968-69 to 258 gm/day during 2009-10.

India is endowed with rich fishery resources which include a 8118 km of coastal line, 2.02 million sq. km Exclusive Economic Zone, continental shelf 0.506 million sq.km, Rivers and canals 0.19 million km, Reservoirs 3.15 million ha, Ponds and tanks2.36 million ha and brackish water area of 1.24 million ha. Fisheries and aquaculture is an important component for national development rendering livelihood security for millions of poor people, income and employment generation, rich nutrition at cheaper price and augmenting national income. Indian fisheries share 4.7% of global fisheries production, 2.5% of global trade, 1.07% of country's GDP, 5.34% of national agriculture GDP, 18% of national agricultural exports and provided Rs.8000 crore as export revenue to the country. The present fisheries production is 7.90 million tonnes out of which 2.91 million tonnes comes from marine fisheries (41%) and 4.82 million tonnes from inland fisheries (59%) (2009-10).

The policy approach to agriculture, particularly in the 1990s has been to secure increased production through subsidies on inputs such as power, water, and fertilizer. It also focused on increasing minimum support prices rather than building new capital assets in irrigation, power, and rural infrastructure or improving the standards of maintenance of existing assets. Despite its past achievements, Indian agriculture continues to face serious challenges because of its ever-increasing population, limited land and water availability, and degradation of natural resources. There are wide gaps in yield potential and national average yields of most commodities are low. Although the Green Revolution increased production and productivity of food crops, improved food security and raised rural incomes, India still has a large poor (27.5 % living below the poverty line in 2008-09) and malnourished population.

The Rashtriya Krishi Vikas Yojana (RKVY), launched in August, 2007 has become the principal instrument for increasing the states' investment in agriculture and allied sectors. Outlay under RKVY for 2010-11 have been substantially increased to Rs.6,722 crore, which includes Rs.400 crore for the special initiative for pulses and oilseeds in dry land areas by origanising 60,000 pulses and oilseeds villages in rainfed areas.

It was decided by the GOUl in June,2004 to double the flow of agriculture credit in three years with reference to base year 2003-04.The flow of agriculture credit since 2003-04 has consistently exceeded the target. From the level of Rs. 86,981 crore credit flow in 2003-04, the agriculture credit disbursed in 2009-10 has touched Rs.3,66,916 crore. The target of credit flow for 2010-11 is Rs.3,75,000 crore. From this year 'onwards credit is available @ 5% rate of interest for those making timely repayment. However, these measures though appear to be substantial, are not adequate to address the present day challenges.

Although the annual growth rate in total GDP has accelerated from below six percent during the initial reform years (starting 1991) to more than eight percent in recent years, agricultural growth has decelerated. While the non-agriculture sector has witnessed rapid growth, there has not been any significant decline in the labour force employed in agriculture and this has created a serious disparity between the agriculture and non-agriculture sectors and urban and rural India. The output growth of the majority of commodities has decelerated after the mid-90s. And currently Indian agriculture is at crossroads.

Addressing these emerging challenges would require a new approach which has to be distinct from the earlier green revolution approach. Agricultural research and extension therefore need to emphasize the following new dimensions in the present era:

- Focus on gene revolution, emphasizing application of biotechnology-tissue culture for multiplication of elite germplasm, GM crops, marker assisted breeding etc
- Emphasize use of bio fertilizers, bio pesticides and bio remediation of ground water
- Address issues like sustainability, resource integration and technology integration as the primary focus

- Apply precision farming and mechanization for optimal use of precious resources and human labour
- Linkage with industry, market driven and export oriented agriculture
- Increase application of cutting edge technologies
- Thrust on post harvest, food processing and value addition technology
- Highlight quality in addition to increase in quantity
- Protect IPR and farmers' rights
- Integrate livestock, fisheries and other allied agro-enterprises with crop production and Exploit advances in information technology.

The working group on agricultural extension for the formulation of XIth Five Year Plan acknowledges the current limitation of public sector extension and emphasizes the following for strengthening agricultural extension. These include:

- Adoption of farming system and farmer participatory approach
- Research —extension farmer and market linkages
- Focus on vulnerable groups
- Up-scaling ATMA to all districts,
- Routing all state and central government extension funds through ATMA
- Support to private extension and public private extension
- Focus on women in agriculture
- Strengthening Kisan Call Centres
- Market led extension
- Promotion of Agri-clinics and agri-business centres

However, these measures though appear to be important, are not adequate to address the present day challenges.

Current Extension Scenario: A Critical Analysis

Extension in this context includes all those agencies in the public, private, NGO and community based initiatives that provide a range of agricultural advisory services and facilitate technology application, transfer and management.. While public sector line departments, mainly the Department of Agriculture was the main agricultural extension agency in the 60's and 70s, the last two decades have witnessed the increasing involvement of private sector, NGOs, community based organisations and media. In the public sector, the extension machinery of the state Department of Agriculture (DoA) reaches down to the block and village level. The village extension workers of the DoA continue to be an important source of information for farmers in India, even though information is clearly targeted at grain production, visits are irregular, and the service is pre-occupied with the implementation of government schemes linked to subsidies and subsidised inputs. With the external support drying up with the end of the T&V (Training and Visit) system of extension in the early 1990s, states have been left to fund their extension machinery and this has led to considerable weakening of public sector extension.

In the case of public sector extension, the major reform in recent years has been the establishment of a district level co-ordinating agency, the ATMA (Agricultural Technology Management Agency), initially tested in 28 pilot districts with the World Bank support is now successfully replicated in 593 districts of the country as a Plan programme of Department of Agriculture and Cooperation. Under ATMA, grass root level extension is mainly channelized through the involvement of BTTs (Block level Technology Teams) and FACs (farmer advisory committees), farmer groups/ farmer interest groups and self help groups.

ATMA is a district level autonomous agency entrusted with the role of agricultural technology management in the district. The district collector/deputy commissioner heads ATMA Governing Body, with members drawn from the line department, KVKs, farmers and NGOs (Fig 1). Based on the experiences gained from the pilot district, the Ministry of Agriculture, Government of India in 2004-05 decided to expand the ATMA model across all the districts in the country. Apart from bringing some additional resources for extension activities to be decided at the district level in consultation with farmer representatives, ATMA is yet to fully address many of the institutional constraints affecting extension performance.

The number of KVKs (Krishi Vigyan Kendras) funded by the ICAR has increased during this period. Presently 588 KVKs are established in the country. The motto is to cover each district with one KVK with a mandate of technology application through OFTs, demonstrations and training. It is an institutional approach and is comprehensive in nature. It functions on farm based model with a built in research-extension linkage through a multi-disciplinary team. It ensures feedback and feed-forward through participatory management. It is the largest research based extension body in the country by the ICAR at the district level. However, the effective reach of these KVKs is marginal mainly due to inadequate linkages with other development agencies. Moreover, their main focus is on technology testing, assessment and application under farmers' condition through conducting on-farm trials, demonstrations and training.

Extension services in the case of animal husbandry and fisheries continue to remain weak. While public sector extension arrangements have weakened, the number and diversity of private extension service providers has increased during last two decades. These include NGOs, producer associations, input agencies, media and agri-business companies. Many provide better and improved services to farmers, but their effective reach is limited and many of the distant and remote areas and poor producers are neither served by the public nor the private sector.

Extension continues to be funded as part of central and state level schemes/programer without much operational freedom at the local level, though the strategic research and extension plans (SREP) under ATMA envisage bottom up planning for extension. While the farmers require a wider range of support to address the emerging challenges, extension mainly functions as an agency for technology dissemination. Most of the organizations including the public sector departments continue to work in isolation.

Marketing extension has been a recent addition but is understood and implemented mostly as provision of output price information in various markets and this is highly inadequate to address the challenges in marketing. Other extension support facilities created include, farmer training centres at the district level; SAMETI (State Agricultural Management Extension and Training Institute) at the state level, EEI (Extension Education Institute) at the regional level; and MANAGE (National Institute for Agricultural Extension Management) at the national level. An analysis of the Extension System functioning in the country is presented in the Table 1.

Table 1 : Analysis of the Extension System functioning in the country

SI No	Extension organisations	Functions/Roles/Capacity	Gap/Limitations
1	ATMA	Aimed at decentralized decision making and bringing convergence among extension providers in a district: Promotion of commodity interest groups; Development of a strategic	No dedicated manpower; Limited resources, Convergence limited to activities undertaken with the specific budget for ATMA.

Contd. ...

SI No	Extension organisations	Functions/Roles/Capacity	Gap/Limitations
		research and extension plan: Provide additional funds to these agencies for key extension activities such as farm schools, demonstrations, exposure visits and trainings	
2	KVK	Technology application (technology assessment and refinement) through on-farm trials, front-line demonstration and training	Limited reach and lack of ownership by host institutes, Several vacant positions and frequent transfers. Inadequate operational funds; Weak linkages with other development agencies in the district; Poor technology and methodological backstopping from the host institutions; Non availability of critical facilities like soil and water testing facility, farming system models, demonstration units- etc in some KVKs
3	State line departments (Agriculture, Animal Husbandry, Fisheries, etc)	Regulatory role; Implementation of development programmes that involve distribution of subsidies and subsidized inputs; Organising extension programmes	Only the Department of Agriculture has staff assigned upto block and village level; Large number of vacancies, especially in remote and distant regions; Grass root level VEWs lack technical competence to deal with emerging challenges: Lack expertise on cutting edge technology, organizational, management and marketing aspects; Implementation of

Contd. ...

SI No	Extension organisations	Functions/Roles/Capacity	Gap/Limitations
			schemes/programmes leave little time for extension. Extension is weak in animal husbandry and fisheries departments
4	FTC	Training farmers on new technologies	Defunct in most places, Acute shortage of funds and manpower
5	SAU (Directorate of extension)	Implement extension programmes of the SAU and oversee activities of KVK	Staff and fund shortage; Weak state support; Inadequate links with development departments; No adequate field *Contd....* presence
6	NGOs	Exhibit wide diversity in terms of reach, credibility and capacity; Have good knowledge and networks with communities in villages they operate; Present in difficult and remote regions; Innovative in their approaches; Can potentially complement approaches of the public sector extension.	Effective reach restricted to select villages in their areas of operation Many of them do not have adequate technical capacity Wide variation in credibility and track record
7	Private Agri-business firms	Agri-input firms mainly involved in product demonstrations; Agro-processing and marketing firms mainly commodity oriented but do provide integrated support (inputs, technical support and marketing) for contract growers;	Present in only select regions; Narrowly focused on business interest; Lack focus on long term capacity development of farmers

Contd. ...

SI No	Extension organisations	Functions/Roles/Capacity	Gap/Limitations
8	Media	Dissemination of information on new technologies	Low literacy levels constrain penetration of print media; Poor content, inadequate coverage and little telecast time through television channels; Potential of media yet to be fully realised due to lack of integration of mass media with other extension programmes/approaches
9	Private consultants	Support large farmers growing cash crops and high value horticulture	Limited to select crops and regions Affordable only by large farmers
10	MANAGE	Training senior and middle level extension managers Conduct studies on extension systems and policies Conducts management educational programes in agriculture Provide Consultancy	Need greater focus on extension systems research and policy initiatives Limited faculty and funding Need to establish international linkages
11	SAMETI	Training middle level extension staff at the state level Conduct studies on extension systems at the state level	Lack of infrastructure, expertise (faculty); Inadequate focus on extension studies; Weak links with actors outside the public sector
12	EEI	Training middle level extension managers at the regional level	Inadequate infrastructure Inadequate training capacity, linkages and autonomy.

In the present day context, agricultural extension needs to assume new challenges and reform itself in terms of content, approach, structure and processes. Adequate focus has to be given on effective technology selection, optimization, application and management. Hitherto the extension efforts were largely influenced by the approaches and models that evolved during the 60's to 80's. These are insufficient to deal with the current concerns emerging out of globalization, sustainability, and other dimensions of agricultural development envisaged to meet the current challenges.

Extension Research and Education

Addressing the emerging challenges would require new approaches which must be supported by extension research and education system. The extension research and education system should focus on the application of cutting edge technologies, thrust on technology integration, use of information communication technology, quality aspects in addition to increase in quantity, thrust on post harvest, food processing and value addition, resource management etc.

Right technology and Right methodology are the two key elements required to be focused by the livestock extension research and education systems. The inventory of right technology and methodology need to be made available by the system. Extension Research and education system must reform itself in terms of content and adequate focus has to be given on effective technology selection, optimization, application, sustainability and management.

The frame work of strengthening the Extension research, education and delivery system includes.

- Assessment of existing extension education system, approaches and organizations against the back drop of changing scenario to come out with practical solutions for strengthening/ restructuring the system.
- Broadening the scope of technical mandate keeping in view the current demand scenario.
- Development of location specific, participatory gender sensitive and customized extension materials and methodologies.
- Emphasis on FSR/E and farmer participatory approach.
- Strengthening the operational linkages and partnership between research, extension, farmer, market and other key stake holders.
- Empowering farmers and organising them into commodity groups/associations and federating them.
- Integrating ICT in the extension research, education and technology development system.
- Focus on issues like IPR, Farmers Rights and DUS guidelines.

The management concepts such as 5-S, Six sigma, JIT, TEM, TQM, LFW, RFD, network analysis, gap and impact analysis etc. needs to be integrated into the extension education system to update the course content. ***The extension education system is conceptually 15 years back and needs to be updated.*** The major issues, responses required, suggested actions need to be analysed and an action plan is to be prepared not as an academic exercise but with a mission mode approach. ***Till date the practice of admiring and adopting approaches evolved and suggested by overseas experts is continued rather than working out an indigenous methodology and framework.*** KVK and ATMA/ SREP are the finest models and approaches put under operations in the country. This needs to be strengthened by professionally managing them as well as working out a convergence between them in their function.

Frame work for strengthening technology delivery system:

The technology delivery system needs to gear up their capacity in terms of manpower, expertise, finance, structure, institutional linkages and the kinds of methods, approaches and delivery system they employ. The issues to be taken up for strengthening the delivery system are;

(i) Farmers empowerment and farmer organizations development
(ii) Institutional linkages, convergence and Partnership issues
(iii) Technology backstopping, application, integration, ICT use and management
(iv) Frontier areas of extension, HRD and skill development
(v) Policy issues.

The above five issues are discussed here under:

I. Farmers empowerment and farmer organizational development

Lack of quality and dedicated man power dedicated in extension is a serious constraint and challenge encountered by the system. An extension system with business and professional approach is a must. This is possible only if the system recruits personnel who can bring in different kinds of expertise. For instance, expertise related to cutting edge technologies, organizational development, market development, legal issues related to farmer rights, IPR, ICT etc. are very much important. This would involve some de-learning of our conventional technology dissemination approach and learning new ways of doing things. Strengthening of natural and regional level training facilities for continuous skill up gradation is a must. Assessment of quality of extension personnel and services, conceiving a natural level mission mode approach is the need of this hour.

Since time immemorial, farmers are under the mercy of 3 Ms, i.e. monsoon, market and money lenders. Now, with the entry of 4th M, i.e. multi nationals Indian farmers are put under the strategically designed and professionally executed cut throat competition by the multinationals. Today, right from seed, plant protection, machinery, post harvest and to market the multinational companies have made their presence and the livestock sector is under tremendous pressure from these external elements.

Addressing the complex challenges of bringing about the food and livelihood security would require solutions which are beyond the decision-making capacities of individual farmers. This would necessitate new forms of collaboration as the agriculture sector in India is dominated by small farms. Collective decisions on resource use and marketing would necessitate evolving new forms of collaboration and this is particularly important as this sector is dominated by small farms often with weak bargaining powers and limited political voice. To deal with the new set of challenges, farmers need a wide range of information and support and these include aspects related to technologies, organizational development, entrepreneurial development and access to appropriate financial services. This would necessitate a reformulation of the objectives of extension as to empower the farmers and their organizations.

Globalisation Issues

The international competitiveness in agricultureproducts will be influenced considerably by our preparedness to take advantage of the opportunities opened up by the new agreements under Uruguay round. Supply and demand forecasts for agro products in India by both International Food Policy Research Institute (IFPRI) and Govt. of India indicate that there will be modest surplus even in 2020 in spite of increase in local demand due to population growth, growing urbanization, increasing affluence and changing life styles, enabling limited but significant international trade. Since the agricultural production system in India is entirely labour intensive, relying on the use of family labour, the local competitiveness depends on low opportunity costs for labour, the value captured from non-food farm outputs like crops residues and manure and the opportunity for capital accumulation in the form of livestock, perennial crops and other immovable farm assets.

The international competitiveness of Indian agro products-particularly milk and milk products assessed on the basis of Nominal Protection Co-efficient (NPC) indicates that ghee is slightly competitive while milk powders are not. This calculation, however, does not take into account market distortions on account of overt and covert subsidies by the major exporting countries. Meat prices in India and their NPC indicate fairly high international competitiveness. However, they lose out on Sanitary and Phyto Sanitary (SPS) standards and have only limited markets confined to West, South and South East Asia.

Ensuring quality standards and freedom from annual epidemics are pre-conditions of the guaranteed market access. Small holder production system often faces difficulties in capturing the economics of scale in marketing, input supply and services delivery. Agricultural extension education plays an important role in this context to make the our agro products quality specific and cost effective. The role of agricultural extension education in this direction is :

- Training of farmers and entrepreneurs on export quality standards and phytosanitary requirements.
- Training on Fair Average Quality Standards (FAQ) for livestock products.
- Market intelligence through information technology and cyber extension.
- Sensitization training to middle level extension functionaries to improve their technical and professional knowledge and skills.
- Educating the farming community and the industry, the anticipated implications of the WTO agreement and lend a helping hand in building confidence and converting the challenges into opportunities in global trade.
- Conducting Livestock Extension Education programmes on Good Agricultural Practices (GAP), Good Laboratory Practices (GLP) Good Manufacturing Practices (GMP).
- Capacity building in the areas of understanding WTO, SPS, legal issues of SPS, food safety, risk analysis, disease risk analysis, diagnosis etc are the prime areas of importance for the livestock extension education system.
- Facilitating adoption of improved knowledge on production, value addition and marketing by farmers.
- Farmer organizational development including leadership development and supporting farmer organizations to take up new initiatives

(technology application, networking, financing and marketing through supporting producer companies).

- Organise special programmes for women in agriculture to achieve the above two objectives.

II. Linkages, Convergence, Partnership and Institutional Issues

Extension faces several institutional challenges. These include: rigid hierarchy and Centralized modes of planning; a tradition of assessing performance in terms of technology adoption; history of working independently; a mistrust of other agencies; and a tradition of upward accountability for resource utilization rather than output achievement and client satisfaction. For instance, lack of dedicated manpower, functional autonomy and attitudinal barriers at all levels have constrained ATMA in its effective functioning. Still most of the line department and extension functionaries are not clear about the approach and ways of integrating extension activities through ATMA.

The KVKs are 100% funded as a centrally sponsored programme under plan budget. The functioning of KVKs is mostly limited to the central grant only as there is no corresponding matching grant either from the state or the host organisation. This has led to lack of ownership of KVKs and lack of accountability resulting in improper manpower deployment, delay in sanction and utilization of the budget, and non-establishment of required infrastructure. These institutional issues currently constrain integrated delivery of support and services at the district and block levels. ICAR has recently suggested a framework for technology development and delivery system (Fig 2). It illustrates the roles of different organizations and the functional linkages among them and is worth considering in this context.

Among other things, we need a uniform national standard for quality specifications with respect to agro products. In line with Telecommunication Regulatory Authority (TERAI), a national food quality standard regulatory authority enforcing one quality standard will help the agriculturalsector in a great way. Presently, so many standards like Agmark, FPO and BSI/ ISI standards need to be confirmed with by the producers. Business environment therefore must be conducive for domestic producers in facing the competition across the border. Strengthening is required right from cold chain facilities at grass root level to handling facilities in the port for facilitating our Indian products to reach overseas consumers and international markets.

Privatization and partnership issues

Promoting private extension and privatization of extension services are two approaches used by national governments worldwide to improve the delivery of extension services. Private extension is not a single entity, but includes a wide range of service providers. The first type is entirely private and they use their own revenues to promote technologies, inputs and services. Most of the private profit oriented actors belong to this category. The second type consists of organizations that receive funds from government and other donors for implementing extension programmes and they are mostly of the "not-for profit" type. The third type consists of membership organizations that raise some resources from members (either as membership fee or service fees) for providing services.

There is a general belief that the private sector is more efficient and cost effective than the government in running certain public services. Private companies are normally smaller than government bureaucracies, which helps them to take action more quickly when needed. They also have necessary resources and use better technology to provide quality and timely service at a lower cost.. Areas like agricultural extension, which involves long term investment in human behavioral change, is not as attractive to the private sector as are those which give quick and positive returns on their investments. Still, there are components within agricultural extension such as the sale of farm inputs like seed, fertilizer and pesticide and advisory services for the same, which could promise profit.

Through the process of privatisation, extension effectiveness is expected to improve by:

- reorienting public sector extension with limited and well focussed functions.
- more number of extension providers (institutional pluralism) resulting from active encouragement by the public sector to initiate, operate and expand.
- more private participation leading to the availability of specialised services hitherto not available from the public system.
- user contributions to extension leading to improved financial sustainability, and support and control by clients leading to client orientation.

Whether privatisation is the only means to achieve overall effectiveness and efficiency in extension can be further debated. Some have questioned the distributional impacts; the dependence on private providers would result in extension. Sulaiman and Gadewar (1994) based on a review of experiences from privatisation of extension in different countries lists major disadvantages arising out of extension privatisation as follows:

i. contradictory message flow.
ii. negative impact on sustainability.
iii. sidelined educational role.
iv. lesser contact between farmers and extension.
v. high cost of technologies.
vi. increase in regional imbalances.

It is worth mentioning here that the public sector extension is not totally free from many of these limitations. Katz (2002) noted that reforms in public extension organisations such as decentralisation, transformation to independently functioning units, or the introduction of payment for services in the private interest and other cost-sharing agreements, coupled with capacity building for personnel, appear in general just as promising as privatisation.

Based on a review of private extension initiatives in India, Sulaiman (2003) has identified the following lessons on private extension.

- The private extension offers much scope for supplementing and complementing public sector extension.
- Crop/commodity focussed extension of private sector though very useful, is narrow in one sense as they do not engage with other related issues such as farmer organisation development or those issues related to sustainability of resource use.
- Remote areas and poor producers (especially those grouting low value crops and having little marketable surplus) are poorly served by both private as well as public sector extension.
- While public funding would remain important, the delivery of all kinds of services need not necessarily be though the public sector. Several of these services could be either contracted by the public sector to other private extension providers such as NGOs and private consultancy firms for delivering specific services in select regions, and client groups or jointly funded and implemented by public and private agencies.

- Public funds also could be utilised to fund farmer organisations to help them contract services from other service providets including public sector. However, efforts should be made to strengthen the capacity of farmer organisations to prioritise, demand, contract and monitor services.
- Private extension is not a substitute for public extension and there is a need for significant public funding for extension in the years to come.
- As farmers are also willing to pay for value-added services, the challenge is to create quality services so that partial cost recovery can commence.
- Financial participation needs to be seen more as an accountability enhancing mechanism and not purely as a mechanism to reduce costs.

Despite repeated emphasis on collaborative extension efforts involving public and private agencies, this approach is yet to get adequate attention. There are very few successful partnerships in the country. Some of the critical constraints related to establishment of successful public private partnerships (PPPs) include:

- bureaucratic hurdles,
- delays in decision making,
- hoarding of information/technologies,
- fear of operational compatibility;
- lack of a common platform to get into an operational MoU among partners,
- lack of initiatives and mission mode approach
- unwillingness to share credit among partners and
- reluctance for investments from private players.

Suitable partnership among national and regional players involving commodity boards, research institutes, farmer organizations and business houses will certainly prove to be successful, provided such partnership arrangements are made on professional terms and conditions, centered around teams, free from conventional bureaucratic control with incorporation of inbuilt project planning, implementation and monitoring arrangements. For example, a viable partnership among NDRI (National Dairy Research Institute), NDDB (National Dairy Development Board), Dairy co-operatives and the state Department of Animal Husbandry in the field of dairying can potentially lead to a successful technology generation, support, transfer and application for dairy development.

Similar arrangement could be worked out in the area of production and marketing of agricultural commodities. Big corporate players like Reliance, ITC, Pepsico and Bharati, have entered the business of marketing agricultural commodities. Partnership arrangements with the public sector for technological support in the areas of production, value addition, cold chain management etc at the farmer end can ensure farmers from receiving wholistic benefit from such partnerships. Lack of a common platform to broker such alliances has led to non-operationalisation of PPP in practice. Under present circumstances, a suitably conceived and operated PPP arrangement will certainly will help the country to address the second green revolution challenges, promote client centered extension, and lessen the burden on government exchequer.

- Explore options for PPP in district plans and SREPs
- Senior staff to be made accountable for operationalising PPPs
- Establish a separate cell for promoting industry and market driven extension
- Orient the staff at various levels on the advantages of PPPs and lessons from experiences so far
- Address bureaucratic delays that constrain partnerships.

Convergence of Extension Services

There are many extension service providers in the field, providing different kinds of useful services like information and service support to farmers. They are state, central government agencies, agribusiness companies, agri-preneurs, input dealers, manufacturing firms, NGOs, farmers organisations and progressive farmers. There is duplication of efforts with multiplicity of agents attending extension work without convergence. There should be a coordinated attempt to synergise and converge these efforts at the district level and below to improve the performance of various stake holders. One such frame work for Technology Development and Delivery System is given at Fig. 2 Some of the issues related to the convergence of extension services are:

- What type of institutional arrangement could be made keeping in view of the operational convenience, for achieving an effective convergence?
- What is the frame work of reference and strategies for such convergence?

- Identification of need for convergence, drawing a convergence model, operational steps for convergence and coordinating the identified activities for effective convergence.
- What is the operational working plan for such convergence?

The critical areas in which such convergence effort is required are:

- Fanner empowerment and farmer organisational development
- Technology backstopping and management
- Public private partnership's
- Frontier areas for extension, HRD and skill development

The details of areas of capacity building exercise in terms of training, method, or procedure of capacity building, infrastructure required need to be worked out. It is beyond doubt that combination of human performance with proper resource structure lead to development. An effective convergence of method/ procedure, Agency/ infrastructure no doubt will ensure proper use of scarce resources, time and energy. An exercise has been made to indicate the possible convergence in the above critical areas of extension and presented hereunder. The details of the areas of capacity building, method/ procedure adopted and agency/ infrastructural convergence are given here under at Table 2.

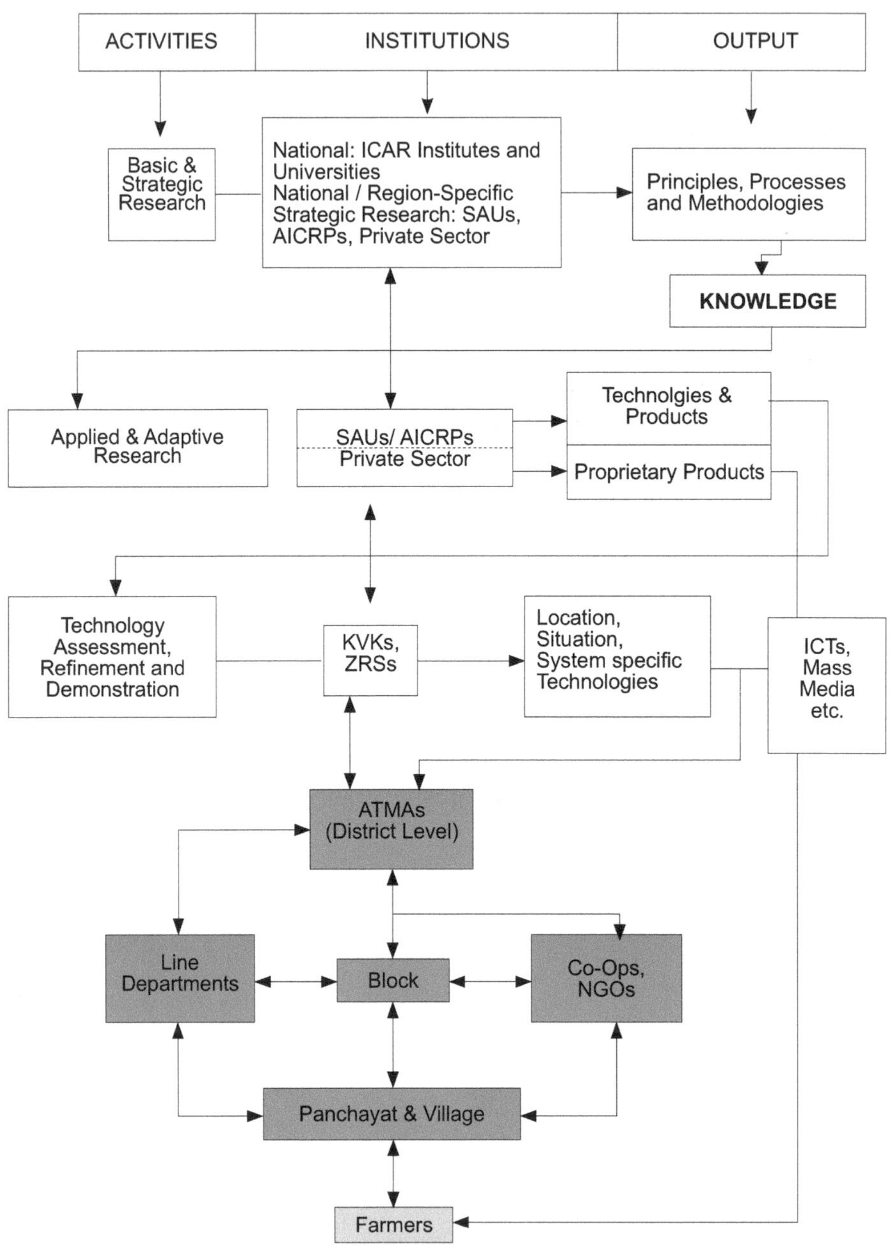

Outcome

Enhanced Profitability, Productivity, Sustainability, Livelihood Security, Employment generation, Competitiveness and Food, Nutrition & Environmental Security

Fig. 1 : A Framework for Technology Development and Delivery System

Table 2 : Infrastructural convergence at Grass root Level

SI No.	Areas of capacity building	Method / Procedure	Agency/ Infrastructure/ convergence
1. Farmers knowledge, skill and attitude			
	i) Breeding ii) Feeding iii) Health care and disease prevention iv) Value addition and marketing	i) Training/ Demonstration ii) Exhibition & farmers fair iii) Extension literature iv) Mass Media v) Use of ICT and cyber extension vi) Field campaign	KVKs, SIRD, NIRD, IIE, ICAR & SAU Extension system. State level livestock development agency and Department of Veterinary and Animal Husbandry.
2. Leadership, communication, skill and managerial development			
	i)Strengthening of village level leadership ii) Developing interpersonal communication iii) Managerial skills such as planning, organizing, coordination etc.	i) Training/ Role play ii) Success stories and cases iii) Management games	KVKs, EEI, IIE, MANAGE, NIRD & SIRD
3. Organisational Skills			
	i) Organisation of farmers groups. ii) Organisation of producer/ cooperatives/ societies/ union or federations	i) Training in leadership ii) Training in group dynamics & group formations iii) Performance linked specialized training iv) Record keeping v)Financial management	i) Cooperative training institute. ii) IIE iii) NIRD/ SIRD iv) EEl
4. Marketing and Business Skills			
	i) Market Analysis ii) Demand and supply Forecasting iii)Supply chain, Cold chain and networking iv) Retail marketing and creation of market network	i) Lectures and skills training by practical ii) Exposure visits to progressive states and leading co-operative dairies	i) Cooperative training institutes ii) MANAGE iii) IIE /EE1 iv)N1RD/SIRD v) Leading management institutes in Rural development.
5. Establishment of grass root level infrastructural facilities			
	i) Al Centres ii) Mobile unit iii) Fodder demonstration units iv) Milk Collection centres v) Bulk Coolers vi) Chilling centres vii) Rural Marketing Network and centres viii) Establishment of Milk Processing centres at regional level ix) Value addition and manufacturing facilities at district level	i) Analysis of the types of breed requirement and quantity of semen requirement ii) Ensuring quality semen supply and availability of skilled insemination iii) Training of rural youths for paid insemination services. iv) Preparation and submission of suitable projects for the establishment of milk colicction centres, chilling centres and rural marketing centres through funding agencies.	i) Department of Animal Husbandry and Dairying ii) State Milk Federation iii) NDDB iv) NEDFI v) NABARD

Cases of Technology Backstopping and Convergence

Technological backstopping to the Livestock Extension System and other centrally sponsored schemes such as NREGS and SGSY is a must at the ground level leading to sustainable development in rural areas. The appropriate technological support will help to create grass root level assets/ infrastructure through the extension programmes and schemes which in turn will help as a local resource in their command to undertake sustainable agriculture. Such an attempt ensures self reliance and less external input dependence agriculture. Some of the successful relevant cases of convergence are being presented here for discussion. They are: (i) Quality Hybrid Maize production, (ii) Mithun microchipping for identification, (iii)SGSY — SHGs — Technology- Financial institutions linkages (iv) Convergence of MANREGA (NREGA) and ICAR, KVK, (v) FSR/E in tribal Areas of E- Godavari and (vi). Banana Fibre Extractor Machine

CASE —1

Hybrid Maize Production Programme in NE Region- A successful case of convergence

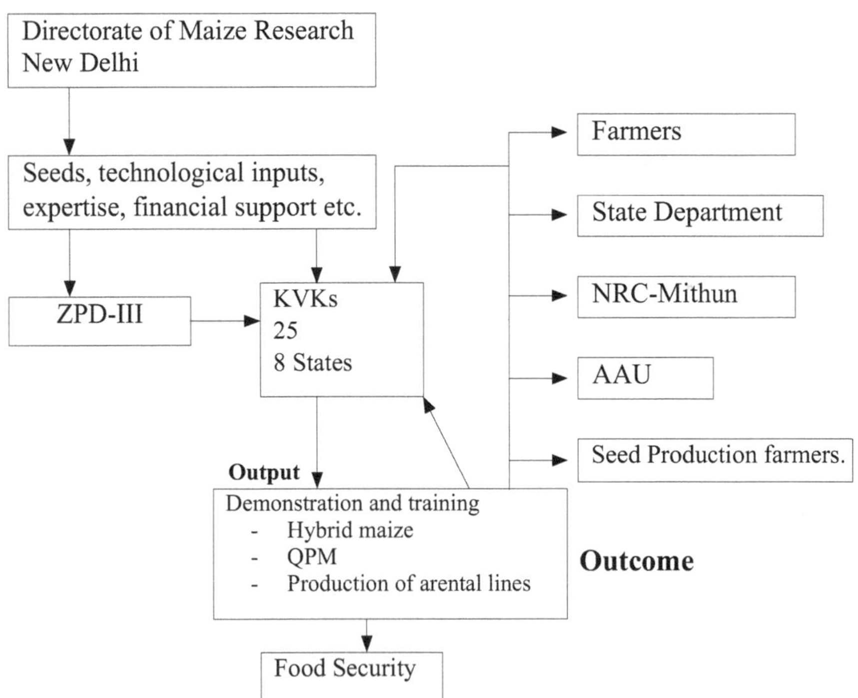

One of the successful convergence attempt made in the NE Region which is now extended to the other parts of the country is the hybrid maize seed production programme where in a successful convergence of various stake holders is being achieved. The partners are PD Maize, 27 KVKs distributed in 8 states, NRC-Mithun, AAU, State Department of Agriculture of respective States, farmers' organisation, NGOs etc. lead to food, nutritional and livelihood security in NE Region. Based on the success the above programme is now being implemented in the whole of the country involving 160 KVKs, eight ZPDs, PD, Maize and Division of Agricultural Extension, ICAR.through this programme the NE region was made self sufficient in parent seed production, hybrid quality protein maize production and baby corn. The out come of the programme is food, nutritional and livelihood security. As a horizontal expansion of the programme, an ambitious mass demonstration by involving 4000 maize farmers in each of the eight NE states is planned and under implementation.

CASE — 2

Mithun Identification Using Microchip Installation : Arunachal Pradesh

Traditionally the mithuns are identified with the identification markings made through ear cut, branding, tattooing etc. since these methods can be easily be manipulated, many times farmers are disputed over the ownership. Theft and ownership dispute is a common problem due to the

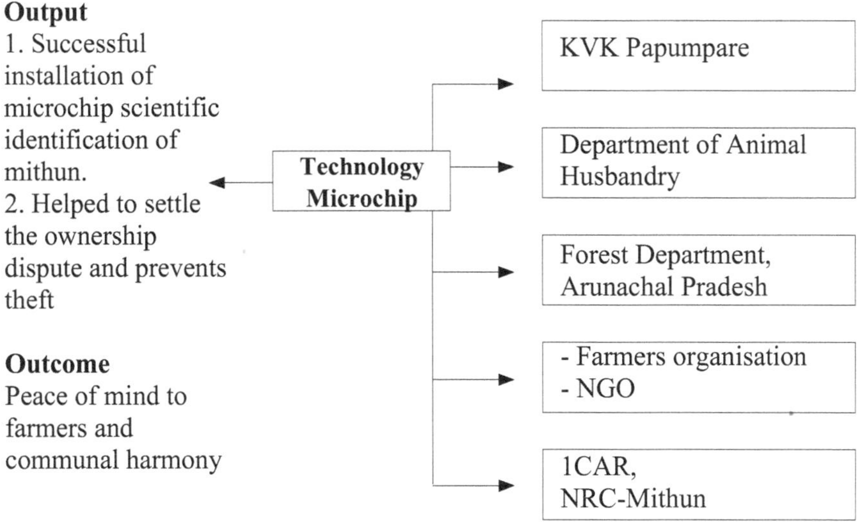

lack of full proof identification practice adopted by the farmers. KVK Papumpare district has come out with the solution of installing microchip installation on the Mithuns for the scientific identification. This programme was initiated with the help of forest department of Arunachal Pradesh for technical collaboration and animal husbandry department for the field level support. The farmers organisation, local NGOs were also actively involved in the above project. The details of the convergence, output and outcome of the programmes are given here under.

CASE — 3

Swarnjayanti Gram Swarojgar Yojana (SGSY) — SHGs — Technnology — Financial Institutions — Linkages

The SGSY provides credit cum subsidy for various income generating activities of SHGs including those related to irrigation and land development, horticulture, animal husbandry and dairy development, fisheries, village and agro based industries, rural handicrafts, handlooms etc. The key issue under SGSY is a social mobilization of rural poor into SHGs; setting up of sustainable micro enterprises by selection of key economic activities depending on available resources, occupational skills, appropriate technology and; financial assistance through a mix of Bank Credit and Government Subsidy; infrastructure, technology and marketing support with forward and backward linkages.

The appropriate technologies developed by ICAR and KVKs for these sectors need to be demonstrated and disseminated through SHGs. One such example of convergence of efforts by KVK, ICAR Institute, DRDA and SHGs in developing entrepreneurship development in East Godavari district, Andhra Pradesh is given here under.

A total number of 293 SHG groups with 2158 beneficiaries from 143 villages established 514 household units. Each individual is earning about Rs. 1000 to Rs. 3000/ month depending upon the activity they have taken up. The KVK CTRI, Rajamundry developed very good linkages with various financial institutions and rural development agencies of state, central and NGOs in a more effective way for economic viability and sustainability of these units in rural areas. These organisations besides acting as a resource centres and programme sponsors, provided financial support, scheme and subsidy components. The details of linkages is given in the figure presented below:

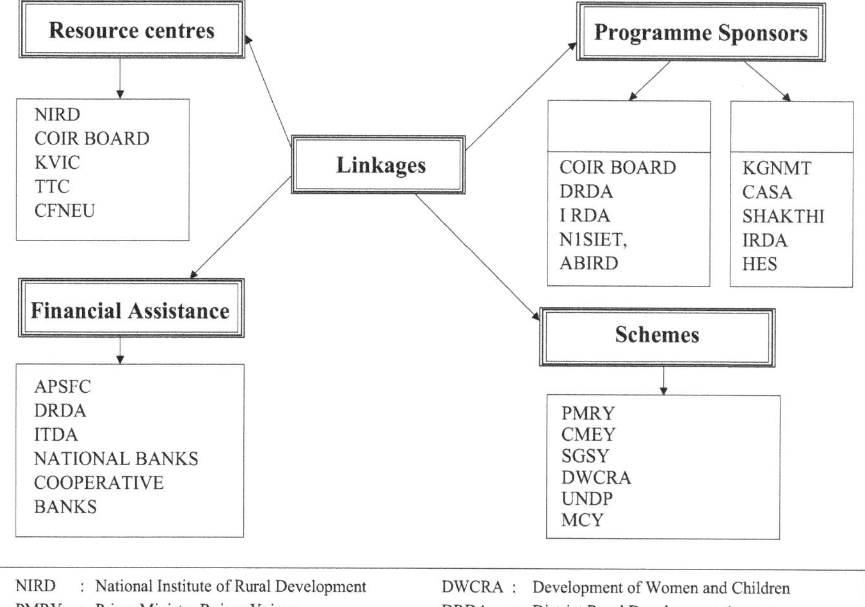

NIRD	:	National Institute of Rural Development	DWCRA	:	Development of Women and Children
PMRY	:	Prime Minister Rojgar Yojana	DRDA	:	District Rural Development Agency
CMEY	:	Chief Minister Employment Yojana	UNDP	:	United Nations Development Programme
KVIC	:	Khadi Village Industries Commission	IRDA	:	Integrated Rural Development Agency
TrC	:	Trainers Training Centre	MCY	:	Mahila Coir Yojana
SGSY	:	Swamajayanthi Gram Swarojgar Yojana	NISIET	:	National Institute of Self Employment Training
CFNEU	:	Community Food and Nutrition Extension Unit	HES	:	Harsha Educational Society

CASE — 4

Convergence of NREGA and ICAR- KVK

Ministry of Rural Development implements various schemes for employment generation and allevation of rural poverty and infrastructure development in the rural areas. The major schemes being implemented by the ministry are the National Rural Employment Guarantee Scheme (NREGS) and the Swarnjayanti Gram Swarojgar Yojana (SGYS). After a series of high level interface meetings held between Ministry of Rural Development and Indian Council of Agricultural Research (ICAR), it has been decided that appropriate technological backstopping to the schemes of NREGS and SGSY would be provided by the Krishi Vigyan Kendras (KVKs) of ICAR at the ground level leading to sustainable development in the rural areas. The KVKs have developed appropriate technologies for on farm and off farm activities. Initially, 50 districts have been identified on a pilot basis for technological interventions by KVKs based on the requirements of the districts.

The detail guidelines mentioning the need for convergence, parameters for convergence, strategies for NREGA (MARD) and ICAR

(MOA) convergence, have been worked out. Initially, the process of convergence with ICAR is to begin with those areas of natural resource management where the KVKs have developed technical experience. The concern for quality is central to NREGA and KVK is expected to promote appropriate technologies for NREGA works. An illustrative list of NRM activities where the expertise of KVKs can be used in conjunction with the choice of works under NREGA was prepared and provided. The list includes, water conservation, ground water recharge, drought proofing, development of irrigation facility, land development, flood control and protection works etc. In addition, the value adding activities on NREGA work also proposed to be attempted through convergence with ICAR technologies.

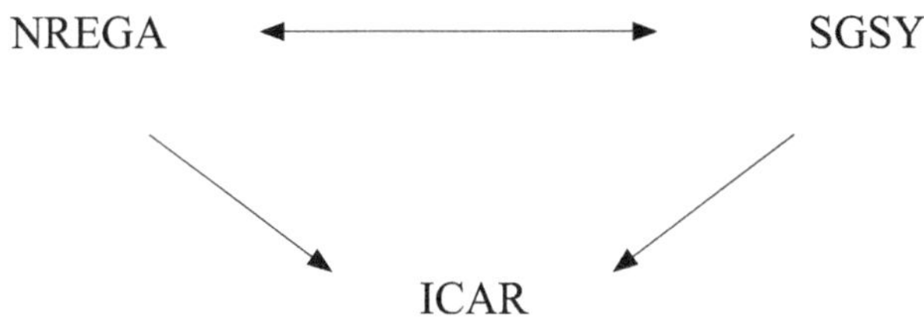

The efforts of different development agencies need to be successfully channelised by means of arriving at a common understanding and working out suitable implementation strategy for achieving the common goal. Our country provides wide opportunities for fruitful convergence of various extension systems operation with a common goal. Such efforts not only save our precious resources but also ensure effective utilization of various facilities created at the grass root level including the human resources and time required for achieving the target. Creating an inbuilt coordination mechanism in these agencies will help to work out fruitful convergence in the possible areas as well as its effective implementation.

CASE — 5

Assuring a Livelihood Security to Tribals in East Godavari District-Convergence of KVK-CTRI, ITDA, AP Forest Department, SAU and NGO

An inter institutional project was undertaken by CTRI, KVK-CTRI and Integrated Tribal Development Agency, East Godavari District towards farming system analysis and agricultural development in the tribal areas. It was revealed from the findings that tamarind collection and marketing was the main source of income to the tribal family. The findings further revealed that the tribals were facing the problem of declining yield of tamarind and so their income was constantly declining inspite of good demand for tamarind in the market. The senile plantation and cutting of tamarind trees were the main cause of yield decline and it was estimated that around 25,000 fresh tamarind trees to save the tribals from the approaching livelihood crisis due to declining forest produce of tamarind.

A joint action was conceived by the KVK-CTRI, ITDA, AP Forest Department, SAU and local tribal NGO to plant fresh tamarind tree saplings so that after 6 to 8 years, tamarind fruits will be'available. With the help of forest department, College of Forestry and ITDA about 25,000 tamarind saplings were procured and with the community participation through local NGOs in the agency area, the afforestation drive with PK-Variety of tamarind was initiated. Within a span of three months before monsoon, all saplings were planted and at the end of third year it was reported by the forest department that about 16,000 saplings were alive and grown well. It is now beyond doubt the convergence efforts of these organisation have resulted in assuring the tribals with their livelihood source.

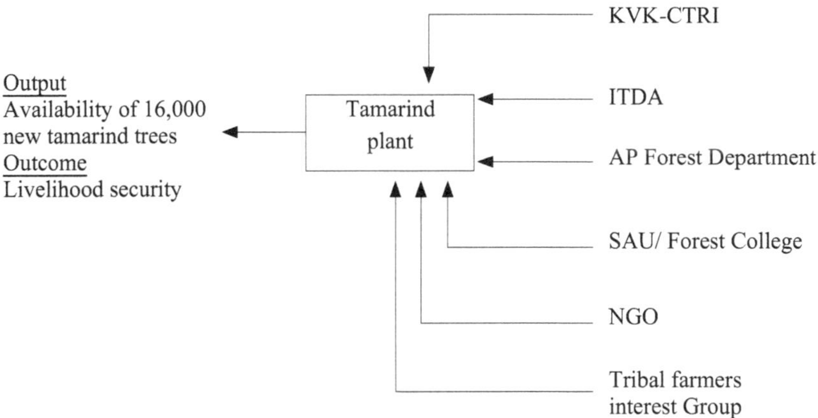

Similarly, more than 500 families in the tribal area were covered with back yard poultry units by introducing the nicoboro fowls brought from A&N Islands. Nicobari eggs were bought from CARL Port Blair and hatched at Rajahmundry for establishing a parent line at KVK- CTRI. From the parent line , eggs were produced and the chicks/ eggs were distributed to selected tribal families with a condition that the beneficiary should adopt five more beneficiaries in their village and support them with a minimum of 10 eggs/ five chicks. In that manner a network of 500 families could get the benefit of nicobari chicks in their back yard. The nicobari chicks are capable of producing a minimum of 145-100 eggs under no management condition. These birds are also sturdy and tolerant to many of the local diseases. The convergence effort of KVK-CARI, CARI Portblair, ITDA and tribal associations helped in establishing more than 500 backyard poultry units in the agency area ensuring nutritional security and income generating opportunity.

CASE — 6

Banana Fibre Extractor: A case of successful convergence

The manual fibre extraction process from banana stem is a cumbersome process. In the manual process an expert person can hardly produce a maximum quantity of 500 to 600 grams of dry fibre in eight hours. It is a tedious process involving patience, drudgery by means of straining palms of the person who extracts the fibre. Blackening of nail ends, finger tips and nail ulcers are some of the common problems associated with the manual extraction process. In addition it also creates a poor working condition due to the spillage of juice and waste pit in and around the extractor. Due to this cumbersome process with less economic output the extraction of fibre from the pseudo stems of banana has not receive desired attention and therefore no commercial extraction of the fibre is made from the pseudo stem of banana even though good quality of fibre material can be extracted and used for commercial purposes.

The survey conducted by the KVK- CTRI confirmed the above problems associated with the banana fibre extraction process. Therefore, it was felt that a suitable user friendly fibre extraction device which is highly essential to solve the above problem is essential. An inter institutional project team consisting of V. Venkatasubramanian, R. Sudhakar, K.Deo Singh from CTRI, Hyderabad designed and developed the need based user friendly machine "Banana Fibre Extractor" for the commercial exploitation of unutilized Banana wastes such as pseudo stems, peduncles and leaf stalks.

Machine highlights

- Reduces drudgery
- 50 times increase in fibre production compared to manual process
- User friendly and economic
- Less maintenance cost and safe to operate
- Clean work atmosphere and clean hands
- 25 kgs of fibre production/ day against 500 grms through manual operation.
- Superior quality fibre in terms of length , softness, strength and colour.

The machine helps banana cultivators to get an additional income of Rs. 2500/ per acre @ Rs. 5/ per plant with an average of 500 plants in an acre.

How convergence happened?

- Andhra Pradesh Agro industries ltd, Hyderabad purchased the commercial manufacturing rights of the machine.
- Andhra Pradesh announced 50% subsidy for the machine and ensured the training and distribution of the mission through DRDA and SGSY.
- KVK-CTRI offered the training/ technology backstopping demonstration of the machine.
- NGOs like Abhyodhay, Harsek Parshed, Amma Society, MSSRF, etc. were involved in the popularization of the machine in their respective operational areas.
- SHGs taken up the fibre production activities.
- DST, GOI selected the machine for meritorious invention award and popularized the machine through their channels/ exhibition.
- Machine was imported by the South Asian Centres and recently to Trinidad under UNDP programme for livelihood assurance project.
- In NE Region alone, more than 60 machines are under operation and recently all the home science SMS of the KVKs — NE region were trained by KVK CTRL sponsored by Zonal Coordinating Unit, Zone- III.

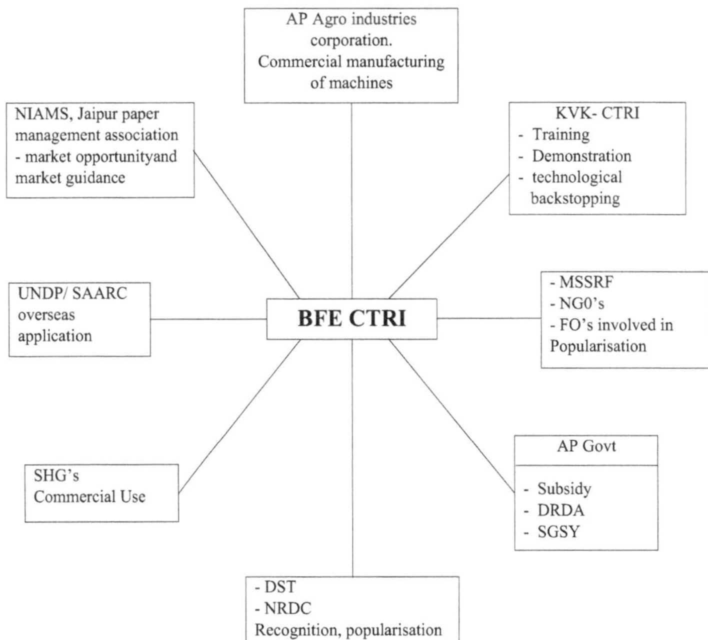

Thus, the BFE technology has been successfully commercialized for the benefit of farming community and rural women for their income generation activity using the locally available materials which are otherwise known as wastes.

III. Technology backstopping, Application, integration, ICT and management

The most important and neglected area of the present extension system is its limited ability to integrate different kinds of technologies. As a result, farmers adopt such technologies in isolation and blames extension and research system for its failures. This has resulted in confusion at the field level and erosion in extension's credibility on technological aspects. Therefore, an arrangement has to be evolved by the extension system for selection and integration of technologies. Promoting sustainable agriculture involves adopting an integrated approach keeping in view the nutrient status, water availability, crop animal combinations, pest and disease build-up, labour, markets, and availability of other support. Approaches such as INM (integrated nutrient management), IPM (integrated pest management), integrated soil and water management, etc involves integration of a wide range of technologies relevant to each area. Extension therefore needs to take up the challenge of integrating suitable technologies focusing on specific issue of agricultural development. The potential approaches are as follows:

- Decide on the problem and what needs to be improved
- measure the current status against the desired status
- analyse the root causes of the problem
- analyse the merits and demerits of potential solutions
- test different combinations through on-farm trials and demonstrations
- implement solutions and monitor long term improvements on sustai nab ility parameters.

ICT and Knowledge Dissemination

The most important role of ICT in Agricultural development is fostering a knowledge intensive sustainable livelihood security system in rural areas. Since ICT can enable us to reach the unreached and include the excluded information, knowledge and skill empowerment communication and information hold the key to the 21st century. An inclusive knowledge society requires the effective harnessing ICT to combat poverty and foster development. The issues of importance in the information led knowledge dissemination are: (i) access (ii) content and (iii) Capacity Building.

Access: The access to information and knowledge is impeded for much of our population die to poverty, illiteracy and violation. Linkages among professional partners are essential to reach those who are unreached and especially those who are under greatest risk of being left out of the knowledge societies.

Content: The farmers need locally relevant information, in the right language, to meet their immediate needs, and it may be more useful to promote more information sharing between local institutions than bringing in new information from outside. It is therefore important to promote information as a catalyst for community initiatives and encourage the adaptation of new technologies within decentralized and locally owned processes.

Capacity Building: It is needed at all levels to equip farmers and extension functionaries towards effective use of ICT in acquiring and dissemination of knowledge. Support is to be provided in terms of training for the use of ICT, establishment of rural knowledge centres, appropriate linkages with research institutes for continuous content updating. Training in information collection, storage and dissemination including the use of innovative formats based on the local culture is a must.

In the future, market determined production is likely to be basis of all agricultural operations. In such a scenario, real time and up to date information regarding market prices, insurance, logistics, warehousing, commodity trading, pesticide and other allied activities and resources, becomes indispensable to the farmers. The market led agriculture is possible through information based extension support system provided through information communication technologies at his door steps.

Information led Knowledge Management

The information content is classified into static and dynamic parts. The static content provides the information on the technologies, package of practices etc. whereas the dynamic information, are weather forecasting, market information, pest and disease outbreaks, in an area. The management of information comprises of a range of practices used to identify, classify and organize the information to facilitate its dissemination to the users. The information based extension support helps the users to get trustworthy information on selection of crops, variety, time and method of plantation, nutrient management, inter-cultivation, pest and disease control, special operation, method and time to harvest. The market and weather information helps the farmers to plan their operation according to the local developments to minimize their risks. Some of the successful information led extension initiatives are e-Choupal by ITC, I-Kisan of Nagarjuna Fertilizers; AKSHAYA by Kerala Govt., Gyandoot Project, Warana Wired Project, Vyapar, ICT based technology dissemination approaches of KVK, Ahmednagar and KVK, Pune, Agri Knowledge Portal of TNAU, Coimbatore and E-Connectivity of KVKs.

Establishment of ICT centres involving corporate sectors under Corporate Social Responsibility (CSR) schemes can be thought of to popularize ICT mode extension initiatives. These centres however have to be linked to the state agricultural universities and research institutes for technology and methodological backstopping. Content development for each technological area is a must. ICT centres need to be supported by content, technological inventories and methodological issues by the research partners. For instance, the ITC's *e-chaupal* initiative has got a strong technological and methodological support from Central Tobacco Research Institute (CTRI/ICAR) and ILTD Division of ITC with respect to tobacco production technologies. It is one of the strong successful cases of public-private partnership in agricultural extension through ICTs. Similar arrangements could be thought of in other sectors too. For instance, NDRI-NDDB-Dairy co-operatives can come together to establish ICT networks for the benefit of dairy farmers. Similarly partnerships between commodity boards and farmer association can establish ICTs in their respective areas of operation. Sustained technological and methodological backstopping arrangement among partners having similar interest is a pre-requisite for the successful implementation of 1CT based agricultural extension. As discussed earlier, clear cut assignment of responsibility and freedom from bureaucratic hurdles are necessary for bringing dynamism in its operation.

Agri-India Knowledge Portal (AIKs)

There is need for establishment of "Agri-India Knowledge Portal' to serve as a single electronic gateway for providing information to all those engaged in the task of development of agriculture. The information on improving agricultural productivity of crops, livestock and natural resources and coping up with disasters is focal to the rural farming communities. This would require access to a wide range of information on new technologies; alternate cropping systems, improved breeds of livestock, information on soil and water quality, pesticide, farm implements, animal health, weather and other related aspects. Besides, non-farm activities also constitute sources of income and livelihood for the rural farmers. Therefore information is also required on the means to incfrease income through processing, value addition and marketing.

ICAR-ERNET network consists of 274 Institutions including National Agricultural Institutes, State Agricultural Universities including their colleges and Research stations, National Research Centres and Project Directorates. Besides, a VSAT-based e- connectivity and networking of 196 Krishi Vigyan Kendras and 8 ZPDs in the country have been successfully completed. Both these networks are being linked through a high capacity leased line to make a unified high speed secured Intranet. Mobile based agro advisory services have been launched in different parts of the country by different organisations also need to be strengthened.

The intranet network among NARS institutions facilitate the development of Central Agricultural Data Warehouse, Content Development and Validation System, Agricultural Best Practices, Expert System for users, Multi-lingual content development and convertibility. The portal will also serve as a platform for facilitation of interaction among researchers and extension workers in KVKs through high speed secure intranet.

Various applications and services of the knowledge portal need to be accessible to the users including farmers, entrepreneurs, private sector organisations, other participating government and non-government organisations, extension workers, etc. through intranet. The services include Agro-met advisory services, Market information/ advice, package of practices, Agri-business consultancy, e-learning, knowledge on indigenous farming practices, and agriculture related FAQ. Besides, it will also act as information gateway for all agriculture related schemes and programmes etc. The Portal will also serve as a gateway for up-to-date websites of various agricultural research and education organizations accessed through Internet. The portal will provide a platform for facilitation of interaction among the end user communities through Internet also.

The Agri-India Knowledge portal will also provide for collaboration among other agencies related with agriculture like Meteorological Dept., Agricultural Markets,. Banks and other financial organzation, Input suppliers, Agricultural equipment and machinery manufactures, General Universities for providing services through Agri-India knowledge portal both through secure and high speed Intranet and Internet. The services proposed to be implemented in the collaborated mode include On-line education, Information on Market Prices, Weather Information including early warning system, Agricultural Input status information, Credit and crop insurance services, land record information, etc.

The AIK Portal is neither intended to replace any agency website nor any other government portal. This Portal will, however, provide additional means by which those sites might be accessed by a possibly wider and more varied client base. The AIK Portal will open opportunities for multiple agencies to participate in web delivery and development opportunities. The Portal would be able to integrate diverse interaction channel at a central point, providing a comprehensive context and an aggregated views across all information related to agriculture including *know-what* or declarative knowledge, *know-how* or procedural knowledge, and *know-why* or usual knowledge and creation of new knowledge.

ICAR as leader of Indian NARS need to have its dedicated transponder with adequate bandwidth. This will facilitate the data exchange among ICAR institutes, their stations, SAUs, CG Centres, General Universities, other science departments, KVKs and private agro enterprises more effectively. This will also permit video conferencing among different stakeholders. Implementation of this would require investment and policy changes in capacity building for ICT, free exchange of personnel between ICT resource institution, private sector in ICT and reorientation in extension education strategy.

Necessary Action is to be taken to:

- Improve the ability of extension organization to integrate technologies, test and develop integrated packages for different areas and monitoring performance
- Improve technological backstopping through partnerships and contracting arrangements
- Use the full potential of ICTs to bring together different types of knowledge and technologies from public and private sources
- Promote distance learning and virtual extension through ICTs
- Exploit the advances in mass media to improve reach and relevance and obtaining feedback.

IV. Frontier areas for Extension, HRD and Skill development

Lack of quality manpower dedicated to the cause of agricultural development is a serious constraint and challenge encountered by the system. Agricultural extension should be on agri-business extension mode and this is possible only if the system recruits personnel who can bring in different kinds of expertise. For instance expertise related to cutting edge technologies, organizational development, market development, legal issues related to farmer rights, IPR etc are crucial for extension. This would also involve some de-learning of its conventional technology dissemination approach and learning new ways of doing things. Extension is weak in animal husbandry and fisheries sectors and this would need considerable strengthening. Strengthening of national and regional level training facilities for continuous skill up gradation of extension professionals is to be taken up immediately. Assessment of the quality of extension personnel is a must and a national level mission mode approach is needed in this regard.

Strengthening the expertise of extension organization includes :

- frontier areas of technology
- resource conservation and management
- market development, linking with markets and export development
- quality and standards
- organic agriculture
- enterprise/entrepreneurship development
- skill development in horticulture, seed and plant material production
- research and extension in response to adaptation to climate change and risk management
- financing and insurance
- extension for resource conservation and management
- extension management techniques (programme/project management-PERT, CPM, log frame, 5-S etc)
- legal and regulatory issues (farmer rights, IPR)
- promotion of public-private partnership
- application of ICTs, content development and updating
- skills related to farmer organizational development and farmer empowerment
- strengthen extension in animal husbandry and fisheries sector
- initiate manpower planning in extension organizations, create new positions and fill existing vacancies based on manpower planning
- Address HRD and skill development in extension organizations through the following strategies:
  - new recruitments,
  - contract arrangements,
  - consultancies,

- staff trainings
- partnering with organizations having expertise.

V. Policy Issues

Several organisations implement extension programmes with very little co-ordination. Co-ordination is lacking even among public sector organisations. Establishment of a national extension authority (similar to the National Rainfed Area Authority), can potentially bring about the much needed integration for effective planning and delivery of extension programmes. Integration of extension activities at the district level also needs policy support. Several initiatives that were successful at the pilot stage had failed when external support was withdrawn. Another major constraint is the declining financial support for extension. Enhanced funding is crucial for improving the ability of extension to deal with the complex challenges of the second green revolution. Extension also needs crucial research backup on new approaches, methodologies and management techniques relevant for different situations.

Policy issues need to be addressed through:

- Enhanced funding
- Strengthen research in extension
- Establish a national extension agency/board
- Developing strategies to improve sustainability of pilot initiatives
- Strengthen mechanisms for district level planning
- National Centre for Agricultural Knowledge management.

Way Forward

Several organization implement extension programmes with very little coordination. Co-ordination lacking even among public sector organizations. Establishment of national extension agency/board (NEA) can potentially bring about much needed integration for objective planning and delivery of programmes. Integration of extension activities at the district level also needs policy support. Several initiatives that were successful at the pilot stage had failed when external support was withdrawn. Another major constraint is the declining financial support for extension. Enhanced funding is crucial for improving the ability of extension to deal with the complex challenges faced by it. It also needs crucial research back up on new approaches, methodologies and management tools and techniques relevant for different situations.

To summarize, among other things, (i) enhanced funding, (ii) strengthening extension research and education, (iii) creation of a national extension authority, (iv) developing technology inventory and methodology modules, (v) developing strategies to improve sustainability of pilot initiatives, (vi) Farming System Approach in Extension along with district resource mapping and planning, (vii) networking and partnership approach, (viii) inclusion of newer concept and methodologies, (ix) Creation of national and regional level knowledge management system for continuous technology and methodology backstopping and (x) Resource and market led approaches for sustainability are the most important areas to be focused upon for developing a frame work for technology development and delivery system.

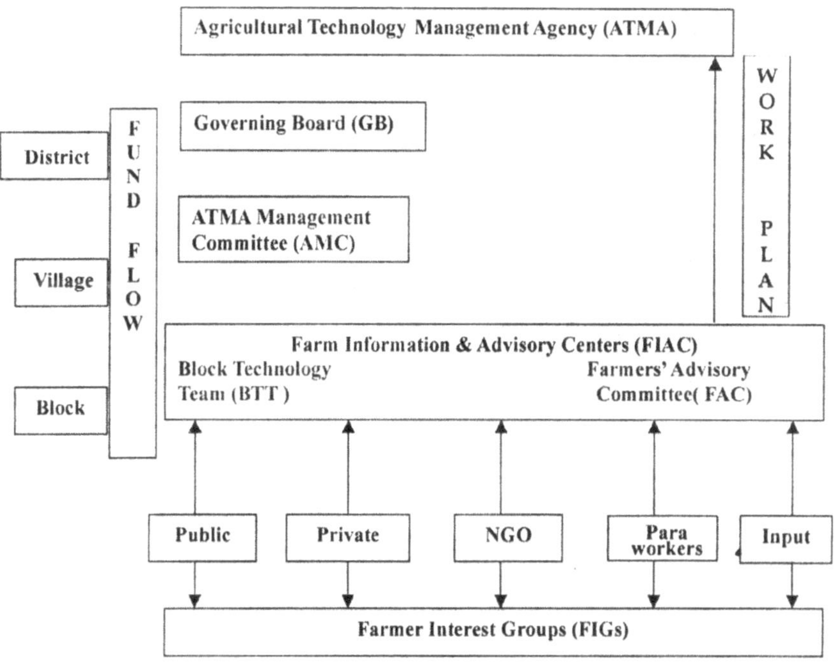

Fig 1 : Organizational Structure of Agricultural Technology Management Agency (ATMA)

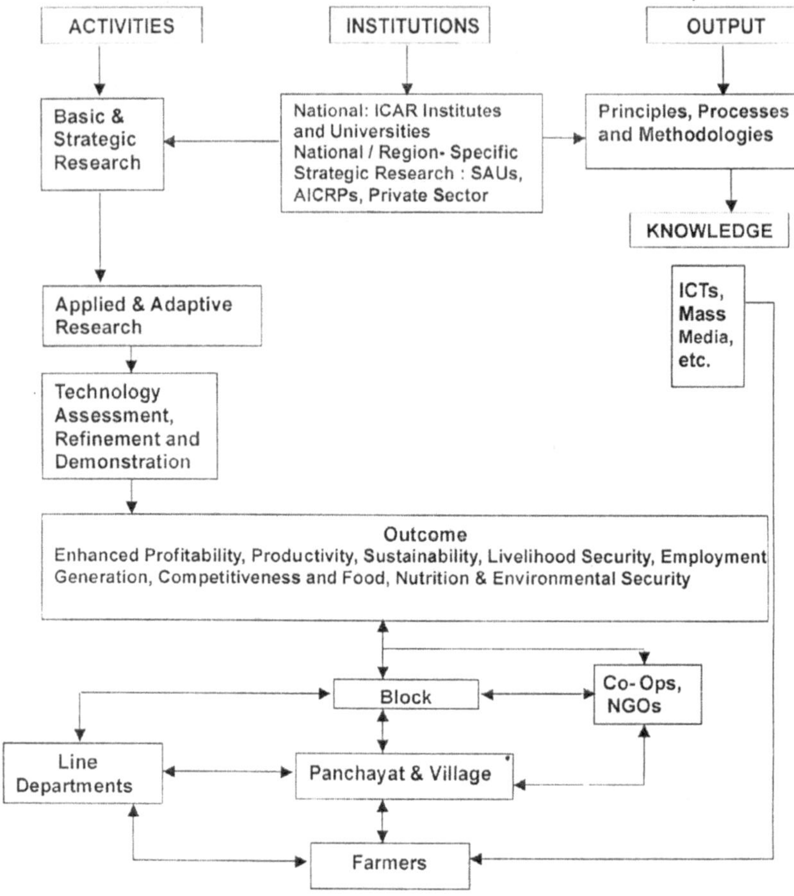

Fig. 2 : A Framework for Technology Development and Delivery System (Source: ICAR)

References

Anonym, (2008). Decision dsupport system in agriculture and allied sector of Papumpare District, Publication of KVK — Papumpare, Directorate of AH and Vet. Arunachal Pradesh, ZC Unit (Zone III) Barapani.

Chandrashekhara, P (2001). Private Extension: India.1 Experiences, MANAGE, Hyderabad.

FAO (2009) Agricultural Extension in Transition worldwide : Policies and Strategies for reform, Training Manual, Food and Agricultural Organisation, Rome.

Katz. L. (2002). Innovative Approaches to financing extension for Agricultural and Natural Resources Management — Conceptual Considerations and Analysis of Experience. Lindau: Swiss Centre for Agricultural Extension.

Ramkumar, S and Rao, SVN. (2001). Wome Self . Help Groups and Cattle Rearing — Preliminary Performance Appraisal, RAGOVAS Pondicherry, India

Sadamate, V.V, Venkatasubramanian V and Sulaiman Rasheed (2008). Technology Transfer and Extension — issues and recommendation, Second Green Revolution summit proceedings, CII, Kolkatta

Subbaiah, Y, Sudakar R, Aruna Kumari and Krishnamurty V, (2008). facilitating entrepreneur development process through SHGs in East Godavari Dist. of Andhra Pradesh. In *Models of technology Delivery mechanism — Experience of KVKs edited by Das. P and Prabukumar,* Division of Agricultural Extension, ICAR, New Delhi.

Sulaiman, V.R. and gadewar, A.U. (1994). Privitisation of Extension Services — Implications in the Indian Context, JRural Reconstruction, 27(2).

Sulaiman, V.R. (2003). Agricultural Extension: involvement of Private sector, Occasional Paper, National Bank for Agricultural and Rural development (NABARD), Mumbai.

Venkatasubramanian, V, Sudakar and Deo Singh, K. (2002). Rural Handicrafts. KVK CTRI publications, Rajamundry.

Venkatasubramanian, V, and Deo Singh, K. (2002). Farming System Analysis and agricultural development in tribal areas.. KVK CTRI publications, Rajamundry.

Venkatasubramanian, V, and Rao, SVN. (2010). Livestock Extension Education, Concept Series-1, Published by Zonal Project Directorate, Zone III, ICAR, Barapani. ISBN. 978-81-910017-2-3.

Venkatasubramanian, V, and Sajeev, M.V. and Singha, A.K. (2010). Concept, Approaches and Methodologies for Technology Application and Transfer, (Second edition, ISBN. 978-81-910017-0-9), published by Zonal Project Directorate, Zone III, ICAR, Barapani.

Venkatasubramanian, V, Sadamate, V.V. and Chandra Gowda, M.J. (2009). Agricultural Systems: Issues and Strategies for Convergence, In Proceedings of National seminar on Agricultural Extension, MoA, Gol, New Delhi: P-P: 43-49.

•••

2

ICT in Krishi Vigyan Kendra - A Transformational Technology for Indian Farmers

S. Prabhu Kumar, D. V. Srinivasa Reddy, C. V. Sairam and B.T. Rayudu
ICAR Zonal Project Directorate, Zone - VIII, Bangalore

INTRODUCTION

Krishi Vigyan Kendras (KVKs) also known as Agricultural Science Centres are the integral component of the National Agricultural Research System, which aims for development and promotion of location specific technology modules in agriculture and its allied enterprises, through Technology Assessment, Refinement and Demonstrations. At present there are 585 KVKs in the country grouped under eight zones and 77 in Zone VIII comprising of Karnataka, Kerala, Tamil Nadu, Goa, Puducherry and Lakshdweep.

The major mandate of the KVKs includes Technology Assessment and Refinement through on Farm testing, popularization of technologies through Frontline Demonstrations to prove their production potentials in the farmers' field. Conducting skill oriented training programmes to farmers and farm women, rural youths and extension personnel mostly focussing the technologies assessed/refined or demonstrated, conducting vocational training programmes for rural youth, popularization of technologies through various extension programmes so as to reach the masses, production and supply of good quality seeds, planting materials and other bio-products to the farming community.

With the advancement in Information and Communication Technology (ICT), technology dissemination can be made by the Krishi Vigyan Kendras using many ICT tools like creation and use of web site, expert systems, video-conferencing, data sharing through net-working and Mobile SMS etc. This paper highlights the Technology Delivery Mechanisms adopted by KVKs of Zone VIII using ICT.

Creation and use of web site

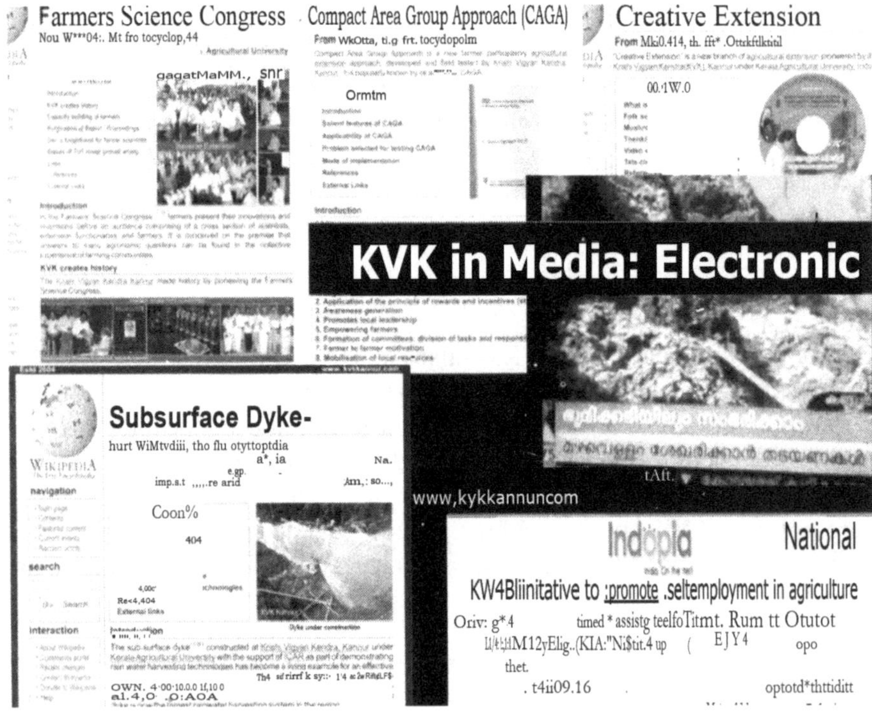

KVKs have their own web site with contents such as host institution profile, infra-structural facilities, district profile and map, mandate, details on core activities of the KVK, thematic area wise action plans and salient achieve-ments, marketing and other related information about development agencies, sales and services, reports and publications and related links. The major advantages are that farmers can get information and knowledge through these sites. The details about their action plans such as training programme dates, venue etc would help them to actively participate and coordinate with the KVKs.

Expert systems

An Expert System is software which would enable the farmers themselves to find solutions to their problems based on their existing input nature. Generally they are user's friendly and prepared based on crop / enterprises or based on various thematic areas. Farmers can decide their choices using their expert system and hence Decision Support System becomes a part of any Expert System. It is defined as a model and associated procedure that exhibits, within a specific domain, a degree of expertise in problem solving that is comparable to that of a human expert. An expert system is a computer system which emulates the decision-making ability of a human expert.

Expert systems generally follow this structure:

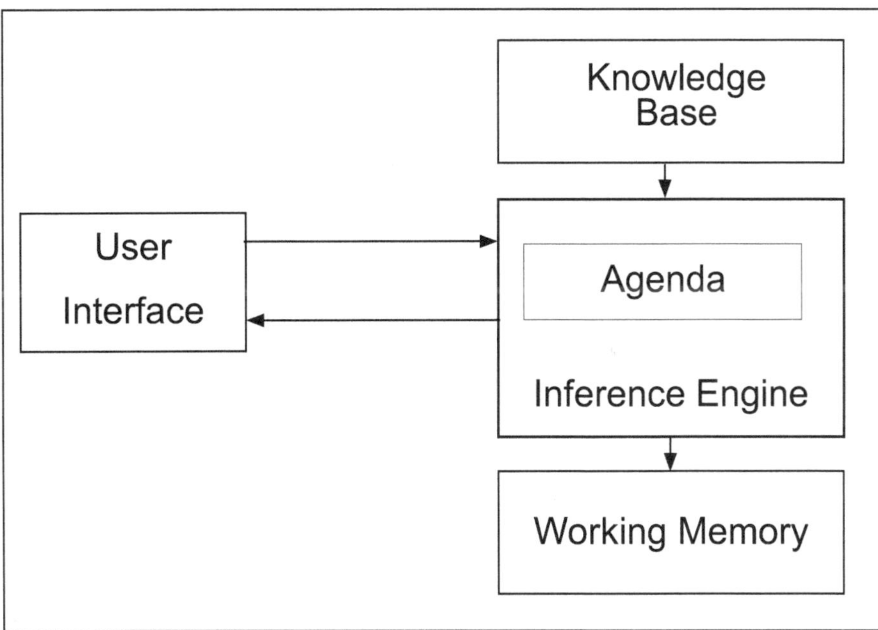

The development step consists of many important tasks. The developer must learn how the expert performs the task (knowledge acquisition) in a variety of cases. There are basically three kinds of cases the developer should discuss with the expert: current, historical, and hypothetical. Current cases can be covered by watching the expert perform a task. Historical cases can be discussed by discussing with the expert a task that was performed in the past. And, hypothetical cases can be covered by having the expert describe how a task should be performed in a hypothetical situation.

The knowledge acquisition process, which started in the specification phase, continues into the development phase. The developer must extract the developer looks for can be grouped into three categories: strategic, judgemental, and factual. Strategic knowledge is used to help create a flow chart of the system. Judgemental knowledge usually helps define the inference process and describes the reasoning process used by the expert. Finally, factual knowledge describes the characteristics and important attributes of objects in the system.

For example Zonal Project Directorate Bangalore in coordination with the Directorate of Extension, TNAU has developed Expert System for coconut. The major content includes planting season and agroclimatic conditions for coconut in Kerala, Karnataka and Tamil Nadu, coconut varieties and hybrids, nursery management, cultivation practices, technologies on crop management, crop protection, and processing. In addition, details on marketing and cost of cultivation are also in this expert system.

Expert system on coconut was Image based Expert System (better than farmers own native language). In this Expert System, high resolution images along with IF-THEN rules and facts were used. All are real time examples, so the farmers can interpret and they can interact accordingly. The front end is VB.NET and the back end is SQL SERVER. As it is query based, most of the data (symptoms, pictures, video clippings, control measures, etc.) comes from the database.

The Home Page consists of i) Planting & Season, ii) Varieties, iii) Nurse Management, iv) Cultivation Practices, v) Nutrition Management, vi) Irrigation Management, vii) Crop Protection, viii) Harvesting Technologies, ix) Coconut Processing, x) Marketing Channels, xi) Farm Implements and xii) FAQ's.

Similarly expert system was also prepared on other crops such as paddy, sugarcane, vegetables and other horticultural crops and on animal husbandry enterprises. The major advantages of the expert system is that farmers can ask simple queries on various aspects of the cultivation and the system would provide appropriate answers to their queries with various extension tools and aids.

Video-Conferencing

Video Conferencing is one among the advanced ICT which can be effectively implemented by KVKs in years to come. Videoconferencing is a futuristic communication tool using which geographically separated people can conduct any kind of face to face meeting. This technological advance makes distance obsolete. Anybody anywhere can see and talk live to the person face to face who could be thousands of miles away. A truly effective meeting involves much more than the exchange of words - it involves visual communication - in the form of body language, facial expressions. Video conferencing is a tool that allows this natural, full communication of ideas & information to take place over distance - without the overwhelming expense of travel.

Whenever any endemic or major field problems are noticed, SMSs from KVK can visit the spot, observe the symptoms and can interact with eminent scientists in the respective fields at ICAR Institutes / SAUs and find appropriate solutions. Another advantage of this system is to that farmer's themselves can interact and get convinced with the suggested. This ICT is at its infant stage of application in KVK system and the same need to be more emphasis during XII Plan Period.

Data sharing through networking

KVKs are creating and maintaining huge database on districts features, results of their achievements, feedback from farmers etc. The database is being systematically collected, collated and documented using simple database software at Zonal Project Directorate level. The same will be upgraded to advanced networking system by which all the 77 KVKs in the Zone will be connected through networking and the data sharing process would be made possible during the XII Plan Period. E-connectivity is being implemented at present in 32 KVKs and the by the end of XII Plan, all the KVKs in the Zone would be linked.

Mobile SMS

At present, the ratio of the farmers to the extension worker is 1000:1, which is really very meagre. Although the appointed Village Local Workers(VLWs) disseminate the information, they hardly accept any accountability. These two issues have created the urgency to help and guide the poor farmers properly. The cost factor in face-to-face information dissemination at the right time, and the difficulties in reaching the target audiences, has also crated the urgency to introduce ICT. It is only by the introduction of ICT that information can also be upgraded at the least cost. There are several models or ICTs in Indian agriculture, which have made a significant difference in the delivery of services in Indian agriculture. Kisan Mobile Phone Advisory Services (KMAS) is one among those and working successfully for dissemination of latest information in the state of Madhya Pradesh, Chhattisgarh and Odisha. This ICT based alternate agricultural information and rural delivery mechanism through Mobile phone was initiated during 2007 in KVK system. KMS model is based on the linear model of communi-cation which involves four major components of communication process i.e., Sender, Messages, Channel and Receiver. This is the unique programme for making linkages between different stake holders who is key player for making Indian agriculture sustainable in the coming future through intensive sue of ICT tools like mobile phone. Short Message Service (SMS) is being provided by KVK Subject Matter Specialists. The Extension Functionary is the user of information while farmers are implementer at field level. In Madhya Pradesh, the data shows that total 6.0 Crore population having 90,00,000 mobile phone which means that every 5 member family is having a mobile phone. Keeping in view the success of KMS, the same was proposed in 300 KVKs which includes 192 e-linked KVKs. All 77 KVKs in Zone VIII were established the facility of mobile SMS service and the content development are in progress.

Proposed Methodology/functioning:-

Major Steps for Implementing Kisan Mobile phone Advisory Services

Phase-I (Planning)

1. Selection of District/Block/Villages for implementation of the KMAS.
2. Identification of KMAS users (Farmers/ Ext. personnel's/NGO).

3. Collect Bill/Invoice from bulk SMS service provider.
4. Decide Sender ID and one mobile No. for Sender ID registration.
5. Sent official letter to bulk SMS service provider for registration. (Life time registration)
6. Discuss with selected KMAS users for different group categorization.
7. Fill up membership form of KMAS for Membership.
8. List of common words user in agriculture should be discussed during KMAS meeting.
9. Paid necessary charges to bulk SMS service provider and received sender ID and password for KMAS execution. (Through net/web site)

Phase-II (Execution)

1. Prepare list of all KMAS users in Microsoft excel (CSV comma delimited) file for proper execution of KMS programme (this format is supported by Net application).
2. Test your Sender ID. and password through bulk SMS service provider's web site (ex-www///kms.com).
3. Change your provided password for your safety and security purpose.
4. After completion of the above steps, finally KMAS could be launched at the district level through KVK for wider publicity and expansion.

Phase-III (Documentation)

1. Prepare following documents for result analysis-
2. Stock register of KMAS.
3. File of KMAS membership.
4. File of received tech information.
5. File of list of selected blocks/villages and users name with address.
6. File of farmers feedback and evaluation form.

Phase-IV (Monitoring & Evaluation)

1. Performance of the Technology with performance indicators feedback.
2. Execution of KMAS as per schedule.
3. Development of content /message.
4. Understanding of the message.
5. Need and time based information.
6. Applicability of the messages.
7. Received feed back from users during personal contact field visit/training etc.

8. Impact of technology (channel).

Expected Output

1. Faster delivery of information to the farmers field.
2. Dissemination of need-based, timely information to the rural masses.
3. Information delivery with minimum information losses.
4. Timely and speedy feedback from farmers, extension personnel and agro-input suppliers.
5. Impact assessment of KMAS.

The details of SMS sent and number of farmers benefitted during 2009-10 is furnished below.

State	No. of SMSs sent	No. of farmers benefited
Karnataka	5192	21752
Tamil Nadu	1022	8499
Kerala	551	4593
Zonal Total	6765	34844

Synergic planning

During XII Five Year Plan, KVKs need to synergize all the possible ICT tools for their Technology Delivery Mechanisms to the farming community. Single approach may lead to limited success, whereas strategic integrated approach with many ICT tools will sustain the system more effectively and would provide long term success to the aims of ICT application in KVK system.

Conclusion

Information and Communication Technology is an effective modern concept which needs to be appropriately utilized for faster and wider range of technology dissemination. Many ICT tools are available nowadays. KVKs need to identify and adopt most suitable and easy approaches preferably by integration of two or more approaches and provide benefits to the farming community on a sustainable basis.

●●●

3

Business Led Extension Services (BLESs) : Global Perspective

V.G. Dhanakumar
Director, Indian Institute of Plantation Management, Bangalore

Introduction

The business models, in general are based on traditional ways of strategy formulation and implementation, leading to incremental and non-disruptive change in the nature of business and agri-industry practices. In today's knowledge environment, we know that agri-businesses cannot survive by just running harder, but rather by running differently and "smarter" than competitors. (Seven et.al, 2005). This article suggests a sense-testing tool for the extensionists and administrators to enable disruptive innovation of extension-led business models through examples and case evidence.

Business is fundamentally concerned with creating value and capturing returns from that value. The term creating and capturing value reflects two fundamental functions that all extension organizations must perform to remain viable over an extended period of time. Successful extension organizations (SEOs) create substantial value by doing things in ways that differentiate them from the competition. The SEOs might develop core competencies, capabilities, and positional advantages that are different from those of competitors. Four important components of Business Led Extension Services (BLESs) is illustrated in Figure 1.

Business-led extension services are competing on a global economy and thus the unit of business analysis in extension is the world, not just a country or a region.

In the new competitive landscape, extension service must shift its existing strategy from production led extension service to meet the market share to., agri-business led extension service to delight the customer and enhance the proportionate market share through effective knowledge and skill in the process of business development.

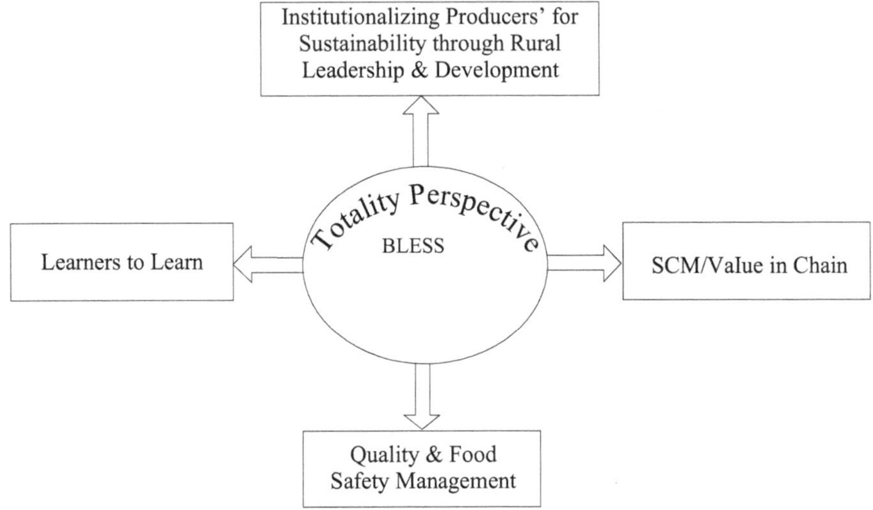

Fig. 1 : Critical Components of BLESS

Business Led Extension Services (BLESs)

BLESS activities involve coordinating the various value-added functions as listed in Figure No. 1. Functional shift in extension is a term commonly used to refer the change in role of activities to decide the best performance within the business process. The existing extension system mostly focused on government centered approach through functional linkage between research, extension and farming community. However, the proposed model of extension service is on agri-business partnership led extension service, whose function will be termed within the core business component namely backward, horizontal and forward integration [#] .

[#] **Backward Integration : The Policy will Promote**

Establishment of a sustained and lasting linkage between the farmers and the processors based on mutual trust and benefits by utilizing the existing infrastructure of cooperative, village panchayats and such other institutions. Development of Futures Market in the best interest of both the farmers and the processors ensuring a minimum price stability to the farmer and a sustained supply of raw material to the processor. Mechanism to reduce the gap between the farm gate price of agro-produce and the final price paid by the consumer. Setting up of an Equalization Fund to ensure sustained supply of raw materials at the particular price level and at the same time to plough back the savings occurring in the eventuality of lower price to make the Fund self-regenerative.

Therefore, the fundamental shift in extension service would be aligned to its integration with research and farming community as backward extension integration, its association for establishment of institutional network as horizontal integration and customer led service as forward integration within the integrated concepts of agri-business management (ABM) and Supply Chain Management (SCM). The extension vision and mission sets the direction for business led extension service, till recently most extension service organized their people and manage their activities through functional groupings i.e., production, input supply, technology transfer, etc. A primary goal of extension approach in the past was to develop efficient way to prosecute their work. In return, personnel in-charge of these functions become experts and seek to achieve better performance, which is termed in supply chain literature as functional excellence as measured by high productivity, maximum area covered, high rate of technology adoption, achievement in target, etc.

Interestingly, global extension services are finding that its functional excellence does not equate with agribusiness performance, which is achieved through high target coordination. A theory of organization development suggests that in a stable environment (Pre-WTO era) the functional excellence lead to business excellence. However, virtually all extension function in dynamic agri-business environment induces a greater uncertainty of demand, shorter product life cycle, etc. Therefore, extension service needs to bring about functional integration in order to succeed in the volatile environment (Post-WTO era). The road to competitive advantage will begin, if extension service strategy shifted from managing function to managing business processes.

A profound shift in extension service should be a process led extension service within the dimension of business to achieve excellence in agribusiness than agricultural development. A process led extension would always focus on holistic dimensions of agribusiness i.e., extension

---

Forward Integration : The Policy will Promote

Establishment of a strong linkage between the processor and the market to effect cost economies by elimination of avoidable intermediaries. Establishments of marketing network with an apex body to ensure proper marketing of processed products. Development of marketing capabilities both with regard to infrastructure and quality in order to promote competitive capabilities to face not only the WTO challenge but to undertake exports in a big way.

Horizontal Integration

Establishment of institutional linkage between forward and backward integration and fonnmlation of effective strategy for agri-business performance and development of local organizations to empower farmers.

education and service towards supply, purchase, production control, distribution, marketing and customer delight. Thus, BLESS encompasses all the traditional agribusiness function, their coordination within individual farming and across farming community in the process of supply chain and its value for agriculture.

(1) Supply Chain Management (SCM) and Its Value in Chain

The Council of Logistics Management, the "preeminent worldwide professional association of logistics personnel" (CLM, 2003), defines logistics as *that part of supply chain process that plans, implements, and controls the efficient forward and reverse flow and storage of goods, services, and related information between the point of origin and the point of consumption in order to meet customers' requirements.* This definition tells us that logistics, manages all the movement and storage activities that are associated with product and service flows. It is focused on what we call the "focal extension", that is, on managing extension's inbound and outbound flows of goods, services, and related information. Information on marketing plans, effective diffusion of information, pricing structure, product management status, ownership, finance, etc. as the systematic, strategic coordination of the traditional functions within a farming community, for the purposes of improving the long-term performance of the individual farm and the supply chain as a whole.

There are three conceptual models presented and the first model (Phase 1) is denominated as independent supply chain entities where characterized as inventory push i.e. physical distribution management. Production is handled in isolation and its output is pushed down to the finished warehouses (or) market. The second approach aimed at internal integration within sales, procurement, production, warehousing, distribution and transportation. The final phase of supply chain management extends the scope to link external partner like suppliers, vendors, distributors, Third & Fourth Party Logistic partnerships for value added activities and customers. It advocates managing relationships, information and material flow across-agri-business borders. The details related to three basic models are illustrated in Figure No.2.

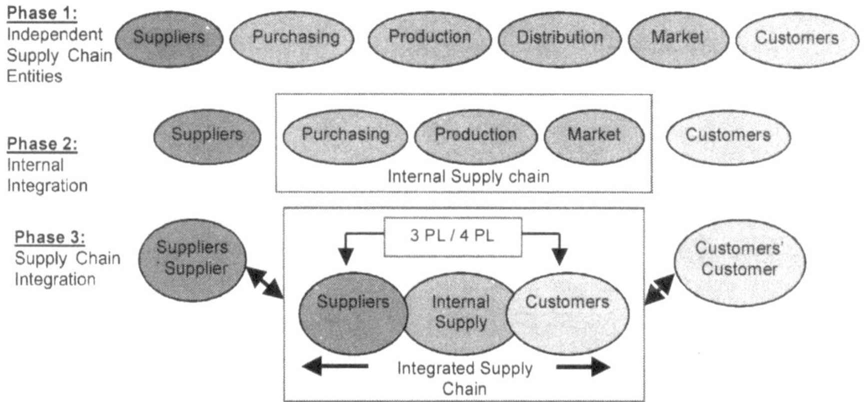

Fig. 2 : Supply Chain Modules

In conclusion, SCM is a management process that deals with inbound and outbound flows, from the perspective of the agri-sector, its suppliers, and its customers. This means a fundamental aspect of supply chain management is the consideration of not just the cost and profit goals of agri-sector, but of all the stakeholders involved in managing the value in supply chain.

(2) Institutionalizing Producers' for Sustainability through Rural Leadership

The concept of rural leadership strengthens its communities by preparing leaders as catalyst to effectively deal with today's changing world. The social and economic issues facing communities require leadership. Communities deserve leaders who are visionary, yet pragmatic; bound by public needs and non-political favours. In order to meet challenges and recognized opportunities, our communities need leaders who show initiative, listen intently, understand the breadth of perspective on issues, assume responsibility and exercise sound decision-making within institutional dimensions of BLESS network as illustrated in Figure No.3. Rural leaders often referred as link man/ woman and support leaders of the society. An attempt to develop rural leadership program at par with the WISCONSIN Rural Leadership Program (WRLP), is essential for the Indian agri-business society.

In order to implement the first component of BLESS i.e. SCM and its value in service, the institution building approach must focus at *'micro'*, *'meso'* and *'macro'* levels. Existing traditional community structures, traditional institutions, organisation and community welfare organisations are to be effectively utilised to build a trust and confidence at the micro-level. Mobilisation of the traditional leaders gets them involved in community development. At the same time, community based self-help structures are introduced and developed from the base up. Both participation and mainstreaming are taken as essential components for building internal development networks. Institution building at the meso-level would help the community reach the district development structure and external support systems for institution building support as well as articulation of need and influencing policy and decision making. The third component of institution process is at macro level, which facilitates the development activities by government, commodity boards and private sectors to jointly initiate the business process both for micro and meso sub-system. The three levels of development concept for the institution building processes are illustrated in Figure No. 3A and 3B.

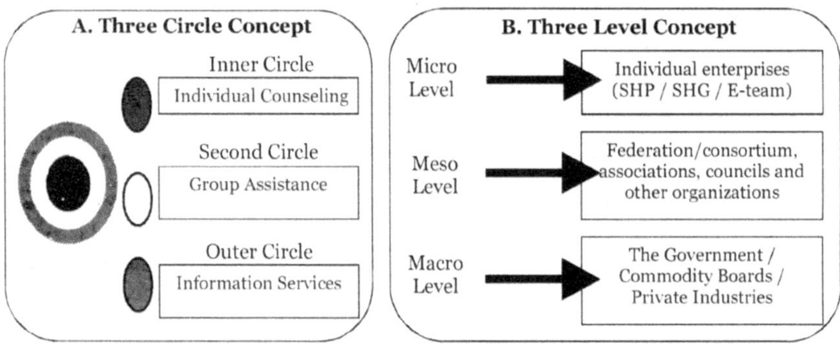

Fig. 3 A, B The Three Levels of Development Concept for the Institution Building Processes

The conceptual components further explains on how a business development programme at the grassroots must plan and develop micro, meso and macro level institutions in different phases and how to merge them together with sectoral systems and organisational structure to form a representative sector. Understanding the importance of developing representative sector is not difficult; what makes the assignment hard is the commitment to have a clear implementing strategy and a set of phasing out time. Only being *local* is not the answer. The grassroots institution developed must be *representative* and the resultant model or representative sector shall have a wider geographic coverage so that integrated planning and implementation could be possible for different sectoral programmes. It means that developing one representative sector that covers a village or two would never work. There should be Regional Development Institutions (representative sector) working as intermediaries between the central coordinating unit and the local at the micro-level, all linked and netted together through proper systems and well defined structures.

The Figure No. 4A and 4B gives a holistic framework of institution building processes and shows how traditional, modern and local & district level structures can function together in a perfect harmony within commodity and related agri-business sector.

The three-stage process of institution building at meso-level starts with the simple approach of community organisation according to the participatory approach followed by many other development programmes at the micro-level. However, organising the communities around a "common interest" alone is not enough, because more often than not, the "common interest" is the package that the rural extension development programme offers and this interest vanishes away with the phasing out of the external programme, leaving no structure, no system and no organisation behind. Secondly, these organisations need to be formed and developed by the external support programme in a systematic manner.

Third party organizations (other than commodity representation) to do social mobilisation and forming village organisations might lose not only the professional touch required to groom these organisations, but may also result in diverse out put obtained in a costly manner. Moreover, the objective is to form meso-level, representative organisations on top of the local organizations at the micro-level. If local, unrepresentative of sector are involved to work as contractors for social mobilisation at the grassroots, they might stand nowhere at the end of the process as their constitutions and bylaws might not allow them to become representative organisations. Nor would they like to be replaced by another set of local organisations, more focused and more capable. Involvement of the already present commodity representatives can only of some advantage, if they are willing to work on a continuous advisory role. The basis of all institutional promotion is the formation of local organisations that are established on a particular region or area.

The focus should be on building resources, capacities and lompetencies of the community members. Mobilisation starts from the micro-level while decentralisation is done from the macro-level. At the meso-level institutional capacities are built for self-help promotion, participatory and integrated action planning. We need to mainstream functions and upscale capacities of the members from the micro-level to upwards.

The interface and the involved actors between the micro, meso and macro level need to involve in visioning, need identification, articulation and prioritisation. The bottom-up development network at the micro-level need flexibility that shall be provided through multi-sectoral and pluralistic development structure.

For example, "Agriculture remains one of the major industries in the NNY region. The vitality and vibrancy of the agricultural industry has a direct relationship to the vitality of the rural communities in this region. Local food production and the increase or strengthening of the production is a vehicle for economic development in the region."

Over the past few years, the Direct-to-Consumer agriculture sector in the Adirondack North Country Region of the state has experienced a 22.3% increase in the number of farms involved in selling their products directly to the consumer. This increase has occurred even though the overall number of farms in the region has decreased by 6.6% due to effective institutionalization processes.

Local Foods is a term used to describe both foods produced and consumed locally and those activities that support efforts to make connections between local foods producers and consumers. Local foods are about relationships between people: producers - those growing and marketing the food consumers - those purchasing and consuming the food community developers - those helping to make or support connections between producers and consumers.

There are unique efforts in all parts of the state which are geared toward increasing the production and consumption of local foods. In the North Country, production agriculture has transitioned to take advantage of the local food movement and it is making a difference for farmers, consumers and communities in that region.

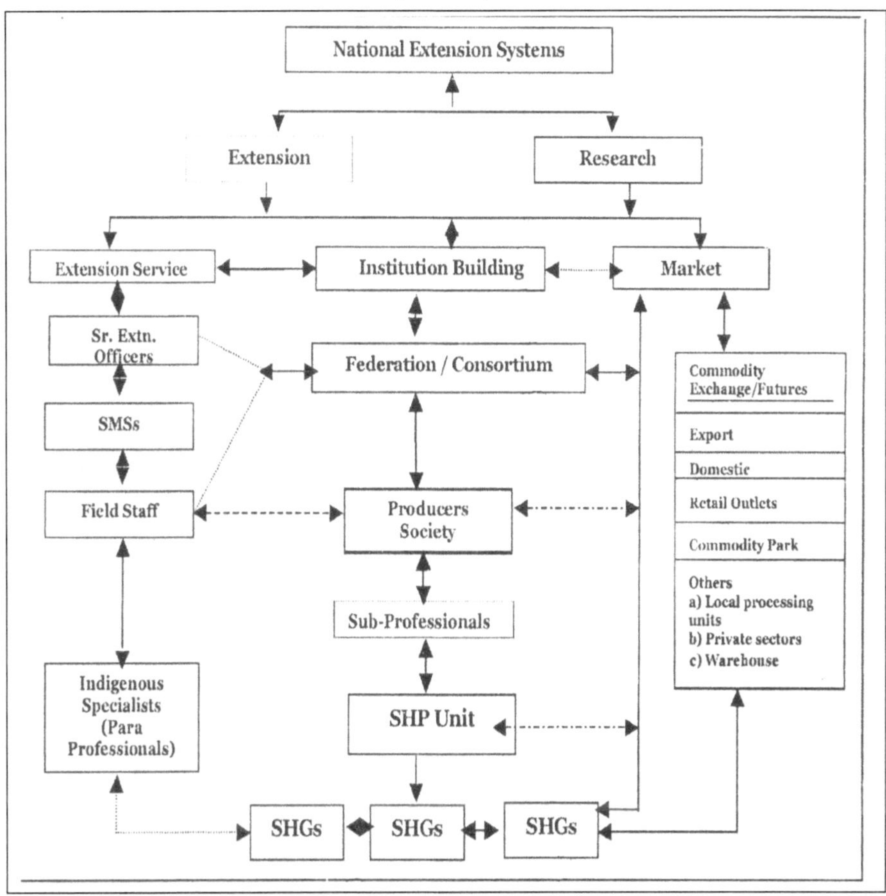

Fig. 4A : Structural Dimensions of Grassroots Institution Building for Sustainability (GIS)

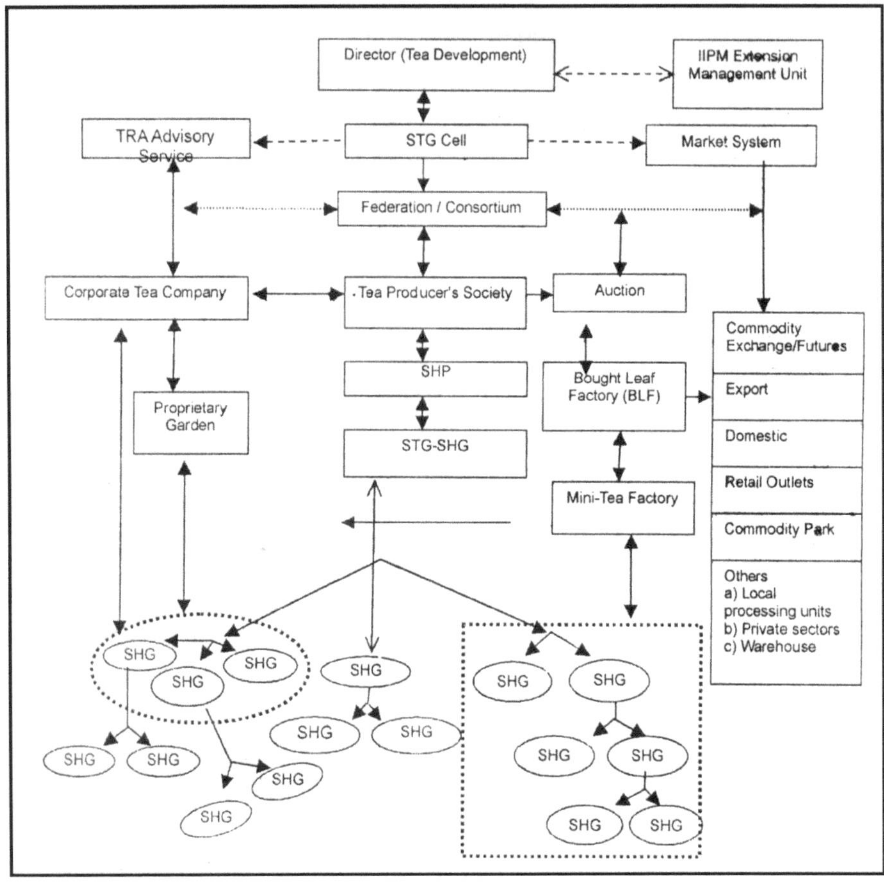

Fig. 4B: Module on Institutionalizing STGs in Tea Sector

(3) Quality and Food Safety Management System

Access to good quality, safe and nutritious food is considered a basic right of the people. Consumption of unsafe, contaminated food leads to food-borne diseases, which cause considerable morbidity and mortality. Agri-business sector and food processing industry is widely recognized as a 'sunrise industry' in India having a huge potential for uplifting agribusiness economy, creation of large scale manufacturing and food chain facilities. India has enormous growth potential from its current status of being the world's second largest food producer to the world's number one producer. However, there are several bottlenecks that need to overcome for achieving such position. It is very important that processed food available is safe, hygiene and free from contamination, intoxication and adulteration. The unholy connection between an unsanitary, pest-infested environment and ill health amongst the masses has been recognized since very early ages. Sanitary practices now form an integral part of production process in the agri-food industry "(Roday, 1999)". Major changes in the fields of agri-business sector in recent years, including technological acceleration in food production, economic and market development, have brought to the fore a variety of issues which are relevant to food qualsafe and techno-risk management. But the elimination of pests by use of pesticides requires an extreme caution and the producer should not go over-board in this quest. For this reason, the landmark publication "Silent Spring" by Carson was a timely warning to those who ignored the impact of scientific developments on qualsafe and techno-risk management (QTM) with respect to environment and human health. For these very concerns, the FAO/WHO/ CODEX have recognized the international food security as a priority area of focus in agriculture with special reference to Globalgap, HACCP/ISO 22000; ISO 9000/14000, SQF, etc. The QTM project is aimed at creating awareness regarding the harmful and inefficient agricultural practices followed by the Indian farming community, which not only makes farming unprofitable for farmers but also poses risks to sustainable development in agriculture and health of the consumers.

(4) Learners to Learn Strategy for BLESs

Therefore, the importance of learner-centered learning for sustainable development has led development agencies to make institutional sustainability in the design of planning process. *Institution,* is a subtle concept and therefore subject to confusion. It refers to stable, valued, recurring pattern of behaviour. Institutions thus include rules or procedures that shape how people act, and role or organizations that have attained special status or legitimacy. *Sustainable institution* in a strict sense the term is redundant since institutions are by definition, sustained ways that people interact. But in development circles the conventional meaning of this refers to consciously designed organizations that do one or more of the following:

They survive over time as identifiable units, 2. They recover some or even all their costs and 3. They supply a continuing stream of benefits to the community. Focuses on generic framework for understanding institutional sustainability framework viz., a) systems that function in relationship of their environments; b) organized and managed entities whose organizational structures and procedures must match the tasks, products, people, resources, multipl associations and context they deal with and c) settings intimately concerned with the exchange of resources and knowledge across the nation through learner-centered learning mode.

Institutionalizing India's Tea Planters, as Learner-to-Learn Group Project Experience

The tea industry occupies an important place in India not only because of the significant level of production of tea, but also it relates to the livelihood concerns of a large section of its population. India is also the world's largest consumer of tea, and given its large population and the present high levels of economic growth, it is expected to play a significant role in determining global demand.

Currently, the focus in India is on improving all segments of the tea value chain and its several efforts on all dimensions. This includes efforts towards improving field productivity, business performance, labour efficiency through HRD initiatives & standards, Cost Minimization (CM) and Profit Maximization (PM), post harvest processing, value addition, encouraging internationally accepted quality certification systems, environmental management, improving and upgrading infrastructure viz. extension and R&D system, packaging, marketing, branding and ultimate customer delight.

The pursuit of excellence in Total Factor Productivity (TFP) has long been a hallmark of IIPM thought its Tea Board sponsored HRD programs to the sector. Since its founding in 1993, the Institute has endeavoured to assemble promising tea planters and their stakeholders in tea commodity designed to stimulate, even inspire, its members to develop their talents to the fullest. Figure No.5 provides the prime factors and its inter-linkages with Total Factor Productivity (TFP) components.

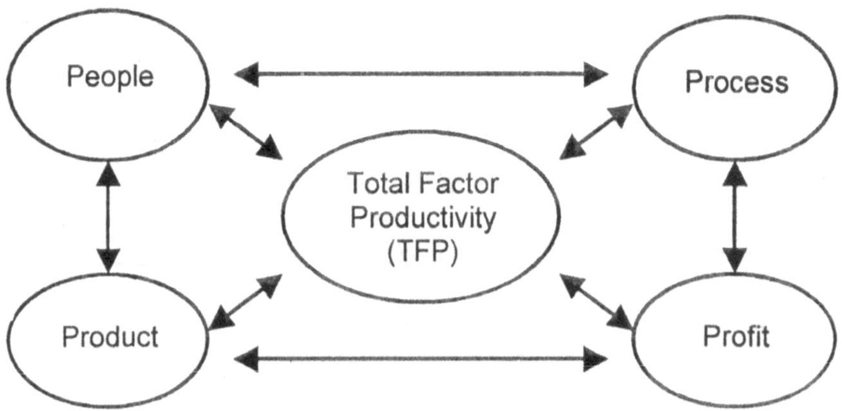

Fig. 5 : Tea Planters' Productivity Council for TFP

There is an immediate need to supplement the ongoing capacity building programs by providing further a planters' forum with intellectual and practical inputs on a regular basis. Therefore, ***Tea Planters' Productivity Council (TPC)*** with an objective of learner-to-learn and provide a forum with intellectual and practical inputs on a regular basis, under the sponsorship and financial assistance of the Tea Board of India.

In today's competitive environment, the tea sector depends on ***"High Performance Teams"*** to deliver value faster. This bring a cross-functional set of planters together and what offers individual performers or groups from teams is the ability to share knowledge and exchange ideas. TPC approach is where team dynamics come into effect. TPC is a community of learning that provides extra-ordinary opportunities to talented planters and their stakeholders from around the globe. TPC explores how members in tea plantations perceive and deal with change by describing the learners-to-learn team building process through forming, storming, norming and performing through TPC, the council members could find strengths and opportunities for overall improvements.

Conclusion & Implications

Several key managerial elements and lessons emerge in applying the BLESS tool for the development of disruptive extension led new business model. A systematic approach helps extensionist to grasp the concept of adapting and creating an extension led business models for strategic infliction. These tools also assist extension system in discerning change, if it is possible for the extension system to create or revert its business model within its dimensions. It is a matter of fact that most disruptive ideas, whether products, services, processes or industrial extension units, NGO led extension services rather than public/government incumbency. The larger incumbents such as government extension service mostly lack kind of thinking and flexibility to incorporate business flair and need to launch business led incubators and experimental venture within extension networks. The norms of doing business led extension service in agri-sector have been reinvented with a new basis for competition besides the conventional mode of operation. The question is what steps the conventional extension system should focus from switching the production led extension model to that of business led extension service model, when it proves to be successful. Why does the way of doing current extension model usually not becoming the model for the whole sector? On the other hand, extension groups might find it quite easy to distinguish themselves from new competitors by their own way of doing things and benefit from the increased demand.

Therefore, the extension survival in today's fast changing world depend on being a head in business model thinking and adaptation. The key dimension of a future reinvented extension led business model will be within the component such as supply chain management orientation, institutionalizing producers, learners to learn and rural leadership & management, Quality & Safety Management System, identifying right customer and developing competitive strategy, especially disruptively changing the way of doing extension service. The major lesson from this paper is to be able to perceive that business led extension service model change opportunity and to be able to grasp it while it is still possible to do achieve requirements of production.

References

Buffa, S.E. (1994). Modern Production/Operations Management. John Wiley & Sons: Toranto.

Dhanakumar, V. G., (2009). "Institutionalizing Tea Planter Centered Capacity Building: India's Tea Planters' Productivity Council (TPC) Perspective". International Journal of Extension Education, 5: 27-32.

Dhanakumar, V. G., (2006). "Integrating a Value in Supply Chain Management (SCM) for Agri-Business Led Extension Services (ABES) in India". International Journal of Extension Education, 2: 21-33.

Dhanakumar, V. G, (1999). "New Techniques for Plantation Management: TQM Perspective". The Planter's Chronicle, February, Pp.87-93.

Dhanakumar, V.G., (1995). "A Comparative Analysis of Institution Building for ARD in the Developing Countries". Quarterly Journal of International Agriculture, Berlin.

Dhanakumar, V. G., (1994). Staff Decision Making Patterns, Village Leadership Performance, and Local Institutionalization Processes in Agricultural and Rural Development Programs. Journal of Agriculture Food and Human Values Society,, 11 (2&3): 67-74.

Dickinson L (1987). Self-instruction in Language Learning. Cambridge: CUP.

Esman, M.J, (1980). Paraprofessionals and Rural Development. Cornell University. Centre for International Studies. Rural Development Committee.

Garten E. J, (2000). World View: Global Strategy for the New Economy. A Harvard Review Business Book: Boston.

Holec H (1988). "Student Autonomy — Learner Training and Self-Directed Learning". In S. Nicholls (Ed.), Current Issues in Teaching English as a Second Language to Adults. UK: Edward Arnold.

Indian Institute of Plantation Management (2009). "Annual Report on Tea Planters' Productivity Council (TPC). March.

Leibod et.al., (2002). Strategic Management in the Knowledge Economy: New Approaches and Business Applications, Wiley: New York.

Seven, Voelpel, et.al ., (2005). European Management Journal. 23 (1) : 37-49.

Wenden, A. (1991). Learner Strategies for Learner Autonomy: Planning and Implementing Learner Training for Language Learners. Hemel Hempstead and Englewood Cliffs, NJ: Prentice Hall.

White W. F. (1981). Participatory Approach to Agricultural Development. Cornell University. Centre for International Studies Report.

●●●

4

Role of Natural Resource Management in Livelihood with Special Reference to Non Timber Forest Products (NTFPs)

K. N. Krishnakumar* and Ganesh Yadav**
**Professor and **Program Manager, IIFM, Bhopal*

INTRODUCTION

Natural resources are the untamed natural essence of the environment which contain a significant amount of material as well as aesthetic values. Natural resources are broadly categorized into renewable, flow and non-renewable sources. While renewable natural resources primarily involve living sources such as forests, crops fish, reindeer, coffee, etc. and non-living sources like water and soil. Natural resource management refers to the management of natural resources such as land, water, soil, plants and animals, with a particular focus on how management affects the quality of life for both present and future generations. Natural resource management is congruent with the concept of sustainable development. It is a scientific principle that forms a basis for sustainable global land management and environmental governance to conserve and preserve natural resources (Wikipedia, the free encyclopedia, 2009). Though forest, water, agriculture land and crops, minerals, livestocks and many more constitute natural resources in India but forest represents the second-largest land use in India after agriculture, covering about 641,130 square kilometers, or 22 percent of the total land base (FAO 2005). Forestry is the second largest land use after agriculture and accounts for about 1.5 % of the nation's GDP (World Bank, 2006) in India. On one hand, forests support the livelihood of the poor, especially the tribals, and on the other hand, it is remarkable to see that the poorest people of our country live in the rich forest and natural resource base (Narain, 2008).

In pre-history, forest was believed to cover more than half of the global land area and benefited human life a great deal through hunting and collecting food, and over 90 per cent of the demand for forest resources was for fuelwood. After industrialization in the seventeenth century, however, much wood was used for shipbuilding, house-building. and other industrial products.In India, forest has always played an important role in providing sustenance to millions of rural people especially the tribal living in and around the forest areas. It has been estimated that about 275 million poor rural people in India depend on forests for at least part of their subsistence and cash livelihoods.As tangible output, forest produces two major products i.e timber and Non Timber Forest Products (NTFPs). The extraction of various NTFPs is very widespread, both within and outside protected forest areas.

Moreover, with increasing emphasis on participatory models of forest conservation, such as joint forest management and community-based conservation, in which NTFP extraction constitutes an attractive economic incentive for local people, the role of NTFP in forest conservation and livelihoods has gained significant impetus. The total present value of the NTFPs and services from tropical deciduous forest in India, based on the use option and existence values is estimated to vary from Rs.1,21,059 to Rs.1,99,870 per hectare (Chopra, 1993).

NTFPs and Livelihoods

According to the United Nations Food and Agriculture Organization (FAO 1997), it has been estimated that 80 per cent of the population of the developing world use NTFPs to meet some of their health and nutritional needs and several million households worldwide depend heavily on NTFP products for income. The example of India indicates that NTFPs contribute over 75 percent of total forest export revenue and add significantly to the income of about 30 percent of rural people, i.e. about 250 million people (ITTO 2006).However, the level of benefits that NWFPs can provide is location-specific and boom-andbust cycles for specific products impact income generation significantly. An estimated 67% - 90% of the tribal income is earned from collection of NTFPs (Bhattacharya and Faiz, 2004).Over 50% of forest revenue and 70% of forest export income come from NTFPs and the major source of both self-employment and indirect employment in forestry is in their collection, processing, and sale. Timber products have overshadowed

NTFPs as major commodities in modern times. However, the important contribution of NTFPS to food and resource security and to financial well-being is gaining increasing recognition. In fact, in some areas, the financial impact of NTFPs may be even greater than that of forestry. For example, a study in Zimbabwe revealed that small-scale NTFP-based enterprises employed 237,000 people, compared to only 16,000 employed in conventional forestry and forest industries in the same year (FAO, 1995).The annual value of direct contribution of NTFPs in India is expected to be about US$27 billion compared to about US$ 17 billion for wood products. NTFPS account for 70% of the forest based export earnings (US$ 500 million) and provides 55% of the total employment in the primary forest sector. Different policies, legislations and programmes of Government of India has empowered the communities the ownership rights of the NTFPs for collection, trade, value addition, marketing through national level legislation which is a milestone for poverty alleviation of 350-400 million people living around the forests. India has national level legislation for ensuring benefits arising out of the traditional forest related knowledge (TFRK). The GOI is now focusing on forest certification as well as certification facility to forest dependent communities for quality control of value added NTFP to explore better market. Around 10 million people are dependent on forests products for household use and cash income. Campbell *et al.* (2003) found that in 1986 NTFPs accounted for almost 40% of forest department revenues, 75% of net export earnings from forest sector revenues and 75% of net export earnings from forest produce. The opportunity for self-employment which NTFPs-based enterprises provide to the forest dwellers is estimated to be 3.3 million person years (Campbell *et al.* 2003a). Surayya (2000) found that in Andhra Pradesh as much as 39% of income for FD comes from NTFPs including fuel wood. Mishra et.al (2002) found that an average of 43.7 percent of FD received their subsistence income from trade in NTFP. Over 50 million people are dependent on NWFPS or NTFPs for their subsistence and cash income The contribution of NWFPs and eco tourism to the forestry sectors gross is valued at Rs. 259.85 billion.1.6 million person years of employment in India are from NWFPs.India exports a large number of NWFPS products to other countries earning foreign exchange revenue to the tune of Rs 10 billion (US$ 384 million) annually. People below poverty line often sustain their livelihood on these forests by collecting a number of forest products for their living. In tribal and marginalised agricultural areas, 70% of household earning depending on NTFP based resource. The World Bank (2001) estimated that 1.6 billion depend to varying degrees on forests for their livelihoods, with 350 million living in or near dense forests depending on hem "to a high degree". The importance of forest product income is usually more in the way it fills gaps and complements other income, than in its absolute magnitude or share of overall household income" (Byron and Arnold, 1999). In India, over half of its forest revenue and about 70% of export income is contributed by NTFP (Sekhar *et al.*, 1993). Among the NTFPs), *tendu* leaves (leaves of *Diospyros Melanoxylon*) used as wrapper for making *bidis* (country cigarettes) are the most important. The enterprise supports about 10 million people in the cottage industry of rolling the final product. For example, it is claimed that 1.6 million person years of employment in India are from NTFP while the forestry sector in total provides 2.3 million person years of employment (Shiva and Mathur, 1996).

NTFPs: Definition and Classification

Definition

Non Timber Forest Products (NTFPs) refers to a wide array of economic or subsistence materials that come from forests, excluding timber. Similar terms include "non wood," "minor," "secondary and "special" or "specialty" forest products. In simple term NTFPs represent income opportunities from forests and forestry that do not involve cutting down trees for wood products. In forests with low timber production potential, NTFPs represent the major actual or potential source of income.FAO adopted term non-wood forest products which eliminates forest services and which came from other wooded lands and trees outside forests. Define NWFP consisting of goods of biological origin other than wood, derived from forests, other wooded lands and ToFs.

The National Commission on Agriculture (NCA, 1976) has classified MFPs or NTFPs as follows:

1. Fibres and Flosses,
2. Grasses (other than oil producing), Bamboo canes,
3. Essential oils,
4. Oil seeds,
5. Tans and Dyes,
6. Gums, Resin, and Oleoresins,
7. Drugs, Spices and Insecticides,
8. Leaves,
9. Edible products,
10. Lac and its products and
11. Miscellaneous products,

National Importance of NTFPs

NTFPs are not only locally important but also nationally because of its economic, social and religious value, besides ecological importance. It can be noted that 40% of state revenues and 70% of net export earnings for India comes from forest produce and it generate income for 2 million person years of employment. Economic significance can be seen from the facts that 0.17 million forest dependent villages use NTFPs, it is source of major livelihood 147 million people and 160 million people are partly dependent for livelihood need — (Fuel, Fodder, Food), (Medicinal plant, grazing), (small timber, Bamboo), (leaves-cottage industries product).Owing to the increasing recognition of the importance of NTFPs as a means to sustain livelihoods of the rural poor, the Government of India (GoI), various state government, funding agencies, Non Governmental Organizations and research institution are trying to mainstream NTFP policies in a right manner which can support the livelihoods.

Annual Production of NTFPs in India

Altogether the annual total production of these NTFPs is estimated to be 13450.11 tons. Out of these, 4 NTFPs, i.e., bamboo (4717.6 tons), mahua seeds (697.6 tons), sal seeds (7097 tons), and tendu leaves (360 tons) constitute the bulk of annual production. The production of these is 12872.2 tons (95.7%). Annexure 3 gives a detail of the state-wise annual production of NTFPs in India. Table 4.8 gives the annual production of a selected number of 22 NTFPs.

Table 1 : Annual Production of NTFPs in India

SI. No.	Products	Annual Production (Tons)
1.	**Bamboo**	**4717.600**
2.	Eucalyptus oil	0.150
3.	Fibres and flosses	15.000.
4.	Ghatti and other gums	3.5000
5.	Grasses other than oil producing	80.000
6.	Kadaya gum	15.000
7.	Katha	5.750
8.	Lac	30.000
9.	Lemon grass oil	0.950
10.	Mahua oil	28.750
11.	***Mahua* seeds**	**697.600**

12	Myrobalans	132.250
13	Neem seeds	115.000
14	Other seeds	57.500
15	Others	0.300
16	Palmarosa oil	1.600
17	Resin from pines	45.000
18	Rosin	24.000
19	Sal oil	23.000
20	***Sal seeds***	**7097.000**
21	Sandalwood oil	0.160
22	***Tendu leaves IBidi leaves***	**360.000**
Total		**13450.11**

Source : Gupta and Guleria, 1982, pp. 133-134

Study conducted by Gupta and Guleria (1982) in their study found that NTFPs are under harvested and though have great potential of production and employment if proper strategies are adopted (Table 2).

Table 2 : Actual and Potential Production and Employment in Collection of the NTFPs in India

S. No.	NTFP Items	Production (tons)		Employment Man Years	
		Actual (in 1980)	Potential	Actual (in 1980)	Potential
1.	Fibres and flosses	5.5	49.5	14.4	94.0
2.	Grasses	350.0	525.0	1200.0	1800
3.	Bamboos and canes	1946.0	4330.0	49.0	110.0
4.	Essential oils	1.7	102.6	27.2	140.8
5.	Non-edible oilseeds	419.2	6670.8	109.0	1440.9
6.	Tans and dyes	187.2	290.0	21.2	33.2
7.	Gums and resins	91.2	175.5	87.0	145.7
8.	Lacs and tasar silk	232.3	334.9	83.7	127.5
9.	Drugs, spices and insecticides	2.6	4.6	55.6	102.7
Total		**3235.7**	**12482.9**	**1647.1**	**3994.8**

Source : Gupta and Guleria, 1982, pp. 133-134

Values of unrecorded contribution of NTFPs

NTFPs are collected by collectors from the forest , area as and when available. In most of the cases, collection goes unrecorded. Thus,value of NTFPs are left un enumerated due to lack of any systemic documentation of collection, storage and marketing. A study conducted Forest Divisions of Godhra, Baria, Junagadh and the Dangs in the state of Gujarat found that the average recorded contribution of the NTFPs for the period of five years from 2000-2001 to 2004-2005 to the Gross State Domestic Products was Rs.726 lakh at current prices. The unrecorded value of NTFPs was Rs.366/ha and the net benefit was Rs.46/ha (Table 3). The projected value of unrecorded NTFPs would be Rs.6940/- lakh for the state of Gujarat with forest area of 1896200 ha. It is interesting to note that the value of unrecorded contribution was nine times more than the value of recorded contribution.

Table 3 : Estimates of NTFPs collected free of charge by people and their value

Name of Forest Division	**Forest area (ha**	**Name of NTFPs**	**Households (HH) benefited**		**Value (Rs./ HH)**	**Total value of NTFPs**	**Cost of collection of NTFPs**
			%	**Nos.**			
Godhra	110591	*Madhuca indica* (Mahua) Flowers and seeds and Gum	30	59102	490	28959980	8865300
Baria,	88363	*Madhuca indica* (Mahua), Gum of *Anogeissus latifolia* (Dhaora), Acacia *catechu* (Khair) and *Acacia nilotica* (Babool), fruits of *Carissa carandus* (Karonda) and *Zyzyphus mauritiana* (Ber) and *Butea monosperma* (Palas) leaves	40	51288	513	26310744	15386400
Junagadh	29254	*Tamarindus indica* (Tamarind), *Manilkara hexandra* (Rayan), *Syzygium cumini* (Jamun), *Emblica officinalis* (Aonla), *Carissa carandus* (Karonda), *Momordica cochinchinensis* (Kantola/spiny gourd), *Zizyphus mauritania* (Ber), *Aegle marmelos* (Bel, *Terminalia bellerica* (Behada), *Sapindus trifoliatus*	8	12658	5122	64834276	8860600

		(Soap nut /Aritha) and bel leaves etc.,					
The Dangs	103535	*Madhuca indica* (Mahua) flower and fruit, *Terminalia bellerica* (Behada), *T chebula* (Harda), *Syzygium cumini* (Jamun) fruit, and Gum of *Sterculia urens, Anogeissus latifolia* and Acacia (etc)	15	5595	252	1409940	1678500
Total	**331743**				**1594***	**121514940**	**34790800**

(* Average of four selected forest division, Based on primary survey)

NTFP on an average contributes to as much as 25-30% of the cash income, while adding to the food security as well Satpura and Vindhyan natural regions region According to a rough estimate, the Satpura region produces about 1200,000 Quintal NTFP (excluding fuel wood, grasses and bamboo) valued at Rs 150 crore (Workshop Proceeding "Conserving Biodiversity in Satpuranchal for Enhancing Peoples Livelihood", August 2003). As an estimate , NTFPs contribute to the livelihoods of NTFPs collectors in Madhya Pradesh as in Boxl (HDR 2000).

Box 1 : Livelihoods of Non Timber Forest Produce (NTFP) Collectors

NTFP Contribution in Panna, Bundelkhand

A study of 53 households living on the outskirts of the Panna Reserve Forest in Panna district, revealed that the villagers earned a total annual income of Rs. 3.23 lakh which were sold to tendu patta co-operatives, collected Mahua leaves and nuts worth Rs. 98,000 which were either sold in haats or to traders in barter and sold fuel wood to the Majhgaon Diamond Mine Employees Colony worth Rs. 65,000 annually. While these households also had other income sources, the NTFP income of the households accounted for the 85% of the total income of 46 households in the village. On an average 2.7 persons from 46 families were engaged in NTFP Collection and marketing. Each family on an average earned Rs. 9450 from the forests and total incomes accruing to them annually was Rs. 11,000/-

Forest Produce Gathering in Mahakaushal

Collection of Non-timber Forest produce or NTFP. includes food, fuel, fodder, tendu leave, Mahua flowers and seeds as well as chironji or Achar seeds. Fruits like mangoes, custard apples, guavas and small berries are also abundant in this area. Some other forest produce gathered are the tuberous roots of indigenous trees which are used for eating/ consumption, some green leaves and herbs used as vegetables, leaves of Khakhra (Chevle/dhak), used by the ginger 'cultivators to cover the tender plants of ginger and for making plates and bowls by sticking the leaves together with tiny wood pieces or scraping. For those people who have access to forest and NTFP, say, typically, a household of seven persons, with four of them engaged in NTFP collection, is able to collect tendu leaves worth 200 rupees per day for seven to ten days in a par i.e., an annual income of fourteen hundred to two thousand rupees. "Similarly, for mahua seeds and flowers, the collectors get about the same amount if money over 30-40 days in a year. On an average, they collect mahua flowers and seeds from about 4- 5 trees and each mahua tree yields about Rs. 500/- worth of NTFPs per year. Hence Rs. 2000 — 2500 is the average income from mahua. Other NTFPs and fuel wood, fodder and foods fetch a household with four collectors yield worth about Rs. 1000/- per year. Hence on, an average, a household with four collectors collects NTFP, worth Rs. 4400 to 5500 per year. It needs to be stressed that all households in a villages do not collect NTFPs. It is the people at the subsistence level of livelihood (generally SCs and STs), who collect NTFPs.

Source : HDR 2002

Some Examples of NTFPs Based Enterprises

Local people earn income which are often known as primary processing, secondary processing based enterprises. Some of the prevalent enterprises are Bamboo based sticks and furniture making ,Mahua selling through the SHGs, Nursery Development, Mushroom Jam. Jelly/Achar and Papad making, Plantation of the forest based species viz. Subabbul, eucalyptus, Lantana furniture making, Nicanthaus basket making, Rope making, Natural Dye making, Sisal Processing, Fibre Spinning weaving, Medicinal herbs processing unit, Lac processing, Mango pickle, Char seed decorticato, Leaf Cup and plates maker, Grinding /pulverizing of herbs and other materials- Ber powder, Satawar powder, Aonla powder, Triphala etc., Aromatic oil extraction, Bio fuel Expellers Mosquito Repellent Unit based on Nirgundi (Vitex negundo) and Mahua Oil cake, Sericulture Fibre spinning, Honey processing etc.

Strength, Weakness, Opportunity and Threats

NTFPs as natural resource have strengths, weaknesses, opportunities and threats as for as its management is concerned.

Strengths

NTFPs are a traditional and important part of rural livelihoods with which the majority of people are familiar (acceptability) and have ability to generate both rural employment and income. It is capable of providing access to raw materials independent of private landholding. There is national market demand greater than production at present. NTFPs products fits with labour availability for small scale fanners as well contributes to diet, health and income needs. It is not only income generating but also act as buffer during famine and household crisis..

Weakness

Though strengths are enormous ,NTFPs suffer from weaknesses. There is difficulty in developing bye-back arrangements without minimum support price. NTFPs lack resource inventory and product potential in different areas, suitable storage facilities at village level. Burden of taxes on NTFP products (VAT and mandi tax) are there.Most of the products are harvested unscientifically leading to degradation of resources. In comparison with other category of products , the sector has got insufficient promotion and advertising . The dealing agencies like forst department, Minor Forest Produce Federation and others s are lacking policies, procedures, criteria, etc. that can help guide them as to how to utilize their resources and people for maximum effect and efficiency .

Opportunities

In spite of the several weaknesses and drawbacks, the NTFPs have f local level information centers to inform collectors of market prices and link buyers with collectors which could also provide training in improved harvesting/ cultivation techniques and advice through experts. There is a database of raw material and finished product availability. There is opportunities for building up the institutional capacity of local NGOs and entrepreneurs. There exists a large number of JFM committees through which to work. Recently, due to expansion of aurveda and yoga therapy and herbal cosmetics, herbal markets are increasingly getting importance worldwide.

Threats

Productivity of NTFPs are determine by many factors. Therefore, yield varies from year to year, due to climatic factors, which increases insecurity in income level. Introduction of large processing units could suppress rural-based micro-enterprise development initiatives. Despite of several policy reforms in favour of community say in decision amking, strong control still exercised by Forest Department within forest areas from which the majority of NTFPs are source.Industries are by and large strongly economically motivated and exploit collectors. Productivity of NTFPs are declining due to over harvesting and inappropriate management and development of natural habitats.

Need for Extension efforts for Sustainable NTFP based Livelihoods

NTFPs based livelihoods development requires specific approach and well defined strategies for extension. It has been seen that the extension approach which is site specific, addresses local solutions and involves all stakeholders (participatory) and integrate social, economic, ecological and technology and skills are most effective. Strategies are most important at implementation level.The strategies which have included awareness generation among community members/local traders/local entrepreneurs using personal approach,mass campaign and PRA / Focused group discussions methods have been far effective than any other conventional methods. For successful program implementation, trainings to field foresters/ Community leaders / Local traders/ SHGs / JFMCs members using demos, case studies,f ield visits have far reaching impavts as far as skill development and its applications are concerned. At Planning & Strategy Level regional workshops, facilitating decision making — providing data access through IT, supply of multimedia kits, Training of Trainers, Training on sustainable NTFP management techniques, Training on Extension techniques / PRA are found to be useful in forest reach area conservation and NTFPs based activities.

Conclusion

NTFPs are more valuable than timber in terms of its economic, social, ecological and economic significance worldwide. The wild origin of NTFPs are free available to local poor collector for household use and sell in the local market. Semi processed/processed products are exported which contributes to the foreign money resource base and also add to the national GDP. Thus it not only provide income and sustenance to the local people living in the remotest area , but also provide raw materials to the big industries and pharmaceuiticals. In spite of the many advantages it has, the NTFPs resources are degrading day by day due to over exploitation and improper management strategies. To reverse the process of over exploitation and unsustainable harvesting, there is felt need of suitable extension approach and strategies for its management and livelihoods development.

References

Arnold, J.E.M. & Stewart, W.C., 1991. Common Property Resource Management in India. Tropical Forestry Papers No 24. Oxford, Oxford Forestry Institute.

Arnold, J.E.M., 2001. Forestry, Poverty and Aid. CIFOR Occasional Paper No. 33. Center for International Forestry Research (CIFOR), Bogor, Indonesia.

Bhattacharya P, Hayat S. F (2004). Sustainable NTFP management for rural development: a case from Madhya Pradesh. International Forestry Review, 6(2):167-8.

Campbell, J Y (1994). Putting people first. NWFPs and the challenge of managing forests to enhance local income. J Non Timber Forest Products. 11-2: 102-107. As cited by CAMPBELL et.al 2003 Non-Timber Forest Products: Availability, production, consumption, management and marketing in Eastern India on line cited on 21st April 2005, available at www.frp.uk.com/dissemination_ documents/FTR.pdf.

Chopra, K., 1993, The Value of Non-Timber Forest Products: An Estimation for Tropical Deciduous Forests in India, Economic Botany 47 (3): pp 251-257.

FA0.1997.State of the World's Forests 1997. FAO, Rome, Italy.

Food and Agriculture Organisation (FAO) 1995. *Non-wood forest products* for rural income and sustainable forestry. Non-wood Forest Products 7: 1-2, Rome.

FAO. 2005. Global forest resources assessment 2005. FAO Forestry Paper 147. /Rome, FAO.

Gupta, T. and Guleria, A. 1982. Non-wood Forest Products in India. Economic Potentials. CMA Monograph,No 87. Oxford and IBH Publishing Co. Pvt Ltd. New Delhi.

HDR, Madhya Pradesh.2000.

International Tropical Timber Organization (I1170). 2006. *Status of tropical forest management* 2005. Yokohama, ITTO.

Mishra , M., Surayya, T., Mishra. R. 2002. Sustainable harvesting, Value Addition and Marketing of selected Non-Timber Forest Products A case study of Koraput and Malkangiri Districts, Orissa State, funded by RCNAEB, Bhopal.

Narayan, D., Chambers, R., Shah, M.K., Petesch, P. 2000. Voices of the Poor. Crying Out for Change. Oxford University Press. New York.

Narain, S., Singh, S., Panwar, H.S., Gadgil, M. 2005. The Report of Tiger Task Force. Joining the Dots Published by Project Tiger, Union Ministry of Environment and Forests, GoI.

National Commission on Agriculture (NCA, 1976).

Sekhar, C., Rai, R.S.V. and Surendra, C.1993. Price Regime Analysis, Marketing and Trade of Minor Forest Produce. Dehradun: CMFP.

Shiva, M. P. and Mathur, R.B.,1996. Management of Minor Forest Produce for Sustainability, Oxford and IBH Publishing Co. Pvt. Ltd., New Delhi.

Surayya, T. 2000. Dependence of Forest Dwellers on Fuel- wood and Non-wood products for their survival and pertinent Marketing Issues. In: JHA, L.K., RAMANUJAM. S.N., LALRAMNGHINGLOVA, H., and SINGH, L.N. (ed.) Agro-Forestry and Forest Products. Monograph ISBN-8186129, Volume 1.233 pp.

Wikipedia (2009). The free encyclopedia.

World Bank.2001.

•••

5

Women Empowerment — Rhetorics and Reality

K. Thangamani
Professor and HOD, Department of Home Science Extension Education, Avinashilingam University for Women, Coimbatore

Introduction

The status of women in India has been subject to many great changes over the past few millennia. From equal status with men in ancient times through the low points of the medieval period, to the promotion of equal rights by many reformers. The history of women in India has been eventful. In modern India, women have adored high offices in India including that of the President, Prime Minister, Speaker of Lok Sabha, leaders of opposition etc. The incumbent president of India is a woman but the same vast majority of the women suffer from the great discrimination and even the basic human rights are denied to them.

History

With regard to status of women in Indian society at large, no nation has held their women in such a high esteem than India. Perhaps no other literatures have presented more admirable types of women character than Sita, Maitreiyi and others. It dates back to Indian mythology of thousands of years to witness this status of women during the vedic period which was honorable and respectable. The marriage was regarded as sacrosanct and the family ideal was high. Attainment of women in the intellectual field is to be inferred from the fact that some of the hymns were attributed to female Rishis. In fact Vedic age women had sufficient freedom to attend fairs, festivals and assemblies. There was no mention of Purdah System.

During the post Vedic period, women started loosing the status in the society. She lost her independence. She became the subject of protection. It was thought that women were pre-ordained for procreation and they had no other function. During the Mugal Period the socio-economic status of women was very much lower and she had to depend on male for every activity. Social evil like Pardah system came into existence, Child marriage was prevalent.

Lack of education, early marriage, non-existence of employment opportunities, absence of absolute property right were the main causes of inequality of sex in the socio economic field. Incidence of female infanticide and customs of Sati came into existence.

Status of Women during Freedom Struggle

It was Mahatma Gandhi who recognized women as the nation's most precious human resource and paved the way for women empowerment in the independent India with his vision on role of women in nation building. He recognized the role of women individually in the family and as a group in the society could bring changes in the socio economic and psychological well being of the people. As early as 1950 under the Community Development Programme the Mahila mandals were formed in rural areas and which were pioneer to todays. SHG movement in the country. Poor women in India first began to draw into a reform movement by Mahatma Gandhi. Under the movement for freedom, Mahila Mandals and Mahila Samajams were formed all over the country. After independence, with the launching of Community Development Programmes Mahila Mandals and Samajams assumed significance in the process of development.

Status of Women after Independence

After Independence, the framers of the constitution realized the importance of giving equal status to women and gender equality is enshrined in the constitution in its Preamble, Fundamental Rights, Fundamental Duties and Directive Principles. The constitution not only guarantees equality of women but also empowers the state to adopt measures of positive discrimination in favor of women.

The five year plans in India laid special emphasis on the development of women through different approaches for development of women through health, education/literacy, housing, gender budget etc. Since 1980s, the Government of India has shown increasing concern for women issues through a variety of legislation promoting the education and political participation of women, International organization like the World bank and United Nations have focus on women's issue especially the empowerment of poor women in rural areas. Since the 1990s women have been identified as a key agent of sustainable development and women's equality and empowerment is seen as central to a more holistic approach towards establishing new patterns and proCiess of development that are sustainable. The World Bank has suggested that empowerment of women should be a key aspect of all social development programmes.

Women Development in Five Year Plans

- First Five Year Plan: Women as mothers and women's welfare.
- Second Five Year Plan: Hygiene and health care of women and the problems of women workers, equal pay for equal work and expansion of opportunity for part-time employment.
- Third & Fourth Five Year Plan: Education of women.
- Fifth Five Year Plan: Integration of welfare with socio economic development services, family planning and nutritional care of expectant and nursing mothers.
- Six Five Year Plan: Women as a target for welfare; women and development approach; Indira Awas Yojana; 40 percent reservation of women in all anti-poverty programmes; exclusive credit schemes and skill training programme for women's employment.
- Seventh Five Year Plan: The concept of women empowerment and establishment of Women Development Corporation; Women Study Centers; elementary education of rural girls; technical and vocational education.
- Eighth Five Year Plan: Economic liberalization; draw women into political process through constitutional amendment; mainstreaming of women.
- Ninth Five Year Plan: Empowerment of women; focusing on growth, equity and participation; women component plan and gender mainstreaming through women specific policies and programmes.
- Tenth Five Year Plan: Poverty reduction; primary education; National Rural Health Mission; PURA (Providing Urban Facilities in Rural Area); IFAD (International Fund for Agricultural Development); micro- credit.

homogeneous categories; they belong to diverse castes, classes, communities, economic groups, and are located within a range of geographic and development zones. Consequently, some groups are more vulnerable than others. Mapping and addressing the specific deprivation that arise from this multiple location is essential for the success of plan interventions. Thus apart from the general programme interventions, special targeted intervention catering to the differential needs of these groups is undertaken during the Eleventh Plan.

In short, it is claimed that the emphasis on the status of women has shifted from welfare in the 50s to development in 70s and, now to empowerment.

SHG Movement

Outside the frame of Five Year Plans, the SHG movement started as a pilot project in 1980s contributed greatly to empowering women and making them aware of certain fundamental civil rights. It transformed into a variety of governmental and non governmental schemes of rural development. The policies, programmes and schemes undertaken by different sectors of development have today largely contributed to empowering women for better participation in development. SHG approach is a powerful instrument of women empowerment and acts as a catalyze to social change this has been proved by many critically analyzed studies. SHGs have been recognized as a useful institutional strategy to help the poor and as an alternative mechanism to meet the urgent credit needs of poor through thrift.

National Policy for Empowerment of Women 2001.

However, it was realized that there existed a wide gap between the goals enunciated in the plans on the one hand, and the situational reality of the status of women on the other. There was a need for a specific policy document to fix time bound goals and monitor them effectively. So, a National Policy for Empowerment of Women was formulated in the year 2001. It fixed goals under the various heads like judicial and legal system, education, health and nutrition, political participation, decision making and economic empowerment.

All Central and State ministries have to draw time bound action plans for translating the policies into set concrete goals through participatory process of consultation with state/central departments. The Policy significantly included the following:

1. Measurable goals to be achieved by 2010.
2. Identification and commitment of resources.
3. Responsibilities for implementation of action plans.
4. Structure and mechanism to ensure efficient monitoring.
5. Introduction of gender perspective in budgeting process.

It is also reiterated the following socio demographic goals set in Population Policy 2000.

1. To reduce the infant mortality below 30/1000 live births by 2010.
2. To reduce the maternal mortality below 100/1, 00,000 live births.
3. To attain 80 percent of institutional delivery.
4. To promote delayed marriage of girls, not earlier 18 years, preferably after 20 years of age.

Framework for Empowerment

Empower' means "to invest with legal power" and "to enable". The concept of empowerment is from the bottom up approach to development. It is a process involving collective decision making, collective actions and popular participation. Empowerment refers to increasing the spiritual, political, social or economic strength of individuals and communities. It often involves the empowered women developing confidence in their own capacities.

Empowerment is a multi dimensional social process that helps the people to gain control over their own lives. It is a process that fosters power in people to use in their own lives, their communities and in their society by acting on issues they feel as important.

Women empowerment concept was introduced at the International Women's Conference in 1985 at Nairobi. Women empowerment principally aims at enhancing their social functioning by quantitative and qualitative change particularly in the fields of education, health and employment, which will bring the desired level of change. The government and voluntary organizations are providing many income-generating schemes to raise economic status of women.

Social empowerment of women is one of the essential aspects of women's development programme. It envisages equal status, participation and decision making power for women at the household level and also at the community and village levels, by overcoming of social, cultural and religious barriers in their day to day affairs and in matters concerning them.

Globally, as well as locally, the position of women in the family, society and in the place of employment is gradually becoming an issue of concern. At late, there is a growing awareness for the development of women in general who constitute roughly half of the population, both at national and international levels. In the context of national development, women's participation in economic activity is of great importance.

To enable the detailed examination of the fact relating to the concept of women empowerment, a scientific approach- is necessary. Women empowerment was viewed as a myth in the male dominated society and welfare approach was considered as an appropriate means of achieving the objectives. Empowerment was identified or confused with women's liberation of the west and was thought to be not fitting in with the culture and traditions of India. Empowerment was considered as a slippery concept to define and measure and different perceptions were stated by research and development practitioners. In an attempt to conceptualize empowerment, empowerment means giving opportunities to people for developing a sense of autonomy and self confidence in managing their lives. Empowerment is the process by which women groups become more independent, particularly in economic matters and have increased power for the decision making in the family well being as well as that of society. The rural women perceived empowerment as enabling them playing more effective role in economic production without detrimental to reproductive and community management roles by meeting their practical gender needs which included opportunities for income generation, better living amenities and improved social conditions. The index of empowerment consists of eight different parameters such as self perception, perception of role of women in the society, economic independence, decision making, and innovativeness, attitude towards group action, communication and desire to improve living conditions.

All efforts discussed above, have contributed in one way or the other, to empowerment of women. The myth of or stereotyping the roles for women no longer exist among the educated sections. Women's safety has been an important concern providing them opportunities for growth. Maternal and child health along with drudgery reduction is high in nation's development agenda.

Having discussed the policy statements and goals set and also about the framework of NPEW it is necessary to look into, how far we have been successful to realize the goals and understand the gap.

Achievements

Socio Demographic

The Maternal Mortality Rate came down from 667 per 1,00,000 live birth in 1980 to 450 in 2010.

The life expectancy for female rose from 40.6 years in 1970-71 to 67.17 years in 2010. The total fertility rate came down from 5.97 in 1951-61 to 2.72 in 2010. The female literacy rate rose from 8 % in 1947 to 54.16% in 2008.

The female work participation increased from 14.2% in 1971 to 35% in 2008. The relationship between educational level of women and employment showed a 'U' shaped curve, suggesting high level of participation is seen among the illiterates and highly educated women; low level of participation for the primary and middle educated women.

Political participation

The 73rd & 74th Constitutional Amendment are the milestones on the road to empowerment of women. There are 76,200 women chairpersons at the local governance; 65 women in senior positions in the Indian Foreign Service, one of the premier civil service. The highest post in the Foreign Service, the External Affairs Secretary, is occupied by a woman.

Among the rural women also, perceptible changes are happening due to SHG movement. In one of the studies conducted by IFAD, it has been reported that

- 15% of the women group members had visited new places and travelled long distances.
- 94% had experienced new interaction with the staff of institutions such as banks, district and block officials and NGOs.
- There seemed to be an improvement in women development in the household decision making of male headed households on issues such as credit, creation and disposal of assets, children's health, education and marriage.

Reality

So far, we have seen the goals set and some achievements made in women development. However, if you look at the realistic situation, there exists a wide gap between what is said in policy pronouncement and what is achievement. We have a long way to go, to realize, what, we near in rhetoric.

Though we fixed a target of reducing the IMR, by 30/1000, in 2010, still it remains as high as 53/1000 live births. Here again there is a gender variation. IMR for female child is 52.4 and male is 49.3. This indicates that there is a bias towards the male child. The sex ratio, *i.e* the number of females per thousand male remain as 933, which is unfavorable 'to women. Preference for male child and neglect of female are the primary reasons for the lower sex ratio. Sex selective abortion using even the ultra sound prenatal test devices have also contributed to this malady.

The MMR indicates the health and nutritional status of women. In India, it is still as high as 453, which is one of the highest in the world. Lack of appropriate care during pregnancy, in child birth, utilizing the services of untrained mid wife, lack of motivation on the part of the family members to take medical advise during pregnancy are the important reasons for higher mortality rate.

Literacy

Though the literacy rate increased from 8.86 to 54.1% in 2006, still half of the women are illiterate.

Decision-making

Even in political empowerment also, though women have started playing a major at the local governance, still, the act to pass 33% of women reservation in parliament and assemblies is being debated for more than 10 years. It is still doubtful; whether the act will be passed at all.

Women over work

It has been estimated that in one of the micro level study conducted, that in one hectare of farm, bullocks work for 1064 hours, a man works for 1212 hours and a woman works for 3485 hours per year. Women's work is less visible because it is non monetized. Women account for 60% of unpaid female family labour in the farm and 98% of domestic work.

Even among the educated career women most of them play a dual role of home maker as well as career women, with a little input from the male head of the household.

Ownership of property

In spite of the entire multiple role-played by women, they owned only less than one per cent of the property. Among all the states, Tamil Nadu was pioneering in conferring equal rights to sons and daughters in the family property. This was followed by a central legislation. It was found in Kerala that women's risk of physical violence from the husband is dramatically reduced, when wife owns a land or a house.

Violence against women

In recent years, there has been alarming news on atrocities against women in India. Every 26 minutes a woman is molested; every 34 minutes a rape takes place; every 42 minutes sexual harassment incident occurs; every 43 minutes a woman is kidnapped and every 93 minutes a woman is beaten to death.

Mainstreaming

As it was already that the development writers often used to repeat the focus of women in India has shifted from "welfare" in 1950s to "development" in 1970s and now to "empowerment". This is hardly borne out in the programme on the ground. There are largely schemes for relief like old age and widowhood rather than mainstreaming the gender in the development programme.

National Institute of Public Finance and Policy undertook the first ever gender budget exercise. It had found the following proportion of expenditure on women development schemes:

- 67% for protective and welfare services.
- 26% for social service, health, education, water and housing.
- 4% for economic resources and self employment training.
- 3% for awareness generation and regulatory service.

Allocation for women directed schemes are pitiful. Only ten Ministries have specially targeted schemes for women in India. No adequate administrative mechanism for execution and monitoring has been evolved.

Negative effect of Globalization on women

In most cases, free market economy has excluded the women from development processes since many of the traditional employment opportunities have been taken over by the mechanized companies. Many of these multi national companies in the export enclaves hire pot:Jr and uneducated women because they are flexible labor and they can be laid off at any time.

Conclusion

Gender Development Index is a composite index like Human Development Index developed by United Nations to measure inequality between man and woman in the development process. India rank 133 out of 157 countries, for which GDI has been worked out. This in short shows that there is a long way to go between the rhetoric and ground reality. We, the extension scientist, do have the responsibility to make the rhetoric become reality.

- How to make our curriculum in higher education include gender issues?
- How to make our research thrusts to focus towards gender mainstreaming?
- How to reach policy makers with our recommendations towards gender mainstreaming through Society of Extension Education?

These are some of the challenges we face and need to work out solution.

References

http://en.wikipedia.org/wiki/women in India.

Empowering rural women through extension education-An action research in a fishing village, Ph.D thesis, Avinashilingam University for Women, Ciombatore,1992.

hppt:// www. fad.org/gender/I earn i ng/rol e/I abour/i n _tam i I .htm

hppt:// www.merinews.com/shareArticle.do?detail=print&articleID=124708

hppt://ezinearticles.com/?Women-Epowerment---Myth-or-Reality&id=916590

Development in Practice, Improving Women's health in India, 1996, the World Bank, Washington, DC, USA.

India, 2002: A reference annual complied and edited by Research, Reference and Training Division, Publications Division, Ministry of Information and Broadcasting,Government of India.

Jean Dreze and Amartya Sen: Economic Development and Social Opportunity, 1996, Oxford University Press. Delhi.

•••

Section II

Extension Approaches

6

Agricultural Education through Internet

H. Philip[1], P.P.Murugan[2] and P.Mooventhan[3]
[1]Professor (Agricultural Extension) and[2] Associate Professor (ATIC), Directorate of Extension Education, Tamil Nadu Agricultural University, Coimbatore and[3] Scientist (Veterinary Extension), National Research Centre on Meat, Hyderabad.

Introduction

Computerization in education is perpetual need of today. Computers store large quantum of data that can be retrieved and used wherever necessary. They can also be used as an interactive media in transfer of technology system, which aids farmers and extension workers to access in formation of their need and to communicate with each other. Positive advantages in web education over distance learning process are that farming community can access it from any part of the world. Web education has also the added advantage of updating its data base on weather conditions, market situation in hourly basis which is not possible distance education as the process need a long time with regard to the availability of human resource and static course material to be developed for a particular time frame. The distance learning process being conventional in nature and also in place for quite a long years has abundance of resources and infra structural facilities however in web education, the professional resources are less in number and still arm journalism is yet to gain popularity among them. So web based interactive knowledge network system in transfer of technology is a present need which has to be addressed with due attention to improve the quality of existing communication system.

Methodology

The research methods adopted in delineating the locale of study, target respondents, sample and sampling procedures, collection of data and farmers need analysis and statistical analysis used were described here.

1. Locale of study

As per the objective of the study, Video production centre, Tamil Nadu Agricultural University, Coimbatore is the apex centre for implementing and monitoring the scheme. With the approval of university, Anbil Dharmalingam Agricultural College & Research Institute, Trichy and Forest College & Research Institute, Mettupalayam were selected as sub-centers.

2. Sample and sampling procedures

a. Selection of Sub centres

Coimbatore and Trichy districts were selected purposively for the study because these sub centres have network and video conference connectivity with Tamil Nadu Agricultural University, Coimbatore, This also facilitates the farmers for further conduct with the Sub centres.

b. Selection of Villages

Lists of villages with computer and interne connectivity were obtained from the collectorate of Trichy and Coimbatore districts. From the list of villages, six villages viz., three villages from Coimbatore district and three villages from Trichy district were selected randomly. Accordingly these villages will be the study area of the project.

c. Selection of Respondents

The lists of farmers from selected villages were obtained from local administrations. About one hundred and twenty farmers from six villages of both the sub centres @ 20 farmers per village were selected at random .

Table 1 : Distribution of Respondents

SI.No.	District	Identified Villages	No. of farmers identified
1.	Trichy	Navallur Kuttapattu	20
		Puthanatham	20

		Anthanallur	20
2.	Mettupalayam	Odanthurai	20
		Maruthur	20
		Chikkarampalayam	20
		Total	**120**

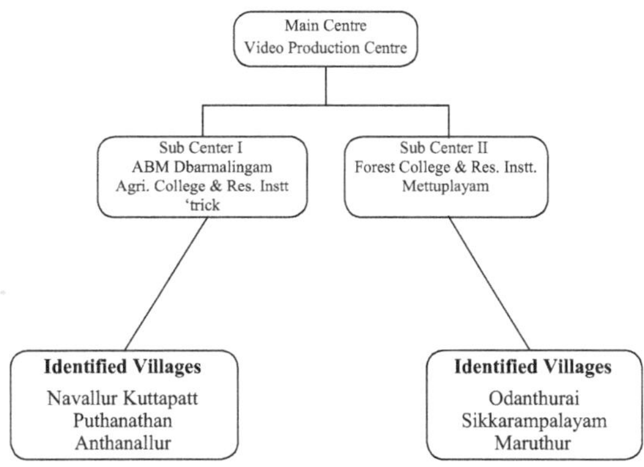

Fig. 1 : Selection of Villages

Findings and Discussion

a. Computer utilisation pattern

The results pertaining to the distribution of the respondents based on Computer Utilisation Pattern reveals that 66.7 per cent of the respondents had not utilized computers so far and the remaining respondents (33.33%) were found to use computers.

b. Awareness of agricultural web sites

Web education awareness was created among the respondents through awareness campaigns in all the selected six villages at Trichy and Coimbatore districts. The results pertaining to the distribution of the respondents with regard to the Awareness of agricultural web sites reveals that majority of the respondents (95.84%) were not aware of any single agricultural web sites and a least percentage of farmers (4.16%) were found with the awareness of agricultural web sites. The Internet can provide a large number of services like E-mail, www etc. A lot of web sites were available for the use of farmers. TNAU has also its own website but least awareness about the agricultural web sites shows that still more computer training needed at the farmers level.

c. Computer training

Established computer facilities with Internet connectivity at the subcentres facilitated the identified group of farmers to access information from the sub-centres and one day computer trainings was given to the farmers. One day computer training was organized in the selected 6 villages of Coimbatore and Trichy districts under this project. The training was provided as part of the project work. Most of the respondents those who have participated revealed a good feedback about the computer (Table 3.). It could be observed that all the farmers gained confidence about computer handling. One important feature could be noted from the finding that farmer participants from Trichy (100%) and Coimbatore (75%) have reported about elimination of computer phobia due to the training programme. The results on the effectiveness of computer training among the farmers of selected districts of Tamil Nadu are presented in the Table 2.

Table 2 : Effectiveness of Computer Training among Farmers of Selected Districts of Tamil Nadu

SI. No.	Particulars	Respondents (Percent)	
		Trichy District	Coimbatore District
1.	Got awareness about computer	80	65
2.	Elimination of Computer phobia	100	75
3.	Learnt about agricultural web sites	95	100
4.	Motivated to learn further through computer	65	75
5.	Gains self confidence in computer handling	100	100

It could be seen from the Table 2 that all the respondents in Trichy and Coimbatore districts have gained self confidence in computer handling. All the respondents (100 %) in Coimbatore district and 95 per cent of the respondents in Trichy district have gained knowledge about agricultural websites as the result of computer training. Almost all the respondents (100%) in Trichy district and three fourth of the respondents (75%) in Coimbatore district have eliminated computer phobia with the help of computer training.

Nearly 80 per cent of the respondents in Trichy district and 65 per cent of the respondents in Coimbatore district have possessed awareness about computer. Further it is seen that, three fourth of the respondents

(75%) in Coimbatore district and 65 per cent of respondents in Trichy district were motivated to learn further through computer. It is interesting to note that, even though the respondents were self confidence in computer handling and eliminated computer phobia through training, further awareness and motivation is needed to implement effective computer training among the farmers.

d. Knowledge gain

The knowledge levels of farmers groups in Odanthurai, Maruthur and Chikkarampalayam of Coimbatore district and Navallurkuttapattu, Puthanatham and Anthanallur of Trichy district were studied who were educated through web based interactive knowledge network system. The knowledge levels of the subjects before and immediately after exposure were assessed to find out the knowledge gain. The result of the selected groups shown that all the farmers group of both the districts gained knowledge. From the result the Maruthur group had highest (80.45%) knowledge gain followed by Anthanallur (78.50%), Chikkarampalayam (76.50%), Odanthurai (61.55%), Punthanatham (60.05%) and Navallurkuttapu (56%). (Table 3).

Table 3 : Mean Knowledge Gain due to Exposure to the Treatments

SI. No.	Groups	Mean knowledge gain		Mean knowledge gain	%age
		Pre-exposure	Post-exposure		
	Coimbatore district				
1.	Odanthurai	9.82	22.12	12.31	61.55
2.	Maruthur	5.12	21.21	16.09	80.45
3.	Ch ikkarampalay am	4.82	20.12	15.30	76.50
	Trichy district				
1.	Navallurkuttapa ttu	10.12	21.32	11.20	56.00
2.	Puthanatham	8.52	22.13	12.01	60.05
3.	Anthanallur	4.32	20.02	15.70	78.50

The post exposure mean knowledge gain in all the villages are more or less same but the Maruthur, Anathanallur and Odanturai got maximum score in the pre exposure knowledge test. It might be due to the influence of agricultural institution located very nearer to the village. This shows the effectiveness of web education among farming community to gain knowledge on latest crop technologies.

5. Difficulties faced by the respondents

Though the one day training was found effective; the farmer respondents expressed some difficulties (Table 4). All the farmers expressed that the duration of the training was insufficient. Since most of them had maiden exposure on computer they took more time to get it familiarised with the computer. Further 75 per cent from Trichy district and 70 per cent from Coimbatore district expressed their difficulties in handling the computer mouse. Fifty per cent of the farmers found difficulties to understand the English language used in computer operation. Forty five percent of the farmers expressed that they found it very difficult in understanding the terminologies used in tool bars. The results on the difficulties expressed by the respondents on computer training among the farmers of selected districts of Tamil Nadu are presented in the Table 4.

Table 4 : Difficulties Expressed by the Respondents

SI. No.	Particulars	Respondents (Percent)	
		Trichy District	Coimbatore District
1.	Difficult to understand English	50	50
2.	Complicated terminology about computer	45	45
3.	Difficult to handle computer mouse	75	70
4.	Insufficient training duration	100	100

A thorough perusal of Table 4 revealed that all the respondents (100 %) in Trichy and Coimbatore districts expressed insufficient training duration as their major hurdle in learning computer. One day training was given to the respondents to get aware of the basic operations in computer. This may be the reason for the respondents to express the constraints regarding the duration. Further to it, about three fourth of the respondents (75 %) in Trichy district and slightly less than that (70%) of the respondents in Coimbatore district expressed difficulty in handling mouse. Half of the respondents (50%) in Trichy district and the same in Coimbatore districts perceived difficulty in understanding English language since majority of the farmers were educated up to middle school level only. Next to this about 45 per cent of the respondents in Trichy and Coimbatore district found the terminologies in computer are complicated and difficult to understand.

Even though the respondents faced difficulty in understanding English language and handling computer mouse, all of the respondents expressed insufficient training duration as their problem. If the training periods extended further, we can expect still more positive changes ih computer learning.

6. Relationship and influence of the independent variables towards knowledge gain

Simple correlation coefficient

In order to understand the association between the selected characteristics of the farmers and their knowledge gain due to web education, simple correlation was computed and the results are furnished in Table 5.

Table 5 : Relationship of Independent Variables towards Knowledge Gain

Variable No.	Variables	`r' value
X_1	Age	-0.915**
X_2	Educational status	0.640**
X_3	Occupation status	0.396NS
X_4	Farm status	-0.218**
X_5	Farming experience	-0.347 NS
X_6	Annual income	1.110 NS
X_7	Mass media exposure	2.748 **
X_8	Computer utilization pattern	-0.864 NS
X_9	Awareness about Agricultural websites	2.400 NS

NS — Non Significant * * Significant at 0.01 Probability leyel

It is observed from the Table 5 that out of nine characteristics studied, two characters were found to be having a positive association with knowledge gain. The characters namely educational status and mass media exposure were found to be positively associated with knowledge gain at 1 per cent level of probability. Age and farm status and farming experience were found to have a negative association with knowledge gain. The other variables namely occupational status, annual income, computer utilization pattern and awareness of agricultural website, were found to have no association with knowledge gain. It is an accepted fact that with higher education and mass media exposure of the respondents to help in increase of knowledge gains. The variable age farm size and farming experience were having negative association with knowledge gain. The increase in age might have decreased the learning capacity of the farmers. A similar trend of increased farm size would have decreased the knowledge gain.

7. Constraints faced by the farmers

It is necessary to study the constraints in any social research. The constraint analysis helps the planners, administrators, development workers, scientists and others to frame policies, and to implement schemes. Hence the constraints faced by the farmers in utilizing web-education technology were studied. The multiple responses of constraints was worked out and the results are presented in Table 6.

Table 6 : Constraints Faced by Web -Education Learners

Sl. No.	Problems	Number	Percentage
1.	Small font size	111	92.50
2.	Dull text color	105	87.50
3.	Less number of photos	85	70.83
4.	Difficulty in handling of computer mouse	95	79.16
5.	Slow downloading and uploading of photos/files	99	82.5
6.	Monotonous font type	77	64.16
7.	Confusion in operating hyperlinked text	62	51.66
8.	Initial operational instructions are in English	42	35.00

The result could be observed from Table 6. The major problems were small font size expressed by the respondents (92.5%), Dull text color (87.5%), Less number of photos (70.83%), difficulty in handling of computer mouse (79.16%), slow downloading and uploading of photos/files (82.5%), Monotonous font type (64.16%), difficulty in operating hyperlinked text (51.66%) and initial operation of the computer are in English (35%). Most of the farmers expressed to have the content typed with bigger fonts and also contrast colour may be due to fact of their readability. Even though broadband connectivity and other highspeed connectivity are provided in cities, the rural sectors are not provided with high-speed Internet facility. Further speed of the computer is also very slow. This may be the possible reason for them to express the problem of down loading.

8. Suggestions expressed by the farmers

Any study would be complete if it gives suggestions for development in the field of investigation. The suggestions rendered by the respondents for effective utilization of web-education technology were studied. The results are presented in Table 7.

Table 7 : Suggestions for effective Utilization of Web -Education Technology

SI. No.	Suggestions	Number	Percentage
1.	Operating instruction may be in Tamil	120	100
2.	Large font size may be used	111	92.50
3.	Contrasting text color may be used	105	87.50
4.	More number of photos may be included	85	70.83
5.	Training to handle computer mouse may be given	95	79.16
6.	Fast downloading and uploading of photos/files may be done	99	82.5
7.	Variations in font type may be included	77	64.16
8.	Training to operate hyperlinked text may be given	62	51.66
9.	Information on other crops may be given	42	35.00

Suggestion rendered by the respondents could be seen from Table 7. Major suggestion were giving information in local language expressed by all the respondents (100.00 %) followed by large font size (92.50 %), contrasting text color (87.50%), fast downloading and uploading of photos/files (82.5%), training to operate computer mouse (79.16%), including more number of photos (70.83%), Variations in font type (64.16%), training to operate hyperlinked text (51.66%), Information on other crops (35.00%).

Conclusion

So far the web has been used mainly among urban sectors. Due to the effort of Govt. of India the ICT has been started piercing into the rural. In this context Educating farmers through Web on new Agricultural Technology has been attempted by Government, Quasi Government and Non profit organizations. So the usage of new media tools in educating farmers is found inevitable.

References

Alan Jollifee 2001. The Online Learning Handbook ,Konkan Page Ltd., London.

Anandaraja, N. 2002. Developing Farmer Friendly Interactive Multimedia Compact Disc and Testing it's Effectiveness in Transfer of Farm Technology, Unpub. Ph.D Thesis, TNAU, Coimbatore.

●●●

7

SWOC Analysis in Mango Pulp Industries in Tamil Nadu

T. Rathakrishnanl[1] and S. R. Padma[2]
[1]Director, Students Welfare and Professor (Extension), [2]Ph.D., Scholar, Tamil Nadu Agricultural University, Coimbatore-3.

Introduction

SWOC analysis is a strategic planning method used to evaluate the Strengths, Weaknesses, Opportunities, and Challenges involved in a project or in a business venture. It involves specifying the objective of the business venture or project and identifying the internal and external factors that are favorable and unfavorable to achieve that objective. A SWOC analysis must first start with defining a desired end state or objective. A SWOC analysis may be incorporated into the strategic planning model.

- Strengths: characteristics of the business or team that give it an advantage over others in the industry.
- **W**eaknesses: are characteristics that place the firm at a disadvantage relative to others.
- **O**pportunities: external chances to make greater sales or profits in the environment.
- **C**hallenges: external elements in the environment that could cause trouble for the business.

The aim of any SWOC analysis is to identify the key internal and external factors that are important in achieving the objective. These come from within the company's unique value chain. SWOC analysis groups key pieces of information into two main categories: Internal factors - The strengths and weaknesses internal to the organization and External factors - The opportunities and Challenges presented by the external environment to the organization.

The internal factors may be viewed as strengths or weaknesses depending upon their impact on the organization's objectives. What may represent strengths with respect to one objective may be weaknesses for another objective. The factors may include all of the 4P's; as well as personnel, finance, manufacturing capabilities, and so on. The external factors may include macroeconomic matters, technological change, legislation, and socio-cultural changes, as well as changes in the marketplace or competitive position. The results are often presented in the form of a matrix.

SWOC analysis is just one method of categorization and has its own weaknesses. For example, it may tend to persuade companies to compile lists rather than think about what is actually important in achieving objectives. It also presents the resulting lists uncritically and without clear prioritization so that, for example, weak opportunities may appear to balance strong challenges.

The overall evaluation of a company's Strengths, Weaknesses, Opportunities and Challenges is called SWOC analysis. A business unit has to monitor Key Macro environment forces (demographic, economic, technological, political-legal and socio-cultural) and significant micro environment factors (customers, competitors, distributors, suppliers) that affect its ability to earn profit. Internal Environment Analysis - In this analysis S - Strength includes, a) Easy accessibility by roads (NH), b) Easy availability of raw materials (mango fruits), c) Availability of sufficient labour is the visible strengths. W- Weakness the important factor includes, a) Dependency on merchant exporters, b) Lack of professional training/for the Managing Directors and managers, c) Lack of sales promotional measures and branding. External Environment Analysis indicates. O - Opportunities, which includes, a) Product monopoly, b) Increasing awareness of processed products among consumers, and C-Challenges was mushroom growth of artificial drinks pose a major threat.

Moreover, this study has also dealt with the important analysis of CINE Matrix - a matrix of Controllable Internal (CI) and Non-controllable External (NCE) factor influencing decision situation. Through these SWOC and CINE Matrix suitable strategies were formulated for the improvement of Mango pulp industries' management pattern. The CINE matrix analysis revealed the Controllable Internal Factors (CIF). Controllable External Factors (CEF), Non-Controllable Internal Factors (NCI) Non-controllable External Factors (NCE) and have been presented else where in this report.

To counter the Non-Controllable External factors enlisted in the CINE Matrix, the management takes all possible efforts to curtail the same and succeeds in the crushing operation. There are 29 mango pulp industries in this district. About 300000 tonnes of mango pulp is produced every year and encouraging 100 crores of foreign exchange through the pulp industries of the district. Once the management has identified the major Challenges and opportunities, it can characterize the business's overall attractiveness. Here, the outcome is higher opportunities accompanied with lower Challenges on the external environment. Hence, it is characterized as an ideal business. Kotler (2001) defined ideal business as follows: "An ideal business is high in major opportunities and low in major challenges". Mango pulp industries are in the category of ideal business. SWOC analysis is a tool for auditing an organization and its environment. SWOC Analysis is a simple but useful framework for analyzing our company's Strengths and Weaknesses, and the Opportunities and Challenges. This helps to focus on our strengths, minimize challenges, and take the greatest possible advantage of opportunities available to you.

Methodology

North-Western zone of Tamil Nadu comprises Dharmapuri, Salem and Namakkal districts; Among the three districts, the Dharmapuri district has witnessed horticulture development to a greater extent and mango in particular. Around 47000 ha mango plantations are in vogue in Dharmapuri district. To use the resources effectively, the mango pulp industries have mushroomed in and around Dharampuri to the tune of 29 in number. Considering the richness of response and the pulp industries, a study was conducted in Dharmapuri to assess the managerial attributes of Managing Directors of pulp industries through a structured questionnaire by way of personal interview. Even after the repeated visits, non-availability of Managing Directors forced the researcher to choose only 20 pulp industries out of 29 available in the zone during the study period. The size of the sample accounted for 69 per cent of the population. The study period is November-December, 2002.

Findings and Discussion

(i) Profile of Mango Pulp Industries

Pochampalli, Kaveripattinum, Burgur, Jgadevi, Choolagiri areas of Dharmapuri district,Vellore, Namakkal, Salem, Dindigul, Them' and Madurai districts of Tamil Nadu, the States of Karnataka and Andhra

Pradesh are the places from where the raw materials required for the factories have been procured. Mango varieties like Totapuri, Alphonso (Kalthar), Pether, Rumani and other varieties are used for pulp production. Among the varieties, Totapuri has been used widely (90%) in all the factories. The worker's potential depends on the work load of individual factories. The office workers have been ranging from 2 to 10 to 150 numbers of skilled labourers (for cutting of mango, machine operation, boiling of pulp, can ceiling and ail staking) and upto 500 numbers of unskilled labourers were engaged per day for various operations of mango pulp industries.

(ii) Procurement of Raw Materials

Mode of procurement of raw materials

All the factories were following the method of direct purchase from the producers and through commission agents. The different methods employed under direct purchase are presented in Table 1. The table reveals that all the factories studied were following the method of direct purchase after harvest (100%) followed by pre-harvest contract (75%). The methods like taking lease of mango garden and purchase from selected mango growers were accounted for 35 per cent and 25 per cent respectively.

The purchase from selected mango growers are practiced because of their loyalty to particular industries and the Managing Directors who have found potential in respect of certain garden, they resort the practice of leasing the garden for a specific period so as to take profit advantage by receiving quality raw material like reasonable price.

Mango Pulp industries

Table 1 : Methods of direct purchase of Mango

SI. No.	Methods	No of Industries reported	Per cent (%)
1.	Pre harvest contract	15	75.00
2.	Taking lease of mango growers	7	35.00
3.	Direct purchase after harvest	20	100.00
4.	Purchase from selected mango growers	5	25.00

Note : Multiple Responses.

(iii) Issues in Procurement

Problems of engaging labourers

Several labour based issues are faced by the factories which inhibit the factory operation of pulp production. Such issues are documented in Table 2. The table revealed that the demand for labourers were acute due to simultaneous operation of factories (100%) which was reported by all the factories followed by demanding high wages by the labourers (95%) during more raw materials are stocked, labour agents got advance amounts from other factories (85%) and labourers will not turn to work without prior intimation (65%) were the major constraints experienced by the factory personnel. The factories were running only for three months (April-June). Based on the export demand the pulp production used to be completed within that period. As the same labourers will be working in shifts in different factories, which reduce the efficiency in production. During the period of pulp production, the labourers will not turn to work without prior intimation. The periodic absences of labourers during pulp production are manned suitably by way of negotiation and personal appraisal with the labour agents. The periodic absence could be noticed only from the local labourers and hence the factory administrators were not depending much on the local labourers.

Table 2 : Labour problems faced by industries

SI. No.	Problems	No. of Industries	Per cent (%)
1.	Labourers not to work without prior intimation	13	65.00
2.	Agent got advance amount from other factories also	17	85.00
3.	When raw materials are more labourers will demand for higher wages	19	95.00
4.	Simultaneous seasonal operation in all factories leads to labour demand	20	100.00

The problems faced by the factory personnel in respect of procurement of mango are presented in Table 3. The table shows that the quality of mango could not be assured being reported by all the respondents accounted for 100 percent, followed by instability in price per unit of mango which is reported by 90 per cent of the respondents. This is the second biggest issue. Other problems reported by the respondents in respect of procurement of mango are the immature mango sent by the growers, damaged fruits could not be identified during the purchase and lack of coordination among the factory personnel to light for the common issues which are accounted for 70 per cent, 60 per cent and 25 per cent respectively. Though there is a federation among the factory personnel, it seems to be defunct. An active participation of members is warranted in this regard to solve the issues amicably.

Table 3 : Problems in the procurement of Mango fruits

SI. No.	Problems	No. of Industries	Per cent (%)
1.	The quality of mango could not be assured	20	100
2.	Immature mango beings sent by growers	11	70.00
3.	No stability in price for verities	18	90.00
4.	Demand fruits could not be identified during purchase	12	60.00
5.	Lack of coordination among factory personnel	5	25.00

The production of quality pulp solely rest with the quality fruits. Fully matured fruits and undamaged fruits alone could satisfy this. Hence, continuous effort is required to get an assessment on quality with innovative approaches.

(iv) Constraints in factory management

The details of constraints in the management of pulp industries are given in Table 4. It may be seen from the table that high investment rate for working capital, labour problem, instability in the rate of raw material supply, instability in the rate of production of mango pulp, lack of investment were emerged as major constraints to the factories in management. These problems normally arouse when the output is in the inventory and results in the investment gets locked up. However the administrators manage the industries by mobilising the fund from their own resources and from the money lenders for a short period, the Managing Directors of pulp industries are keen in selling the output as and when it is introduced. The pulp industry personnel could explore the possibilities of promoting their product with suitable sales promotion measures after branding.

Local sales besides export may also be carried out during the lean season by exploring the potentials for direct consumption through consumer friendly packs with suitable promotional measures. These could facilitate money circulation and the business plan could be further strengthened.

(v) Managerial Aspects of Pulp Industries

Management of any organisations involved in production and sale need to be concentrated on four major areas. They are: (i) Production and operations management: (ii) Materials/inventory management; (iii) Financial management; and (iv) Marketing management.

In respect of operations management, pulp production is grouped under mass production or flow line category. The facilities are arranged according the sequence of operations where the output of one stagebecomes input to the next stages. Here, the main problems will be balancing of production, machine maintenance and raw material supply. Managerial skill observed that the hard work with total dedication, commitment, assertiveness coupled with aggressiveness, keen knowledge in market intelligence of pulp products, team work, round the clock attention during processing period and risk bearing ability were the success attributes for smooth functioning of the factories.

SWOC Analysis

The overall evaluation of a company's strengths, weaknesses, opportunities and threats is called SWOC analysis.

A business unit has to monitor Key Macro environment forces (demographic, economic, technological, political-legal and socio- cultural) and significant micro environment factors (customers, competitors, distributors, suppliers) that affect its ability to earn profit. The business unit should set up a Market Intelligence System (MIS) to track trends and important developments (Kotler, 2001).

Extension excellence interviews were conducted with the administrators to perform SWOC analysis. The observations made have been presented as follows:

Internal Environment Analysis

S — Strength : (i) Easy accessibility by roads
(ii) Easy availability of raw materials (mango fruits)
(iii) Availability of sufficient labour were the visible Strengths

W- Weakness : (i) Dependency on merchaht exporters
(ii) Lack of professional training/for the Managing Directors and managers
(iii) Lack of sales promotional measures and branding

External Environment Analysis

O — Opportunities : (i) Product monopoly
(ii) Increasing awareness of processed products among Consumers

C — Challenges : Mushroom growth of artificial drinks pose a major Challenges

CINE, Matrix - a matrix of Controllable Internal (CI) and Non-controllable External factor influencing decision situation.

Controllable Internal Factors

The CINE matrix analysis revealed the following controllable internal factors. They are:

I. Working hours
2. Pulp quality
3. Job standard and specifications

Controllable External Factors

The controllable external factors as revealed by the firms are:

(i) Neighbouring village relations
(ii) Visitors
(iii)Donation seekers
(iv)Availability of fuel -wood

Non-Controllable Internal Factors

The non-controllable internal factors are:
(i) Nuisance of non-Workers
(ii) Liquor addiction
(iii)Absenteeism during community get together

Non-controllable External Factors

The non-controllable external factors are:

(i) Pressure from political side for several reasons
(ii) Pressure from local population because of unemployment situation
(iii) Hike in the input prices

To counter the non-controllable factors enlisted in the CINE Matrix, the management takes all possible efforts to curtail the same and succeeds in the crushing operation.

Once the management has identified the major Challenges and opportunities, it can characterize the business' overall attractiveness. Here, the outcome is higher opportunities accompanied with lower Challenges on the external environment. Hence, it is characterised as an ideal business. Kotler (2001) defined ideal business as follows: "An ideal business is high in major opportunities and low in major threats". Mango pulp industries are in the category of ideal business.

Conclusion

It is therefore, necessary to develop a uniform system for their modus operandi for achieving the fullest utilization of manpower and crop potential for the development and improvement of better standard of living of the farming community of this district.

The study makes an attempt to focus on the challenges ahead in the managerial techniques on the prospects of pulp industries in Tamil Nadu.

The study has identified various problems faced by the pulp industries. The main problems in procurement were non-assurance of quality mango and sudden absence of labourers during the production season hampering the production process. Besides this, instability in the price of raw materials and its availability also forced the industries in a fix. To solve the problems, strategies have been suggested. Besides, SWOC analysis has also been conducted in the study area and the pulp business has been found as an ideal business.

Strategies

1. Since manual sorting of mango consumes much labour and time, the varietal sorting of mango from the bulk should be done to avoid

damages by inventing a machinery that should identify and remove the damaged from the lot which could save the time and money of the investors.

2. Lethal formulations are the need of the hour to promote uniform fruiting of mango after the harvest so as to meet the uniform raw material input to the industries.
3. The factory owners who went on lease agreement with gardeners should device an euipment which is capable of indicating the maturity of mango so as to avoid harvesting of immature mango.
4. An inbuilt mechanism that is capable of delivering output directly from the input with or without much interventions are the need of the hour to promote the trade abroad.
5. In the absence of good quality natural drinks, the promoters can explore the possibilities of entering into the local markets by designing the consumer packs after branding could promote higher demand for mango pulp.
6. The mango pulp industries should explore the possibilities of entering into product diver sification during the off-season by utilising pine apple, tomato etc., to have multiple turnover benefits.
7. The mango pulp producing firms should have tie-ups with research institutions or by way of creating Research and Development (R&D) wing with them to have more value added products produced out of wastes of mango peel and nuts.

References

Kotler, Phillip (20(11). Marketing Management. Prentice Hall of India Pvt. Ltd., New Delhi, pp. 76-79.

Nesi, 1 (2002), 'Preventing a Near Collapse". The Hindu Survey of Indian Agriculture pp. 175-173.

Nesi, I : "Tor Soil Prosperity and Nutritional Security", The Hindu Survey of Indian. agriculture pp 169-173.

Kannaiyan. S and RamaSamy, C. (2002). Agriculture - 2002, Tamil Nadu Agricultural University. Coimbatore. pp. 2 11-259.

•••

8

Adoption Level of Farmers on Post Tsunami Rice Cultivation Practices

B. Shanmugasundaram[1], K.A. Ponnusamy[2] and H. Philip [3]
[1] Associate Professor (Agri. Extn.) Regional Agricultural Research Station (RARS), Kerala Agricultural University, Pattambi, Kerala.
[2] Professor, Dept of Agricultural Extension and Rural Sociology and
[3] Professor, (Agri. Extn), DOEE, Tamil Nadu Agricultural University, Coimbatore

Introduction

Rice is considered as the most important crop of India both in terms of area as well as value of output. It is estimated to contribute about 2.5 per cent of Gross Domestic Product (GDP) of India and provides employment to a large work force in crop production and in paddy processing.

According to Reinsurance Company 'Munich Re (2001), costs associated with natural disasters has gone up 14 fold since the 1950. Every year from 1991 to 2001, an average of 211 million people in the world were killed or affected by natural disaster. Described as one of the "Worst Disasters" in recent history the Tsunami travelled 4,500 km across the Indian Ocean in a period of about seven hours causing damage to twelve countries including Sri Lanka, India and Thailand most dramatically impacted, The FAO estimates that in the countries hardest hit by the disaster (Indonesia, Sri Lanka, Maldives, India and Thailand), a total of 47,000 ha of agricultural land has been damaged (FAO, 2005).

On December 26, 2004 for the first time in half of a century, India experienced the devastating effects of Tsunami, caused by a series of earthquakes in the Bay of Bengal. Soon after the Tsunami, the government of Tamil Nadu announced an immediate crop compensation of Rs. 2,500 per hectare in cash for standing crops lost. A compensation package of Rs.12,500 per hectare was computed, of which 66 per cent was allocated in the first year, and the balance amount was carried over to the second and third years. The Government of Puducherry has sanctioned an amount of Rs3750/ha as immediate relief to crop loss and a one time relief of Rs 60,000/ha for yield loss and restoration of productivity in Tsunami affected lands.(Dixit, 2006). A holistic package covering desalination on one side and fertility enhancement and sustainability of farming on the other was very much needed and a common package was given by the Department of Agriculture under the Rehabilitation of Tsunami Affected Agricultural Lands (ROTAAL). Keeping the above factors in mind, the present research was designed with the specific objective to find out the extent of adoption of Post Tsunami Rice cultivation practices among Tsunami affected rice farmers.

Methodology

The Tsunami waves caused extensive damage to three states (Andhra Pradesh, Tamil Nadu and Kerala) and two Union Territories (Puducherry and Andaman and Nicobar Islands) in India. However, Tamil Nadu State and Union Territory of Puducherry were selected purposively for the study. In Tamil Nadu the extent of crop area affected was severe in Nagapattinam district followed by Cuddalore district. Hence the two districts were selected .Out of the two affected regions in Puducherry, ie., Puducherry (80ha) and Karaikal (712 ha) the extent of crop area damage was severe in Karaikal region. Hence it was selected. The criteria set for selection of Taluks in Nagapattinam district include the selection of first three Taluks having maximum damage of paddy area. Accordingly, Vedarnayam, Tharangampadi and Sirkali were selected for the study. Based on the same criteria set for Nagapattinam district, Cuddalore taluk was selected for the study. Karaikal region is geographically a small area and hence the selection based on taluk does not arise.

The same criteria adopted for selection of taluk was followed in the selection of villages also. As the extent of paddy area affected was very less (21 ha) the selection of villages was made at random in Cuddalore taluk (Cuddalore district). In Karaikal region about 11 villages were affected due to Tsunami. Out of the 11 villages, four villages were selected at random. Finally 16 villages were selected. The number of villages and respondents selected are given in Table 1. A sample size of 120 paddy affected farmers was considered as optimum.

The term adoption in the study was used to mean the level of use of Post Tsunami rice cultivation practices as recommended by extension agencies. The adoption index for each respondent was calculated using the formula :

$$\text{Adoption index} = \frac{\text{Adoption score obtained by the respondent}}{\text{Possible maximum score}} \times 100$$

The respondent were categorized into low, medium and high adopters by using cumulative frequency method.

Findings and Discussion

The findings and discussion has been carried out by extent of adoption in overall and practice wise adoption of Post Tsunami Rice Cultivation Practices and the results are shown in Table 1.

The study reveals that 54.16 per cent of the respondents were medium level adopters followed by low (30.84%) and high (15.00%) levels of adoption." The fear of raising the crop due to subsequent flood after the tsunami", "complexity in the technology", "technology not need based" and "non availability of relief material in right quantity and right time" were expressed as the factors for low and medium levels of adoption of post tsunami rice cultivation practices.

Practice-wise adoption of farmers on Post tsunami rice cultivation practices

The practice-wise extent of adoption of recommended technologies by the Tsunami affected rice farmer is furnished in Table 1.

Table 1 : Practice wise extent of adoption of farmers on Post Tsunami Rice cultivation practices

(n=120)

Sl. No.	Technologies	FA		PA		NA	
		No	%	No	%	No	%
	Seed selection						
1	Growing salt tolerant rice varieties	29	24	43	36	48	40
	Preliminary cultivation						
2	Levelling of land surface	57	47	50	42	13	11
3	Gypsum application @ 500	14	12	44	37	62	51

Sl. No.	Technologies	FA		PA		NA	
		No	%	No	%	No	%
	Kg/ha						
4	Scrapping and removal of silt	54	45	37	31	29	24
	Seed rate and sowing method						
5	Seed rate @20% extra of normal seed rate	35	29	54	45	31	26
	Transplanting						
6	Planting of 4-6 seedling /hill	10	8	37	31	73	61
7	Planting of aged seedlings (one week more than normal age)	3	2.5	27	22.5	90	75
8	Sowing sprouted seeds 3 DAS under delayed canal water / Excess rainfall	7	6	28	23	85	71
	Fertilizer application						
9	Soil test based fertilizer application	4	3	32	27	84	70
10	Znso4 application @40 Kg/ha	0	0	15	12.5	105	87.5
11	Application of Azospirillum @2Kg/ha	7	6	36	30	77	64
12	Application of 25% Extra N in basal application	6	5	51	42.5	63	52.5
13	Foliar spray of DAP 2% along with 1% urea and 1% Pottash during Panicle initiation stage and 15 days after .	0	0	7	6	113	94
14	Application of 15 Kg P/ha as maintainer dose	0	0	13	11	107	89
	Manuring						
15	Application of FYM @12t/ha	8	6	80	67	32	27
16	Application of green manure crops like Daincha @6.25 t/ha	17	14	50	42	53	44
	Water Management						
17	Leaching soil with good quality water (3-4 times with short intervals)	25	21	49	41	46	38
18	Removal of contaminated water	43	36	42	35	35	29
19	Surface leaching through rainfall and canal water	58	48	49	41	13	11

FA-Fully Adopted ; PA- Partially Adopted ; NA: Not Adopted

From Table 1, it could be observed that 48.00 per cent of the farmers came under "Fully Adopted" category with regard to 'Surface leaching through rainfall and canal water technology' followed by 47.00 per cent in

'Levelling of land surface', 45.00 per cent in 'Scrapping and removal of silt', 36.00 per cent in "Removal of contaminated water', 29.00 per cent in "Following Seed rate @ 20% extra of normal seed rate", 24.00 per cent in 'Growing salt tolerant rice varieties' and 21.00 per cent in 'Leaching of soil with good quality water'. Soil salinity is recognized as a significant problem in the affected areas, although there has been noteworthy progress with respect to its mitigation due to natural rainfall that occurred days after Tsunami. Therefore surface leaching was given due importance by the farmers to restart agriculture. It is also due to the fact that many agencies involved in Tsunami reclamation had extended financial support to restart agriculture in the affected areas. Land leveling and removing of silt is critical in the reclamation of salt affected soils and efficient irrigation management. Hence it is being fully adopted by nearly half of the respondents.

Under "partial adoption", with respect to recommended techniques, 67.00 per cent of the respondents expressed having applied FYM partially. This was followed by 45.00 per cent of respondents "Adopting Seed rate @ 20% extra of normal requirement", 42.50 per cent in "Application of 25% extra N as basal", 42.00 per cent in "Leveling of land surface", 42.00 per cent in "Application of green manure crops like Daincha", 41.00 per cent in "Surface leaching through rainfall and canal water" and Leaching of soil with good quality water. A major constraint immediately after the disaster was the procurement of sufficient quantities of quality seed, availability of appropriate variety of seed suited to prevailing soil conditions and availability of local organic compost/manures and inorganic fertilizers. Poor road infrastructure and remoteness affected transport of relief supplies and impeded the reclamation process.

Under "non adoption", it is seen from the table 23 that 94.00 per cent of the respondents did not adopt "Foliar spray of DAP, urea and Potash". This was followed by 89.00 per cent of farmers not applying Phosphorus as maintainer dose, 87.50 per cent on "Zinc Sulphate application", 75.00 per cent on "planting of aged seedlings", 71.00 per cent on "Sowing sprouted seeds three Days after sowing", 70.00 per cent on "Soil test based fertilizer application", 64.00 per cent on "Application of Azospirillum" and 61.00 per cent on "Planting of 4-6 seedlings/hill". Problems of attracting suitably trained and experienced persons to assist in rehabilitation process was also experienced. Delay in distribution of relief material in right quantity, mismatch in aid with real needs of individuals and communities often being overlooked, fear of another Tsunami, lack of compatability of technology and complexity of the technology resulted in farmers not adopting the Post Tsunami Rice Cultivation Practices.

Conclusion

As far as overall adoption of post tsunami rice cultivation practices was concerned 54.16 per cent of the respondents were found in the medium category. From the study it can be concluded that Tsunami had only brought into the limelight the vulnerable status of Agricultural community that had been reeling under the impact of Alternate floods and droughts. It is also seen that with the support of different agencies alone it was possible to restart agriculture in the affected regions

References

Dixit, Sreeenath.(2006). Brief report on participation in the Consultation on Post-Disaster Management and Prevention strategies a MSSRF, Chennai and follow up visit to Tsunami affected areas of Nagapattinam district, Tamil Nadu. Retrieved July 2006, from http://www.vasat.org/research/tsunami3 htm.

FAO (2005). Country report on impact of the Tsunami on soil and water resources In: *Regional workshop on Salt affected soils from Sea water intrusion: Strategies for Rehabiitation and Management ,* Thailand, FAO.

MunichRe (2001). Topics. Annual Review: *Natural Catastrophes* 2001.

●●●

9

Prioritiztion of Research in Selected Crops of Madurai District

***A. Pounraj*[1], *T. Rathakrishnan*[2] *and Rexlin Selvin*[3]**

[1] *Associate Professor, Dept. of Agrl. Extension and Rural Sociology, TNAU, Coimbatore.3,*

[2] *Director, Students Welfare, TNAU, Coimbatore and* [3] *Professor, Dept. Agri. Extension and Rural Sociology, Agrl. College and Research Institute, Madurai-104.*

Introduction

The Horticulture crops and commercial crops like sugarcane have become a key drivers for economic development in many of the states in India, and it contributes 29.5 per cent and 24.80 percent respectively to the GDP of agriculture, which calls for technology-led development. The research priorities for genetic resource enhancement and its utilization, enhancing the efficiency of production and reducing the losses in environment friendly manner and effectively channellising the marketing for the produce. Accordingly the research was taken up to study on the prioritization of research for the crops viz., sugarcane, banana and jasmine. The study deals with the knowledge, adoption and research priorities on selected crops' cultivation which has already been occurred.

Methodology

The survey method of research design was utilized to seek answers to the study objectives. This study was mainly focused on the research priority and the constraints encountered by the cultivars under the crops selected in Madurai district. In Tamilnadu, Madurai district was purposively selected for the study by considering as Madurai is one among the districts where, sugarcane, banana and jasmine are cultivated as major crops and there is a great scope for improving acreage under horticultural crops. Among twelve Blocks available in this district three blocks were selected for the study considering the maximum area under banana, sugarcane and jasmine. Among the total revenue villages only three villages were selected which had more area under selected crops. Sugarcane, banana and jasmine growers were the respondents for the study. A list of farmers under selected crops was collected from the respective BDOs, A sample size of 60 farmers was considered for the study. The study deals with the knowledge, adoption and research priorities on selected crops' cultivation which has already been occurred.

Findings and Discussion

Overall knowledge level of sugarcane, banana and jasmine growers

To assess the overall knowledge level of sugarcane, banana and jasmine growers on cultivation techniques, necessary data were collected and furnished in Table.1

Table 1 : Distribution of respondents according to overall knowledge level on sugarcane, banana and jasmine cultivation

n=20

Sl. No.	Categories	Sugarcane		Banana		Jasmine	
		Number	Per cent	Number	Per cent	Number	Per cent
1.	Low	4	20.00	3	15.00	6	30.00
2.	Low	9	45.00	6	30.00	8	40.00
3.	High	7	35.00	11	55.00	6	30.00

The Table 1 revealed that majority of the respondents possessed medium and high level of knowledge category (80.00%). The low level was found among only 20.00 per cent of respondents. It could be concluded that majority (80.00%) of the respondents was found to possess medium to high level of knowledge. This might be due to the medium to high level access with Research Stations / Research Institutes.

This findings is in line with the findings of Venkattakumar (1997) who reported that the majority of the respondents belonged to medium to high level knowledge category. It is inferred from the Table 1 that more than half of the respondents possessed high level of knowledge (55.00%) followed by medium level category (30.00%). The low level was found among the 15.00 per cent of respondents. It could be concluded that majority (85.00%) of the respondents was found to possess medium to high level of knowledge. This might be due to the interest of the farmers to know the new technologies as they approach and access with Research Stations / Research Institutes. This finding is in line with the findings of Venkattakumar (1997) who reported that the majority of the respondents belonged to medium to high level knowledge category.

The table I stated that majority of the respondents possessed medium to high level of knowledge (70.00%) on jasmine cultivation techniques followed by low level category (30.00%). It could be concluded that majority (70.00%) of the respondents was found to possess medium to high level of knowledge. This might be due to the medium to high level access with Research Stations / Research Institutes. It is because, the respondents were able to contact researchers in time when they are in need. This findings is in line with the findings of Venkattakumar (1997) who reported that the majority of the respondents belonged to medium to high level knowledge category.

Overall adoption behaviour of sugarcane, banana and jasmine, growers

Adoption of innovations by farmers is the fruit aimed at by all the extension agencies. Rogers and Shoemakers (1970) define adoption as a decision to make full use of new ideas as the best course of action available. The information collected to study the adoption behaviour of sugarcane growers on the adoption of improved practices of sugarcane cultivation were presented in the Table 2.

Table 2 : Distribution of respondents according to overall adoption level on sugarcane, banana and jasmine cultivation

n=20

Sl. No.	Categories	Sugarcane		Banana		Jasmine	
		Number	Per cent	Number	Per cent	Number	Per cent
1.	Low	6	30.00	5	25.00	7	35.00
2.	Low	5	25.00	9	45.00	6	30.00
3.	High	9	45.00	6	30.00	7	35.00

The Table 2 showed that two third of the sugarcane growers (65.00%) had medium to high level of adoption. Only one fourth of sugarcane growers had low level of adoption.It could be concluded that majority (70.00%) of the respondents was found to possess medium to high level of adoption. This might be due to the medium to high level access with

Research Stations / Research Institutes. It is because, the respondents were able to contact researchers in time when they are in need. Adoption of innovations by farmers is the fruit aimed at by all the extension agencies. Rogers and Shoemakers (1970) define adoption as a decision to make full use of new ideas as the best course of action available. The information collected to study the adoption behaviour of banana growers on the adoption of improved practices of banana cultivation were presented in the table. The Table 2 showed that the three fourth of the banana growers (75.00%) had medium to high level of adoption. Only one fourth of banana growers had low level of adoption. It could be concluded that majority (75.00%) of the respondents was found to possess medium to high level of knowledge. This might be due to the medium to high level access with Research Stations / Research Institutes. It is because, the respondents were able to contact researchers in time when they are in need. The table 2 showed that just below two third of the jasmine growers (65.00%) had medium to high level of adoption. Above one third (35.00%) of jasmine growers had low level of adoption. It could be concluded that majority (65.00%) of the respondents was found to possess medium to high level of knowledge. This might be due to the medium to high level access with Research Stations / Research Institutes. It is because, the respondents were able to contact researchers in time when they are in need.

Technology wise knowledge and adoption of sugarcane growers

The practice wise knowledge and adoption of sugarcane growers on fifteen items were assessed and the results were presented in the Table 3.

Table 3 : Technology wise knowledge and adoption of sugarcane growers

Sl. No.	Technology	Knowledge		Adoption	
		No.	Percentage	No.	Percentage
1.	Varieties	91	95.00	17	85.00
2.	Soil	15	75.00	13	65.00
3.	Selection of setts	16	80.00	15	75.00
4.	Treatment of setts	14	70.00	11	55.00
5.	Field preparation	15	90.00	17	85.00
6.	Method of planting	17	85.00	17	85.00
7.	Method, quantity and quality of irrigation	15	75.00	13	65.00
8.	Mannuring	16	80.00	14	70.00
9.	Recommended weeding schedule	13	65.00	12	60.00

10.	Detracing	17	85.00	16	80.00
11.	Earthing up	17	85.00	15	75.00
12.	Control of red rot	13	65.00	11	55.00
13.	Control of early shoot borer	12	60.00	10	50.00
14.	Control of inter node borer	13	65.00	12	60.00
15.	Harvesting	18	90.00	17	85.00

An insight into the extent of adoption on recommended sugarcane practices presented by the respondents, item analysis were carried out and the results were furnished in the Table 3. Most of the (95.00%) sugarcane growers had knowledge about varieties followed by field preparation and harvesting (90.00%). Majority of the sugarcane growers (85.00%) had knowledge on method of planting, detracing and earthing up. The cultivation practices viz.,selection of setts and manuring were well known by the farmers at 80.00 percent level. With regard to other practices viz., Soil, Method, quantity and quality of irrigation, Treatment of setts, Recommended weeding schedule, Control of red rot, Control of inter node borer, Control of early shoot borer the knowledge level of sugarcane growers varied from 75.00 % to 60.00%. From the Table 3 it could be observed that out of fifteen practices the technologies of Varieties, Field preparation method of planting and Harvesting adopted by 85.00 per cent of sugarcane growers followed by 80.00 percent of farmers correctly adopted the technology of detracing. The percentage of farmers ranges from 75.00per cent to 50.00 per cent had adopted the technologies of selection of setts Earthing up, manuring, Soil, Method, quantity and quality of irrigation, Recommended weeding schedule, Control of inter node borer, Treatment of setts, treatment of setts, Control of red rot, Control of early shoot borer. From the above findings, it is observed that most of the sugarcane growers were reported to adopt Varieties, Field preparation, method of planting and Harvesting of sugar cane. They might have adopted the those technologies to get higher yield and to realize the remunerative price for their cane production. Similar adoption trend is noticed in the case of detracing earthing up and selection of setts because, improper management in these practices affect the yield of the sugarcane.

Technology wise knowledge and adoption of banana growers

The practice wise knowledge and adoption of banana growers on eight items were assessed and the results were presented in the Table 4.

Table 4 : Technology wise knowledge and adoption of banana growers

Sl. No.	Technology	Knowledge		Adoption	
		No.	Percentage	No.	Percentage
1.	Varieties	20	100.00	18	90.00
2.	Selection of suckers	19	95.00	18	90.00
3.	Pre treatment of suckers	16	80.00	14	70.00
4.	Spacing	18	90.00	17	85.00
5.	Manuring pits	15	75.00	13	65.00
6.	Desuckering	16	80.00	16	80.00
7.	Nematode control	13	65.00	12	60.00
8.	Panama wilt disease management	13	65.00	11	55.00

An insight into the extent of adoption on recommended banana practices presented by the respondents, item analysis were carried out and the results were furnished in the Table 4. Cent percent of the banana growers had knowledge about suitable varieties, followed by selection of suckers (95.00%), Spacing (90.00%). In the technologies of Pre-treatment of suckers and desuckering 80.00 percent of the farmers have thorough knowledge. Three fourth of the farmers (75.00%) had knowledge on manuring pits correctly. More than two third of the farmers (65.00%) possessed knowledge on nematode control and panama wilt disease management. It could be observed from the Table 4 that vast majority of the farmers adopted the technologies of varieties and selection of suckers (90.00%) followed by spacing and desuckering 85.00 percent and 80.00% respectively. The technologies of pre-treatment of suckers, manuring, pits, nematode control and management of panama wilt disease were adopted by the farmers ranges from 75.00 percent to 55.00 percent. It could be concluded from the above table that majority of the banana growers had adopted the technologies of varieties and selection of suckers to get higher yield and they adopted spacing and desuckering to increase the punches as well as to increase the no in the punches of banana.

Technology wise knowledge and adoption of jasmine growers

The practice wise knowledge and adoption of jasmine growers on eight items were assessed and the results were presented in the Table 5.

Table 5 : Technology wise knowledge and adoption of jasmine growers

Sl. No.	Technology	Knowledge		Adoption	
		No.	Percentage	No.	Percentage
1.	Quantity of farm yard manure recommended per pit	17	85.00	15	75.00
2.	Recommended NPK	15	75.00	13	65.00
3.	Right pruning time	14	70.00	14	70.00
4.	Pruning technique	13	65.00	12	60.00
5.	Control measures for 80.00 budworm	17	85.00	16	80.00
6.	Control measures for nematode	14	70.00	13	65.00
7.	Control measures for iron deficiency	13	65.00	12	60.00
8.	Harvesting	17	85.00	16	80.00

An insight into the extent of adoption on recommended jasmine cultivation practices presented by the respondents, percentage analysis were carried out and the results were furnished in the Table 5. The table clearly revealed that the majority of the jasmine growers had knowledge on Quantity of farm yard manure recommended per pit and Control measures for budworm and harvesting (85.00%), followed by recommended NPK with 75.00%. The technologies of Right pruning time Control measures for nematode Pruning technique and Control measures for iron deficiency were adopted by the farmers with the percentage ranges from 70.00 to 65.00. The table 5 clearly revealed that out of eight practices the technologies of Quantity of farm yard manure recommended per pit, Control measures for budworm and Harvesting were adopted by 85.00 per cent of jasmine growers followed by 75.00 percent of farmers correctly adopted the technology of application of recommended NPK. The farmers ranges from 70.00per cent to 65.00 per cent had adopted the technologies of Right pruning time, Control measures for nematode and Control measures for iron deficiency. From the above findings, it is observed that most of the jasmine growers were reported to adopt Quantity of farm yard manure recommended per pit, Control measures for budworm and Harvesting. They might have adopted the those technologies to get higher yield and to realize the remunerative price for their jasmine production. Similar adoption trend is noticed in the case of application of recommended NPK, Right pruning time, Control measures for nematode and Control measures for iron deficiency because, improper management in these practices affect the yield and of the jasmine.

Constraints faced by the sugarcane growers

Table 6 : Constraints faced by the sugarcane growers

SI. No.	Constraints	Number	Percen-tape
1.	Lack of varieties susceptible for pest and disease	19	95.00
2.	Inadequate credit facilities	14	70.00
3.	Non-availability of labourers during peak season	18	90.00
4.	Fluctuation in market price	16	80.00
5.	Non payment of dues	18	90.00
6.	Illegal reduction of market price	18	90.00
7.	Inadequate information about latest technologies	12	60.00
8.	Inadequacy irrigation water at the time of requirement	13	65.00

The data in the Table 6 shows that nine different constraints expressed by sugarcane growers. The most striking constraints reported by vast majority (95.00%) were Lack of varieties susceptible for pests and diseases, followed by Non-availability of labourers during peak season, Non payment of dues, and illegal reduction of market price as expressed by 90.00 per cent of the sugar cane growers. Fluctuation in market price and Inadequacy of irrigation water at the time of requirement were the constraints expressed by 80.00 and 65.00 per cent of the sugarcane growers respectively. More than half of the respondents expressed (60.00 %) Inadequate information about latest technologies as one of their constraints. These findings inferred that the sugarcane growers are seriously affected by low yield because of high incidence of pests and diseases and in turn they have to spent more moneys towards the plant protection measures. The movement of labourers towards the city for employment which involving less labour as well as Madurai city is nearby and easy accessible by transport conveyance leads to Nonavailability of labourers during peak season. As far as the price of the sugar is concern, the cost of production for the unit of sugar produced within the country is more than that of the unit of sugar imported, hence, the sugar factories are not able to pay the price anticipated by the farmers which resulted to non payment of dues to the farmers, and then the farmers try to sell their products as jaggary in the open market, where the middle men illegally reduce the price of the product which affect the sugarcane growers severely.

Constraints faced by the banana growers

Table 7 : Constraints faced by the banana growers

n=20

SI. No.	Constraints	Number	Percentage
1.	Non availability of quality suckers	20	100.00
2.	More incidences of pest and diseases	20	100.00
3.	Heavy damage by winds	18	90.00
4.	Inadequate credit facilities	17	85.00
5.	Fluctuation in market price	20	100.00
6.	Inadequate extension support	10	50.00
7.	High labour wages	20	100.00
8.	Non- availability of labourers	19	95.00

The banana growers are concern, Non availability of quality suckers, More incidences of pests and diseases, Fluctuation in the market price and High wages for labourers were the constraints as expressed by cent per cent of the farmers as shown in the Table 7. Majority of the banana growers (95.00%) have expressed Non- availability of labourers as one of their major constraints, followed by 90.00 and 85.00 per cent of the farmers expressed Heavy damage by winds and Inadequate credit facilities respectively as their constraints. Only half (50.00%) of the respondent farmers considered Inadequate extension support as their constraints. Banana is highly proned to pests and diseases and involved high cost of plant protection measures, the labourers involved in the work of selecting and taking suckers are having less knowledge on the selection of quality suckers, most of the farmers prefer the nearby city for marketing their produce where the interference of middle men incurring less as well as fluctuation in the price of the product and as mentioned in the constraints faced by the sugarcane growers, the movement of labourers towards the city for employment which involving less labour as well as Mad urai city is nearby and easy accessible by transport conveyance leads to Non-availability of labourers during peak season.

Constraints faced by the jasmine growers

From the Table 8 it could be observed that out of fifteen constraints expressed by the jasmine growers, vast majority (95.00%) of the jasmine growers reported to express Poor keeping quality of flowers as their major constraint and Not convinced with recommended practices, Labour scarcity, High incidence of pest and diseases, Seasonal fluctuation in market and Lack of processing units were the constraints expressed by 90.00 per cent of the jasmine growers, followed by High cost of inputs, High risk 'and uncertainty of returns and Lack of regulated market units were the constraints expressed by 85.00 per cent of the jasmine growers.

Table 8 : Constraints faced by the jasmine growers

n=20

SI. No.	Constraints	Number	Percentage
1.	Not convinced with recommended practices	18	90.00
2.	Lack of credit	16	80.00
3.	High cost of inputs	17	85.00
4.	Labour scarcity	18	90.00
5.	Lack of technical guidance	11	55.00
6.	Lack of irrigation.	14	70.00
7.	High incidence of pest and diseases	18	90.00
8.	Lack of information	10	50.00
9.	High risk and uncertainty of returns	17	85.00
10.	Seasonal fluctuation in market	18	90.00
11.	Lack of regulated market	17	85.00
12.	Undesirable climatic conditions	14	70.00
13.	Lack of processing units	18	90.00
14.	Poor keeping quality of flowers	19	95.00

The percentage of farmers ranges from 80.00 per cent to 50.00 per cent of the jasmine growers reported to express Lack of credit, Lack of irrigation, Lack of information, Lack of technical guidance and Undesirable climatic conditions as their constraints. From the above findings, it is observed that most of the jasmine growers reported to express Poor keeping quality of flowers. If the keeping quality of the flowers increase, then the farmers can sell their flowers for reasonable price, even it will ake more time. Similar trend is noticed in the case of Not convinced with recommended practices, Labour scarcity, High incidence of pest and diseases, Seasonal fluctuation in market and Lack of processing units.

Prioritization of research

The farmers feel that some of the areas of research in crop cultivation are not fulfilled as per their expectations. Hence, the prioritization of research in sugarcane cultivation, banana cultivation and jasmine cultivation were expressed by the farmers so as to evade the loss while cultivating the above crops.

Research priorities expressed by the sugarcane growers

The Study reveals that most of the sugarcane growers (95.00%) had expressed that the priority should be given to conduct research for controlling pests and diseases in sugarcane, followed by the innovation of new machineries for sugarcane field operation as expressed by 85.00 per cent of the sugarcane growers. More than two third of the respondents reported that research should be conducted in the areas of making byproducts, followed by 65.00 per cent of the farmers were in the need of research on new approach on marketing of sugar products. Only half of (50.00%) of sugarcane growers suggested that research should be carried out in new approach on the transfer of technology.

Table 9 : Research priorities expressed by the sugarcane growers

SI.No.	Research Priority	Number	Percentage
1.	Pest and disease	19	95.00
2.	New machineries	17	85.00
3.	By-products	14	70.00
4.	New approach on marketing	13	65.00
5.	New approach on technology transfer	10	50.00

The main reason for the loss of yield in sugarcane is the surveillance of pests and diseases and it should be controlled so as to increase the yield as well as to increase the income of the cane growers. Hence, the farmers might have expressed the priority of doing research to control the surveillance of pests and diseases. Sugarcane cultivation involves heavy labour oriented and the scarcity of labourers leads to the obstacle in the operation of sugarcane cultivation, hence farmers expressed the need of research in , the innovation of new machineries which will reduce the burden of farmers to evade labour scarcity as well as to reduce the cost of cultivation.

Research priorities expressed by the banana growers

Cent per cent of the banana growers (100.00%) had expressed that the priority should be given to conduct research for the production and selection quality suckers, followed by the control of pest and diseases which severely affect the banana as expressed by 95.00 per cent of the banana growers. Majority of the farmers reported to need the research on the varieties prone to damage by winds. More than three fourth of the respondents reported that research should be conducted in the areas of innovation of new machineries in banana cultivation. Marketing system and information system are the fields which needed research priority as expressed by 70.00 and 55.00 per cent of the banana growers respectively

Table 10 : Research priorities expressed by the sugarcane growers

SI.No.	Research Priority	Number	Percentage
1.	Quality suckers	20	100.00
2.	Pest and diseases	19	95.00
3.	Varieties prone to damage by winds	17	85.00
4.	Farm machineries	15	75.00
5.	Marketing system	14	70.00
6.	Information system	11	55.00

The important function in the banana cultivation is production and selection of good quality of suckers which will assure the yield of the anticipation of banana growers. Hence, the farmers might have expressed the priority of doing research for the production and selection quality suckers. The main reason for the loss of yield in banana is the surveillance of pests and diseases and it should be controlled so as to increase the yield as well as to increase the income of the banana growers. Hence, the farmers might have expressed the priority of doing research to control the surveillance of pests and diseases.

Research priorities expressed by the jasmine growers

The Table 11 showed that most of the jasmine growers (90.00%) had expressed that the priority should be given to conduct research on improved cultivation practices, high yielding varieties and for controlling pest and diseases, followed by irrigation management as expressed by 85.00 per cent of the jasmine growers. More than two third of the respondents reported that research should be conducted in the areas of making by-products and processing after harvest, followed by 60.00 per cent of the farmers were in the need of research on new approach on marketing of jasmine..

Table 11 : Research priorities expressed by the sugarcane growers

SI.No.	Research Priority	Number	Percentage
1.	Cultivation practices	18	90.00
2.	Varieties	18	90.00
3.	Pest and diseases	18	90.00
4.	Irrigation management	15	75.00
5.	Processing	14	70.00
6.	Byproducts	14	70.00
7.	Marketing system	12	60.00

The conventional methods of jasmine cultivation will yield lesser than that of the improved cultivation practices. Hence, the farmers might have expressed the priority of doing research for the production and selection quality suckers, high yielding Varieties and to control the pests and diseases of jasmine.

Conclusion

The salient findings of this study are as follows. Majority of the respondents possessed medium and high level of knowledge category (80.00%). Two third of the sugarcane growers (65.00%) had medium to high level of adoption. Three fourth of the banana growers (75.00%) had medium to high level of adoption. Most of the (95.00%) sugarcane growers had knowledge about varieties followed by field preparation and harvesting (90.00%). Cent percent of the banana growers had knowledge about suitable varieties , followed by selection of suckers (95.00%), Spacing (90.00%). Majority of the jasmine growers had knowledge on Quantity of farm yard manure recommended per pit and Control measures for budworm and harvesting (85.00%), followed by recommended NPK with 75.00%.The most striking constraints reported by vast majority (95.00%) were Lack of varieties susceptible for pests and diseases. The banana growers are concern, Non availability of quality suckers, Out of fifteen constraints expressed by the jasmine growers, vast majority (95.00%) of the jasmine growers reported to express Poor keeping quality of flowers as their major constraint and Not convinced with recommended practices, Most of the sugarcane growers (95.00%) had expressed that the priority should be given to conduct research for controlling pests and diseases in sugarcane, Cent per cent of the banana growers (100.00%) had expressed that the priority should be given to conduct research for the production and selection quality suckers,. Most of the jasmine growers (90.00%) had expressed that the priority should be given to conduct research on improved cultivation practices, high yielding varieties.

References

Babu, S. Sezhian. 1990. "Information source utilization knowledge and extent of adoption of sugarcane technologies by registered cane growers", Unpub. M.Sc.(Ag.), Thesis, TNAU, Coimbatore - 3.

Chandrasekaran, G. 1979. A study on consequences of registered sugarcane cultivation". Unpub. M.Sc.(Ag.) Thesis, TNAU, Coimbatore.

Kavitha,M.K.1998.Knowledgeandadoptionofneemasbotanicalinputsinpaddy-an analysis. Unpub. M.Sc.(Ag.) Thesis, TNAU, Coimbatore.

Palaniswamy, A. (1978) "Adoption behaviour of malli and mullai flower growing farmers", Upub.Ph.D. Thesis, AC & RI, TNAU, Coimbatore.

Ratnabai. A.V. 1990. Information and innovation decision behaviour of Registered and non-registered sugarcane growers. Unpub. M.Sc.(Ag.) Thesis, TNAU, Coimbatore.

Venkattakumar, R. 1997. Socio-economic analysis of commercial coconut growers. Unpub. M.Sc.(Ag.) Thesis, TNAU, Coimbatore.

Venkattakumar, R. and K. Nanjaiyan. 1999. Socio-economic analysis of commercial coconut growers. Journal of Extension Edn. TNAU, 10(2): 1150-1152.

Venkatapirabu, J. 1988. A study on the knowledge and adoption of water management practices of paddy, Sugarcane and turmeric. Unpub. M.Sc.(Ag.) Thesis, TNAU, Coimbatore.

●●●

10

Reorientation of Agricultural Extension System in Tamilnadu towards ATMA

S. Parthasarathi [1] and R. Ganesan [2]

[1] *Assistant Professor, PAJANCOA&RI, Karaikal and*

[2] *Dean, Agri. College & Research Institute, Killikulant, Thoothukudi Dt.*

Introduction

The extension approaches followed in agricultural sector during past decade is reoriented in various occasions and new system of approaches were implemented. The present day orientation is shifting the efforts of extension from production orientation to profit orientation. Under these circumstances, the key factor for success will largely depend upon the extension service provider's orientation and farmer's capacity to absorb new knowledge of improved practices and technology. In order to orient officials towards this change HRD is essential. This will help them to impart required competencies in knowledge and skill such as technical, organisational, managerial, communication and business skills.

The present day extension approach i.e. ATMA focusing towards block level planning, farmers participation in planning and implementation of various schemes, group approaches, importance to agri and allied aspects are the major focus of ATMA. The Tamilnadu government in the year 2007 reorganised the extension system in to twotier at district level with the major focus to block level. In each block Agricultural Extension Centres (BAECs) were established which houses for agriculture, horticulture, marketing, agri. engineering departments.

A study to assess the needs of the stakeholder's viz., researchers,

NGOs, Input Dealers, officials of SDA and members of various farmers organisation were conducted with an emphasis to strengthern the BAECs.

Methodology

The study was conducted in Cuddalore District which has different farming situations. The respondents of the present investigation comprised of 62 block level extension officials of the district including ADAs,AOs and AAOs, 33 scientists available in the research stations and KVK. The 17 field staff of three reputed NGOs with agriculture orientation, 52 input dealers and 40 members and office bearers of various farmers' organisations available in the district.

Analysis of organisational needs

The assessment procedure followed by Michel *et al.* (2003) was adopted with modifications for the study. The need analysis was done with three phases as described below.

- Organising focus groups (Phase I)
- Administering a questionnaire (Phase II)
- Conducting feedback sessions (Phase III)

Phase I : Organising focus Groups

The focus group, which ranged from 1 hour to 1 and 1/2 hours, three main issues were asked. They were designed to identify strategic organizational issues, needs, and resources that SDA officials believed would help to serve the farming community in a better manner. Participants were asked to articulate the needs and resources that should be addressed due to restructuring. The out come of the discussions were recorded.

Phase II : Administering questionnaire

Based on the responses from the focus group discussion, a draft questionnaire was reviewed with consultation from extension scientists, officials. The major issues include organisational needs, input management needs and TOT needs.

Phase III : Feedback Sessions

The feedback session was carried out with the officials of the blocks in order to render the opportunity to create an organizational improvement plan for a block based on the findings from the questionnaire.

Findings and Discussion

Organizational needs as expressed by Stakeholders

The needs identified by various stakeholder of the study are given below.

Organizational needs expressed by officials of SDA

The organizational needs as expressed by the block level officials of SDA are presented in the Table 1.

Table 1 : Organizational needs expressed by SDA officials

*(n=54)**

SI.No.	Research Priority	Number	Percentage
1.	Well furnished office	51	94.44
2.	Transport facilities	43	79.62
3.	More powers to ADA	48	88.88
4.	Reduction of official work	49	90.74
5.	Posting of ministerial staff in block level	42	77.77
6.	Regular orientation regarding HRD	39	72.22
7.	More Extension — Scientist meet	47	87.03
8.	Training in ICT	48	88.88
9.	Posting adequate staff	50	92.59
10.	More fund for extension activities	36	66.67

* Multiple responses

It is observed from Table 1 that majority (94.44%) of the block level officials expressed that well furnished office followed by posting of adequate staff in the BAECs (92.59%) and reduction of official work (90.74%) as their primary needs. The other need expressed equally by them includes more power to ADAs and training in ICT (88.88%) and more Extension- Scientist meet (87.03%).

The restructuring of the SDA would enhance more farmers to avail the services in BAECs. This could be made effective only by providing the necessary infrastructure facilities like space for all the officials, meeting halls, computer facility etc. Reduction of clerical work and increase the field work, more number of field visits and field related activities have to be included for which transport facilities should be given to the officials to improve their performance level.

In order to carry out the given task and target, sufficient number of staff is essential. The introduction of many schemes needs more of fieldwork. To carry out the regular work and exercise control over subordinates, the ADAs should be given with more power for effective administration. To ease the official procedure and make the communication effective, the application of ICT is essential.

To know the research output, to get clarify the field doubts and implementing new schemes like precision farming which needs close monitoring and researchers support the number of Extension-Scientist meet should be increased to enhance maximum adoption.

Organizational needs expressed by the researchers

The researchers have well established contact with the SDA by providing research output during various occasions like zonal meetings, field trails etc. due reason that they are having high field exposure and research knowledge. The needs identified by them are as follows.

Table 2 : Organizational needs expressed by researchers

(n=30) *

SI. No.	Needs of SDA	Number	Percentage
1.	Timely availability of quality input	26	86.66
2.	Professional graduates at field level	18	60.00
3.	Adequate staff strength	27	90.00
4.	Easy accessing of field staff by farmers	20	66.66
5.	More field trials on modern technologies	28	93.33
6.	Updated knowledge about recent advancements	19	63.33
7.	Fixing responsibilities and critical review	20	66.66
8.	Incentive to better performing staff	14	14.66

* Multiple responses

It could be observed from the Table 2 that the majority (93.33%) of the researchers identified that more number of field trails was the major need followed by maintaining adequate staff strength (90.00%), ensure the timely availability of input (86.66%), supply of quality input (80.00%). Further, they prioritized the needs such as easy access of officials by the farming community (66.00%), fixing responsibilities and critical review (66.66%) and updated knowledge about recent advancements in agriculture (63.33%).

The possible reasons for such high tune of respondents expressed more number of field trails on modern technologies as a primary need due to the fact that such demonstration could only be exhibit the reality of any technology in the given environmental conditions which gives confidence to extension personnel to carryout the extension activities in a concentrated manner in a given area.

To implement location specific schemes adequate staff strength is essential. The researchers identified adequate staff strength was the second important need for the effective function of the SDA. After restructuring the SDA the blocks are given due importance. The staff strength at present is inadequate and difficult to carry out the schemes like IAMWARM, seed village, ISOPAM etc. This would give a set back to the functioning of field staff and also affect the farmers in utilizing the services of the SDA. Hence, the researchers felt this was an important need of SDA to serve for the farming community. The third need felt by the researchers was timely availability of quality inputs. Inputs are essential for transfer of new technologies. The researcher expressed that easy accessing of field staff was another need so that the farmers should contact the officials as and when required.

Organizational needs of SDA expressed by input dealers

Input dealers play a crucial role in Transfer of Technology. They are assuming greater responsibility in issuing inputs to the farming community. The needs expressed by them are given in Table 3.

Table 3 : Organizational needs expressed by input dealers

(n=50) *

SI. No.	Needs of SDA	Number	Percentage
1.	Technical guidance to input dealers periodically	48	96.00
2.	Close monitoring of stocks	34	72.00
3.	Meetings to update laws	47	94.00
4.	More field staff	37	74.00
5.	Display the stock positions in the office	40	80.00
6.	Frequent notification of recent chemicals/ brands	32	64.00

* Multiple responses

From Table 3 it is inferred that nearly cent per cent (96.00%) of the input dealers expressed periodical technical guidance to them was their primary need followed by regular meetings to update the amendments in law about handling of inputs (94.00%), display the stock position in the offices (80.00%), close monitoring of stocks (72.00%), more field staff in SDA (74.00%) and frequent notification of recent chemicals / brands (64.00%).

The farmers frequently contacting the input dealers to purchase inputs, to know about the recent arrivals of inputs and to hear the knowhow of some technologies. If the input dealers had adequate knowledge about the recent technologies it would help them to educate the farmers and thus increase their sales. Hence they are in need of technical guidance from the SDA periodically apart from the representatives of various input companies.

In order to avoid deviation in following the rules and regulations of the government, they need to have frequent update of the amendments in law or any order issued by the government. These meeting could help them to avoid the problem and serve the farming community better.

More field staff during the cropping season was another need expressed by the input dealers. This may be due to the following reasons such as effective transfer of technology, supply of subsidy and inputs, technical guidance, and quick processing of any official request by the input dealers.

Close monitoring of stock would help them to find the misusers and necessary action can be taken. This favours a fair business and increases the confidence of input dealers in the mind of farmers. So they expressed this as a need.

Frequent notification of branded, certified products would ensure the input dealers to place orders accordingly. This would be help them to avoid the purchase of any unbranded products available in the market. This would helpful to the farmers to get knowledge about the original products and use them properly.

Organizational needs as expressed by members of Farmers Organization

FOs are playing a vital role in disseminating the technology. Being a peer group these organisations act as a powerful change mechanism. The present day commercial agriculture also invites group activity. The needs expressed by the members of FOs in the restructured SDA are presented in Table 4.

Table 4 : Organisational needs expressed by members of FOs

(n=40) *

SI. No.	Needs of SDA	Num-ber	Percentage
1.	Adequate field staff	37	82.92
2.	Avoid frequent transfer of officials	32	78.04
3.	Contact with all sect of farmers	31	75.60
4.	Technical guidance from higher officials	26	63.41
5.	Creating awareness about input availability and schemes	40	97.56
6.	Timely availability of quality input	38	92.68
7.	Monitoring the reach of subsidy	32	78.04
8.	Using village panchayat in implementing schemes	26	63.41
9.	Use of modern gadgets for TOT	20	48.78
10.	Frequent meeting of officials of development departments	34	82.92
11.	More village level demonstration	38	95.60
12.	Avoid duplication of work between departments	24	58.53

* Multiple responses

It is inferred from Table 4 that nearly cent per cent (97.56%) of the members expressed that the SDA needs to create awareness about various input availability and schemes in operation. The other important needs expressed by 95.00 per cent of members were conduct more village level meetings and ensure the timely availability of quality and quantity of inputs (92.68%). The other needs include frequent meeting with officials of development departments (82.92%), adequate field staff (82.92%), avoid frequent transfer of officials (78.04%), establish contact with all the farmers of village, guidance from higher officials and using panchayats in implementing schemes (63.41%), avoid duplication of work by line departments (58.53%) and using modem gadgets for TOT (48.78%).

The government introduces many schemes to the welfare of farming community. The major set back in utilizing these schemes was due•to poor awareness. Creating awareness about the schemes through various occasions and media regarding the schemes and input availability would enhance the farmers to use the schemes effectively and reduce the workload of officials, thus both can get benefited.

In order to visualize the success of any technology at village level more number of demonstration plots should be established. This would clarify the doubts and build confidence among villagers to practice the recent technologies in their field. Further, it would also increase the confidence about the extension activities of the SDA. Thus, the members of the FOs expressed this as their second major need.

The availability of inputs during the peak period should be ensured for carrying out the operations without any delay. The SDA should give due consideration for the availability at BAECs in time with the desirable quality.

Frequent meeting with the officials of other development departments by the SDA officials with the focus on agricultural related schemes and farmer development and sharing information could increase the efficiency of these programmes. This could make the farmers to aware these schemes and get benefit out of it. The current position of staff after restructuring was inadequate. The farmers expressed that they were unable to meet the field officers in time and this causes delay in utilizing the government schemes. The "all under one roof concept" becomes reality only when the block level offices are given with adequate staff, as expressed by the members of FOs. The frequent transfer of officials had results a less intensity of operating the schemes. It was difficult for the officials to carry out the schemes and understands the problems of the villages in new environment. The subsidy given to the farmers were not reaching the target group properly. It should be monitored by the officials to ensure the reach of subsidy in a fair manner.

Organizational needs expressed by officials of NGOs

NGOs, the alternative extension service providers to the farming community playing an important role to serve the rural community. They had well established linkage with the people. The needs expressed by the field officials of NGOs about the restructuring of SDA are presented below.

Table 5 : Organizational needs expressed by officials of NGOs

*(n=17)**

SI. No.	Needs of SDA	Number	Percentage
1.	Creating awareness about schemes	17	100.00
2.	Invite for participation in schemes	17	100.00
3.	Use local knowledge	12	70.58
4.	Timely availability of inputs	15	88.23
5.	More field staff	13	76.47
6.	Regular meeting with NGOs	12	70.58
7.	Common place to exhibit the activities of NGOs	12	70.58
8.	Incentive to best performing officials	10	58.82
9.	Demonstration plots	11	64.70

* Multiple response

It could be exonerated from Table 5 that cent per cent field officials of NGOs expressed that creating awareness about the schemes and draw people's participation in developmental schemes were the major need to serve the farming community. The other needs expressed by them include timely availability of inputs (88.23%), adequate field staff (76.47%), common place to exhibit the activities of NGOs and frequent meeting with NGO field officials (70.58%).

The field officials of NGOs expressed that the farmers were not aware many of the schemes. If they were advertised properly they would come forward to use these schemes. The participation of farmers in the field activities of SDA would create confidence in the minds of people about the benefit of the programmes. Thus the programmes can yield better result in a sustainable manner even after the withdrawal of official support.

Time of availability of input should be made known to the farmers. It helps the farmers to avoid frustration about availability and unnecessary travel to the offices. The number of field workers was found to be very low, hence the number should be increased according to the proportionate of area and number of farm families, was yet another need expressed by the NGOs. Due to poor staffing many of the schemes were not reached the beneficiaries. The farmers find difficulty to avail the services of the SDA due to inadequate staff.

The other needs expressed equally by the NGOs officials includes regular meeting with NGOs and use of local resource and knowledge. The NGOs in the village were closely working in villagers especially with the down trodden. They are well versed with the profile of the farmers. The regular meeting of the NGOs officials and SDA officials would be beneficial to exchange the information about village particulars and to carryout common goals of development.

The SDA block offices should provide platform to exhibit the activities of NGOs working in the geographical area. This would help the farmers to know about the activities and use them subsequently. More over it would also help the officials of NGOs and SDA would get benefited by sharing knowledge, resources for the common purpose. This would in turn help the SDA to serve better the farming community.

Conclusion

The role of government is critical for the reconstruction of.agricultural extension, especially for fostering the public good. Only the public sector can effectively and efficiently carry out certain functions and indeed, only government can assume the responsibilities that affect the state as a whole. Only governments can ensure that extension services work for the public good, even if those services are provided by contracting with private sector providers.

This modification has to be taken with due consideration and the views and opinion about this restructuring among the stakeholders the necessary modification can be carried out for effective functioning. It is high time for them to update themselves with various changes taking place locally and globally. Capacity building and institutional strengthening require for qualified service and ensure strong links with and modernization of the various components of the formal and non-formal agricultural extension system.

References

Anonymous. 2000a. Policy Framework for Agricultural Extension, Extension Division, Department of Agriculture and Co-operation, Ministry of Agriculture, Government of India, New Delhi.

Michael Havercamp, Elizabeth Christiansen and Deborah Mitchell.2003. Assessing Extension Internal Organizational Needs through an Action Research and Learning Process. *Journal of Extension*. Available in the web http://www joe.org/ joc/2003october/a2.shtml

11

Strategies to increase the Area, Production and Productivity of Cassava in Salem District in Tamil Nadu

A. Janaki Rani[1] *and P.P. Murugan*[2]

[1] *AssistantProfessor(Agri.Extension)Dept.ofTradeandIntellectualProperty, TNAU, Coimbatoreand*[2]*AssociateProfessor(Agri.Extension), ATIC, TNAU, Coimbatore*

Introduction

Cassava is an important tuber crop in India grown in 2.7 lakh hectare area with a production of 71 lakh tonnes. The average yield of tapioca is 22 tonnes per hectare. Cassava is a richest source of starch (25 to 35%) mainly processed for starch and sago. Cassava is an industrial crop of Tamil Nadu occupies 32.97% of the area and 45.98% of the production in India. (Edison *et al* 2006). In Tamil Nadu major traditional tapioca growing districts are Salem, Namakkal, Erode, Cuddalore, Dharmapuri and Kaniyakumari mostly as rainfed. Among the districts Salem District stands first in area (27000 ha) and with the production of 10.475 lakh MT(Anonymous, 2006).

There are about 900 sago and starch factories in and around Salem, Erode, Namakkal and Dharmapuri districts which depends on cassava tubers. The number of factories in Salem district alone is 650. It is estimated that 60% of the starch produced in India is from Salem district. Because of ease in cultivation, drought tolerance and raise in prices of tubers, the area under cassava is on increase in other District of Tamil Nadu viz., Erode, Trichy, Coimbatore.

Even though the area is increasing the production is not sufficient to meet the demand of the food and textile industries. Even though the area is increasing, the farmers are cultivating tapioca in larger area; the scientific method of cultivation is poor. Scientists are developing lot of varieties, new improved technologies etc., while seeing the knowledge and adoption of recommended practices were found to be least.

It has been well demonstrated by the research system through its out reach programmes that yield of tapioca could be boosted to a level of 35 tonnes per hectare in farmer's field if farmers adopt the recommended practices. But in practical farmers are getting 20-25 t/ha.

For the past three years, particularly in Salem District, the cassava area and production has been reduced. In 2005-2006 the area was about 27,000 ha but this had comedown drastically to 10,564 ha in 2007-2008. It slightly increased to 15,728 ha in 2008-2009. This had created a shortage in the supply of tuber to starch mills and affected the production. Shortage in the supply of ca tubers is posing a serious threat to the survey of tapioca starch manufacturing units in Salem Districts. Due to the low production about 300 factories have already been closed.

Table 1 : Area and production of Cassava over years in Salem District (Statistical report 2008, Salem District of Tamil Nadu)

Year	Area (ha)	Production (lakh mt)
2004-2005	9,908	3.567
2005-2006	27,007	10.475
2006-2007	13,287	4.783
2007-2008	10,564	3.803

Many farmers switching over to short term crops such as sun flower and maize in many parts of Salem and neighboring districts Namakkal and Dharmapuri. If the remaining starch manufacturing units are closed because of shortage of tubers, the livelihood of a large number of people employed in these units will be affected.

Hence, there is an urgent need to channelise our efforts to increase the Cassava yield in farmer's field to achieve the targeted yield. In view of these a study to asses and identify the factors influence for the low yield at farm level and reasons for area shrinkage may be useful to generate'technologies and to design promotional strategies.

Methodology

The study area considered for this project is Salem District in Tamilnadu. There are 9 taluks and 20 blocks in Salem district. Based on the maximum area under Cassava, Attur taluk (9649 ha) was selected purposively. From Attur taluk two blocks namely Attur and Pethanaiken palayarn were selected based on maximum area. From the selected blocks, 3 villages/ block was selected based on maximum area. The selected villages viz., Manjini, Mullaivadi, Ammampalayam, Ramanaiken Palayam, Veeragoundanur, and Chelliyam Palayam. In each village 10 cassava farmers were selected randomly. Totally, 60 farmers were selected as respondents for the study. An interview schedule was constructed, pre tested and finalized and used for collecting data from the respondents. Group meeting also held to identify the constraints. Relevant data collected pertaining to the study was analyzed, interpreted and meaningful conclusions were drawn using percentage analysis and mean.

Findings and Discussion

The constraints faced by the cassava growers were presented in the following tables.

I. Production constraint

Table 2 : Distribution of respondents according to their production constraints

SI. No.	Constraints	No.	Percentage	Rank
	I. Production constraint			
1.	Mosaic virus and tuber rot	50	83.33	I
2.	Labours scarcity	48	80.00	II
3.	Un availability of quality planting materials	44	73.00	III
4.	Lack of short duration varieties with high starch	41	68.33	IV
5.	Decline in area due to the introduction of short duration crops	27	65.00	V
6.	Assured irrigation	39	45.00	VI
		Mean	**69.11**	

It could be observed from Table 2 that in Cassava cultivation mosaic virus and tuber rot are the important diseases in reducing the yield loss upto 40 per cent. This was expressed by majority of the farmers (83.33 per cent). Labour scarcity was another production constraint expressed by most of the respondents (80.00 per cnet). Agricultural labourers are being seasonal, there is always shortage of labour during peak season. The migration of the labour from agricultural to other occupations and to other sectors has also contributed to the labour problem. Unavailability of quality planting material was expressed by 73.00 per cent of the farmers. Cassava is an eight to ten months old crop. So, lack of short duration varieties is one of the major constraints expressed by 68.33 per cent of the farmers. Nowadays, Cassava area is reducing due to the introduction of short duration crops like maize, sunflower etc. They felt that they have to wait one year for harvest of tapioca as tubers and also not fetching good rate in the industries. But the above crops had three months of duration and cultivation of these crops may increase the income within three months was expressed by 65.00 per cent of the farmers. Due to lack of water facilities cassava growers could not able to irrigate their lands. This was expressed by 45.00 of the growers.

II. Processing constraint

Table 3 : Distribution of respondents according to their processing constraints

Sl. No.	Constraints	No.	Percentage	Rank
	II. Processing constraints			
1.	Poor keeping quality	48	80.00	I
2.	Lack of post harvest technology units	43	71.66	II
3.	Lack of Government support for commercialization	38	63.33	III
4.	Lack of technical guidance for value addition	34	56.66	IV
5.	Lack of Government programmes for value addition	31	51.66	V
		Mean	**64.66**	

Due to physiological and pathological processes within the tubers the keeping quality of the cassava tubers gets reduced. This was expressed by 80.00 per cent of farmers. Due to this problem forced to sell their produce even for lower rate to the industries (Lakshmi *et al.,* (1992)

Processing the tubers into sundried chips and other products which have a better shelf life. So lack of Post harvest technology units in the study area was expressed by 71.66 per cent of the farmers. In Tamil Nadu, especially in Salem District no government factory's for commercialization into food, feed and industrial purposes. This was expressed by 63.00 per cent of the farmers. The research system only concentrates on the production aspects. If farmers were trained in value addition, the women tapioca growers will get additional income in their leisure time. More than 200 value added products can be made from it so lack of technical guidance for value addition expressed by 56.66 per cent of the farmers. Extension programmes motivate farmers to know the production practices and value addition to start a venture on their own. So lack of Government programmes for value addition was experienced by 51.66 per cent of the farmers.

III. Marketing constraint

Table 4 : Distribution of respondents according to their marketing constraint

SI. No.	Constraints	No.	Percentage	Rank
	I. Marketing constraint			
1.	Exploitation by middle man	53	88.33	I
2.	Male practices in Point scale fixation	52	86.66	II
3.	Lack of regulated market	50	83.33	III
4.	Low price for tubers	48	80.00	IV
5.	Government factory to fix rate	45	75.00	V
6.	Adulteration by factory personnel by mixing Urea and `K'	35	58.33	VI
		Mean	**78.66**	

It could be observed from the Table 4 that majority of the farmers (88.33 per cent) expressed that the tubers are sold mainly through brokers (More than 90% sales). Private industry personnel fix the rate of tubers/ton. Here brokers are playing major role was felt as a major market constraint.

Cassava is the only crop sold through point scale method. Sometimes the point scale fixation was not done in front of the farmers. They felt that some malpractices were done by factory personnel and was expressed by 86.66 per cent of the farmers. And also they felt if the rate fixed based on weight basis, no need for mal practices.

There is no regulated market or market federation for cassava, which leave the farmers at the mercy of contract merchants or carrying out the job by themselves. So, they need direct procurement and it was expressed by 83.33 per cent of the farmers. Due to low keeping quality interference by middle man and excess flow of tubers the rate of tubers become getting lower. This was expressed by 80.00 per cent of the farmers. In and around Salem district 650 private industries are functioning. But no Government factories are functioning to fix rate for tubers. This was felt by 75.00 per cent of the farmers.

About 58.33 per cent of the farmers expressed that the factory personnel mixing Urea and Potash in the water in order to reduce the point scale value for tubers. By this practice they felt the starch quality and starch content might have reduced.

IV. Socio-economic constraint

Table 5 : Distribution of respondents according to their socio-economic and extension constraint

SI. No.	Constraints	No.	Percentage	Rank
	Socio-economic constraint			
1.	Lack of training programmes	34	56.66	I
2.	Lack of credit facilities	25	41.66	II
3.	High cost of inputs	25	38.33	III
		Mean	**51.00**	

Absence of adequate financial institution like Agricultural Banks, Co-operative Society etc., and rigid rules and regulations to get credit was felt by 41.66 per cent of farmers. They also felt that the inputs are high end expressed as a constraint by 38.33 per cent of the farmers. They expressed that subsidy may be provided to encourage the Cassava cultivation. Only few training programmes were conducted by government and held at distant places. This was expressed by 56.66 per cent of the farmers.

V. Technological constraint

Table 6 : Distribution of respondents according to their technological constraint

SI. No.	Constraints	No.	Percentage	Rank
	Technological constraint			
1.	Top dressing on 90th day	44	73.00	I
2.	Whitefly and tuber rot management	41	68.33	II
3.	Not aware of TNAU varieties	39	65.00	III
4.	Sett treatment methods	38	63.33	IV
5.	Micro nutrient application	34	56.66	V
6.	Lack of awareness on raising nursery	34	56.66	V
		Mean	**78.66**	

Top dressing on 90th day with Urea and Potash is essential for the development of tubers. Due to lack of technical guidance it was not adopted and expressed by 73.00 per cent.

Due to lack of knowledge farmers could not know the control measures for whitefly and tuber rot and it was expressed by 68.33 per cent of the farmers. Most of the farmers in the study area cultivating only the local varieties like Kunguma Rose, White Rose, Burma etc.,. They does not know much about high yielding Tamilnadu Agricultural University varieties like CO2,3,CO(TP)4 and CO(TP)CTCRI-5 and importance in increasing the yield and resistance to CMVD and it was expressed by 65.00 per cent of the farmers. Lack of awareness and knowledge farmers were not follow the sett treatment methods and it was expressed by majority of the farmers 63.33 per cent. Micronutrient application and raising nursery were not known by 56.66 per cent of the farmers.

By working out the mean for different constraints marketing constraint stands first and marketing analysis may be given top most priority for research. Production constraint ranked as second. So training cum demonstration on the cultivation aspects and value addition may be given to increase the production and productivity of Cassava in Salem District.

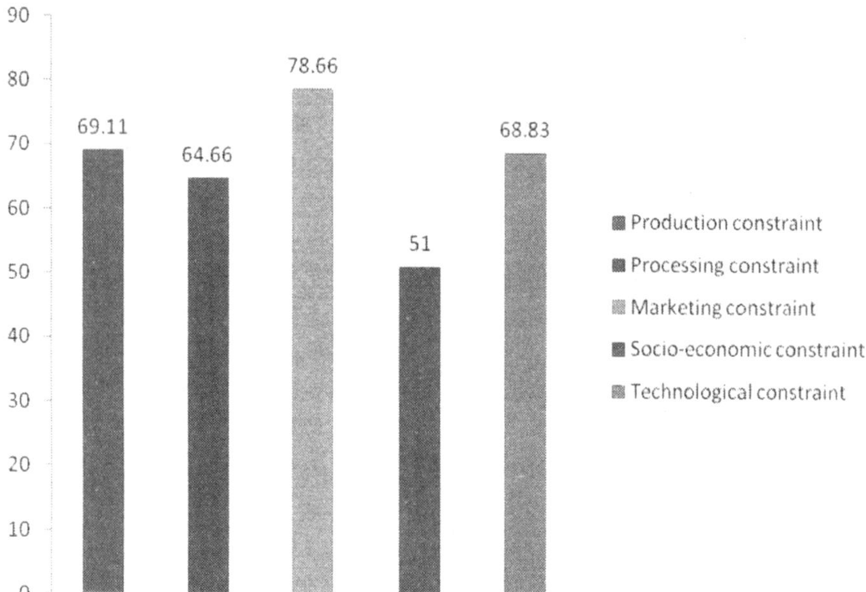

Fig. 1 : Mean of the constraints encountered by the farmers

Strategies needed

Based on the constraint analysis the following strategies may be adopted to increase the production and productivity and also to improve the socio economic status of Cassava growing farmers.

- ✓ Development of Cassava Mosaic Disease Resistant, short duration varieties with high starch content for industrial utilization.
- ✓ Large scale production and distribution of quality planting material of cassava
- ✓ Popularization of TamilNadu Agricultural University varieties through various extension methods.
- ✓ Creation of awareness on recommended technologies like Nursery techniques, Sett treatment methods, importance of top dressing on 90th day, micronutrient application and management of CMVD & Tuber rot.
- ✓ Frequent training cum demonstration programmes on Cassava cultivation and value addition with communication materials.
- ✓ Training for Self Help Groups, NGOs, women farmers on value addition of Cassava and to start their venture on own.
- ✓ Establishment of post harvest technology units for processing and storage, provision of credit facilities, establishment of regulated market, direct procurement from farmers and price fixation in terms of quantity by weight basis/setting up of association for point scale fixation is the need of hour.

Conclusion

It is estimated that currently 3.3 million tones of cassava is being used in human consumption sector, 1.5 million tones of cassava in industrial sector and 0.9 million tones in animal feed sector. The demand and supply projections indicate that by 2020 A. D, the potential demand for cassava will exceed the potential supply by 3.047 and 2.758 million tones (Edison et al 2006) This gap can be bridged by expanding the cassava area in non-traditional areas and productivity improvement through the adoption of improved cultivation practices especially in the industrial zones of Tamil Nadu.

References

Anonymous. 2006 Horticulture Statistical report of Salem District in Tamil Nadu. The Directorate of Horticulture, Tamil Nadu.

Edison. S. Anand, M. and Srinivas, T. (2006) Status of cassava in India. In overall view — published by CTCRI, Trivananthapuram.

Lakshmi, K.R., Anantharaman, M. and Ramanathan, S. 1992. Problems and prospects of cassava in India. Agricultural situation in India. 47(6):527-530.

Lakshmi, K.R. and Anantharaman, M. 1992. Growth and instability of cassava in India, Journal of root crops. 27(2) : 591.

Lakshmi, K.R. and Pal, T.K. 1986. Trend of area, production and productivity of cassava in India. Agricultural situation in India, 41(8):609-614.

12

Market Information Utilization Behaviour of the Farmers Receiving Dynamic Market Information (DMI) through Short Message Service (SMS)

***R. Balasubramaniyam*[1] *and N. Anandaraja*[2]**

[1] *Leaf Superintendent, Agri Business Division - ILTD, ITC Ltd. Mysore, Karnataka, (India) and*

[2] *Assistant Professor (Agri. Extension), e-Extension Centre, Directorate of Extension Education, Tamil Nadu Agricultural University, Coimbatore, Tamil Nadu*

Introduction

India now emerged as the world's second largest producer of fruits, vegetables and tea holding a share of 10 percent. The production base of horticultural crops has been expending since independence (Chadha, 2009). But as a controversy to the previous statement there is a clear indication about the declining farm sector performance-in addition to its deleterious economic ramifications often causing large scale rural distress which often ends up in farmer suicides (Sundaresan, 2009). This distress conveys much on the economic rationale ruling the agri and rural domain for its incompatibility in the changing market environment. The identified major problems faced by farmers in the present scenario were non-availability of information about marketing services including availability of crop specific markets, existing demand of a commodity and daily market price for their produce which can be broadly categorized under lack of market-led extension services.

As an initiative to help the farmers to overcome from all the above mentioned problems Tamil Nadu Agricultural University (TNAU) in collaboration with Centre for Development and Advanced Computing (C-DAC) implemented a pioneer project called Dynamic Market Information (DMI) which aims at providing daily market prices (both wholesale and retail) for 157 selected perishable commodities (67 Vegetables, 36 Fruits, 32 Flowers, 14 Spices and 8 Plantation crops) in thirteen south Indian markets comprising Tamil Nadu, Kerala and Karnataka. The information is uploaded regularly in the DMI internet page linked with the following websites namely www.tnau.ac.in, www.indg.in and www.agritech.tnau.ac.in. The DMI webpage also provides the information about traders in the respective markets and also the previous price database. Mobile phone is known to be the most powerful ICT tool of the present (Yunus, 2008) because of its maximum penetration level in rural India. DMI is being implemented through mobile phones to almost eight thousand subscribers in Tamil Nadu for the past two years, its impact in the perspective of utilization and development is unknown. It becomes important to study the farmers' utilization behaviour of the market information provided by DMI regarding the decision making by farmers in all their crop production and marketing aspects. Hence a research study was designed and conducted with the specific objectives to study the Information utilization behaviour of the farmers of Dynamic Market Information (DMI) through Short Message Service (SMS).

Methodology

The study was conducted in Dharmapuri and Erode districts of Tamil Nadu. Three blocks from each district were selected and two villages were identified from each block summing up to a total of 12 villages. A total of 120 respondents were selected based on systematic random sampling technique as 10 respondents from each village. Data collection was done through interview schedule and focused group discussion methods. The respondents were interviewed personally using a well structured interview schedule to obtain relevant data regarding their profile, socio-psychological characteristics and DMI utilization pattern.

Findings and Discussion

DMI utilization behaviour of farmers

The Dynamic market information utilization behaviour of the respondents was studied with identified ten items namely possession of media, use of media, place of use, purpose of use, frequency of use per day, information preference, decision on information, forwarding informa tion to others, cross checking and satisfaction level of using DMI. Based on the particulars collected, the respondents were categorized into low, medium and high. Majority of the respondent (80.0%) had medium degree of Dynamic market information utilization behaviour followed by high (15%) and low (5%) levels.

DMI project is operated in the selected study area for the past two years. Majority of the respondents were experienced with the utilization of DMI within the range of 1 to 2 years. This might be the reason for the existence of medium degree of Dynamic market information utilization behavior among majority of the respondents. Liu (2002) argued that the economic costs of modern market information services could sometimes be high, and difficult to sustain without support. Also, externallygenerated Marketing Information Services appear still only to have been a minor source of information for many intermediaries, who continued to rely more on personal contacts for their market information.

Pattern of DMI utilization Behaviour of the respondents

The various aspects of Dynamic market information employed to study its utilization behaviour is furnished under Table 1.

Table I : Distribution of respondents based on various aspects of DMI utilization behaviour

(n=120) *

SI.No	Particulars	Number	Percent
a.	Possession of media		
	Mobile	120	100
b.	Use of media		
	General	97	80.83
	Entertainment	23	19.16
c.	Place of Use		
	All (home, farm & market)	120	100
d.	Frequency of use per day		
	Twice	71	59.2
	More than twice	49	40.8
e.	Information preference		
	Price	40	33.3
	Market	30	25.0
	Traders	50	41.7
f.	Decision on information		
	Where to sell	74	61.7
	Where to sell	46	38.3
g.	Forwarding to others		

Sl.No	Particulars	Number	Percent
	Yes	72	60
	No	48	40
h.	Cross checking		
	Yes	87	72.5
	No	33	27.5
i.	Satisfaction		
	Yes	82	68.33
	No	38	31.66

Possession of media

All the respondents (100.00%) possess mobile phones as the media for receiving and forwarding the Dynamic Market information.

The reason for non — possession of computers with Internet for getting to DMI is due to its high cost and it requires knowledge and skill to operate and maintain.

Use of Media

Almost 81 percent of the respondents use the mobile telephones for general communication purpose other than using it for getting SMS from DMI. All the respondents use the media for the purpose of Agricultural activities only. Since their major occupation is agriculture, the sharing of information is predominantly on various aspects of agriculture like information sharing on inputs, crop management practices, communicating with experts in various institution, information sharing on market prices, information gathering on market arrivals and demand etc.

It is also inferred from Table 1 that 19 percent of the respondents use the mobile phone for entertainment purposes like hearing music, watching video clips and playing games installed in the mobile phone. Though many of the respondents are aware of the entertainment and academic utilities of mobile phones, they do not find leisure and interest to use them.

Place of use

Mobile telephones being a handy portable device it is obvious that all the respondents (100%) use the mobile phones at their home, farm and market. This result adds to the reason for higher level of penetration of mobile phones in rural areas.

Frequency of use per day

Majority (59.2%) of the respondents used to view at the SMS sent through Dynamic market information twice a day followed by 40.8 percent of respondents using it for more than twice a day. This result shows that reinforcement and retention of information is highly possible when it is sent through short messaging service (SMS).

Information Preference

It is interpreted from the Table 1 that 41.7 percent respondents prefer to get information about commodity specific Traders in various markets followed by 33.3 percent of respondents preferring for price information for more than one crop and 25 percent of respondents prefer to get information about various markets nearby their locality.

During the study the respondents expressed that the need to know about potential traders in the markets nearby is very important to them so that they may take an attempt to bye pass the exploitative middlemen in the market further they prefer to get market price information for more than one crop because as already discussed all the respondents in the study area cultivate more than one crop at a time and so they need the price information on many crops so that they can fetch better price for all the commodities grown in their farm.

Decision on information

It is known from the Table 1 that 61.7 percent of the respondents make decision on where to sell their produce after getting the information from DMI followed by 38.3 percent deciding on when to sell the produce.

The reason attributing to this result is; based on the price information given by DMI about a commodity in various markets, the respondents make the decision on where (i.e. which market) to sell the produce. Obviously their choice of selecting the market will be based on high prices for the produce, distance of the market, transport, and infrastructure facilities of the market.

Forwarding to others

Table 2 shows that majority (60%) of the respondents used to forward the messages received from DM1 to their fellow farmers and friends who were not registered for DMI or receivers of information about some other crop other than the information forwarded for a crop. It is followed by 40 percent of respondents who do not forward the messages to other for the reasons that all the other known farmers to him are already registered users of DMI and he values the information as a treasure and do not want to share it with others.

Cross checking

Table 1 further reveals 72.5 percent of the respondents used to cross check the information provided by DMI to other sources like traders in various markets, commission agents and other farmers. They expressed that the information provided by DMI is highly credible and useful.

Satisfaction

It is inferred from Table 1 that all the respondents sampled for this study expressed a high level of satisfaction regarding the overall utility of Dynamic market information disseminated through mobile telephone.

Conclusion

It is inferred that 80 percent of the respondents had medium degree of DMI utilization behavior. All the Respondents possessed Mobile Phone as the media for receiving information from DMI and 81 percent of them use the media for General Communication purposes besides using it for DMI followed by 19 percent of the respondents use the mobile media for entertainment purpose. Mobile Phone being a portable and handy device, 100 percent of the respondent uses it in all the places like farm, house and market. All the respondents use the media for agricultural purpose as a major contributor 59.2 percent of the respondents use (or) view the SMS sent through DMI users 41.7 percent prefer to get information about traders in various markets followed by 33.3 percent preferring price information for more than one crop and 25 percent of respondents prefer to get information about many potential markets nearly to them.

Majority (61.7%) of the respondents make decision on where to sell the produce after getting the information from DMI followed by 38.3 percent deciding on when to sell the produce . About 60 percent of the respondent forward the message received by them from DMI to other friends who are need of the information followed by 72.5 percent of the respondents cross check the information provided by DMI with various other sources and 68.33 percent respondents expressed that they are highly satisfied with the overall performance and utility of Dynamic market information disseminated through mobile telephony. and 31.66 percent of the respondents were not satisfied with the overall service provided by DMI because of various constraints experienced by them.

References

Chanda, K.L.2009. Horticulture: The Next in Agriculture, Agriculture Today Yearbook 2009, New Delhi.

Liu, 2002. E-Commerce in Taiwan's Agricultural Marketing in "Bulletin of the Food and Fertilizer Technology Center for the Asian and Pacific Region" (FFTC), Taipei, Taiwan, November 1, 2002.

Nambothripad, P. 2000. A study on commercial vegetable cultivation in Oddanchatram area and its socio-economic impact on the growers, Unpub. M.Sc.(Ag.) Thesis, Ac & RI, TNAU, Coimbatore.

Rosaiah Yeluri, 2002. An analysis of knowledge gap on 1PM technologies of cotton-A system approach, Unpub. M.Sc. (Ag.) Thesis, AC & RI, TNAU, Coimbatore.

Sundaresan, C.S.2009. Sustainable Agri-business, Agriculture Today Yearbook 2009, New Delhi.

Venkatesan, R .2000. Farmers awareness and adoption level of recommended tomato cultivation practices, Unpub. M.Sc. (Ag.) Thesis, AC & RI, TNAU, Coimbatore.

●●●

13

Information Sharing and Dissemination at Uzhavar Udhaviyagam — A Promising Linkage Model for Sustainable Extension Services

R. Sendilkumar [1] ***, C. Cinthia Fernandaz*** [2] ***and S. Parthasarathy*** [3]

[1] *Associate Professor (Agri. Extension), Kerala Agrl University, Thrissur,* [2] *Assistant Professor (Agri. Extension), Tamil Nadu Agricultural University and*

[3] *Assistant Professor (Agri. Extension), PAJANCOA & RI, Karaikal.*

Introduction

Public extension services in the agricultural sector have not kept pace with new challenges and opportunities. Overall, the reform measures initiated by the Government are yet to penetrate agriculture and allied rural sectors. The development of linkage in agriculture is extremely important for the system approach, in order to provide proper knowledge and services to the farmer contributing to the agricultural production. Many service sectors in agriculture are working independently to achieve their organizational objectives. This results in less organizational effectiveness and exhibits poor forward and backward linkages. Public private partnership and NGO partnership has been promising institutional interventions to provide need based information through information and communication technologies (Van Den Ban 1997). Policy frame work for Agricultural Extension (2000) has notified the provision for establishing partnership with private and other public agencies. Uzhavar Udhaviyagam is one of the interventions implemented by Govt of Puduchery since 2000, with an objective of integrating various agro extension services offered by related public sectors, private sectors and

NGOs, cooperative sectors under one umbrella. This is a'so providing ample scope for establishing 3 Ps (Public private/NGOs partnership) model and participatory extension services. Keeping this focus point a study has been formulated with an objective to study the degree of linkage on information sharing and dissemination regarding the extension services rendered by *Uzhavar Udhaviyagam.*

Methodology

The *Uzhavar Udhaviyagam* scheme was launched by the Government of Puducherry only in two regions (Puducherry and Karaikal). The Karaikal District was selected for the study. The district has got 11 Uzhavar Udhaviyagam units headed by an Agricultural Officer and with established partnership of Farmers advisory committee (Pasumai Padai) were considered for the study.

A well structured questionnaire and focus group discussion method were used to study the degree of linkage of *Uzhavar Udhaviyagam* (Farmers helping centre) with other agencies in terms of information sharing and dissemination on the line of rendering different extension services to the farmers. The Primary data was collected from the unit head concern. The responses were obtained on three point continuum scale and mean score was calculated for the purpose of analysis.

Findings and Discussion

The functional linkages of *Uzhavar Udhaviyagam* with other allied organization was studied on the perspectives of information sharing and dissemination with respect to identified functions and services such as weather based farm advisory services, providing information on Govt schemes with respect to agriculture, details on agri-inputs availability, market intelligence, registration facility for seed certification, hiring services of farm machinery & equipments, and providing technical information. The data on these said perspectives were collected and furnished in the Table.l.

Table 1 : Functional linkages of *Uzhavar Udhaviyagam* with other organization

n=11

SI. No.	Functions of *Uzhavar Udhaviyagam*	Linkages with other organisation	Mean Score
1	Weather based farm advisory services	Office of the Additional Director of Agriculture	2.363

Contd. ...

		Farmers advisory committee(Pasumai Padai)	2.363
		Pandit Jawaharlal Nehru college of Agriculture and Research Institute. (PAJANCOA & RI)	1.272
		Meteorology department	1.909
		Farmers association on irrigation (Pasanatharar Sangam)	1.363
2	Regarding the details of Govt schemes with respect to agriculture	Department of Agriculture	2.727
		Farmers advisory committee (Pasumai Padai)	2.363
3	Details on agri-inputs availability	Pondicherry Agricultural Service Industrial Corporation(PASIC)	2.363
		Farmers advisory committee (Pasumai Padai)	2.363
		Private Input agencies	2.000
		PAJANCOA & RI	1.181
		Krishi Vigyan Kendra, Madur	2.090
		Seed depot	2.727
		Krishi Vigyan Kendra, Sikkal	2.090
4	Market intelligence	PAJANCOA & RI	1.878
		Uzhavar sandhai	2.272
		Regulated Uzhavar sandhai Market/ Market committee	2.090
		Farmers advisory committee (Pasumai Padai)	1.190
5	Registration facility for seed certification	Pondicherry seed certification agency	2.727
		Farmers advisory committee (Pasumai Padai)	1.454
6	Hiring services of farm machinery & equipments	Farmers advisory committee (Pasumai Padai)	2.363
		Machinery suppliers	1.181
		Mechanic operator	1.363
		Private owners	1.818
7	Providing Technical information	PAJANCOA & RI	2.363
		AO (Tech)	2.727
		KVK (SMS) Madur	2.181
		Agri portal	2.000
		Farmers advisory committee (Pasumai Padai)	1.818
		KVK (SMS)sikkal	2.181
		NGOs(CEE, Pradan)	1.818

The *Uzhavar Udhaviyagam* have a medium to strong linkage with Pasumai padai (2.363) regarding information dissemination function on weather based advisory information. It has a medium linkage with Meteorology department (1.909) and relatively lesser linkage with Pandit Jawaharlal Nehru College of Agriculture and Research Institute (PAJANCOA&RI) and Pasanatharar Sangam (1.363). The Farmer advisory committee named as Pasumai padai, basically established for the purpose providing mutual linkage on the information sharing and dissemination function. Hence it showed such type of linkage. However, the existence of Indian meteorology department in the district is yet another added advantage in providing data regarding the weather forecasting, also being the contributing factor for such linkage.

With respect to the information providing function on Government schemes, *Uzhavar Udhaviyagam* had established strong linkage with its parental organization namely Department of Agriculture(3.0) being the primary source for all such information. Since, Pasumai Padai is also acting as vehicle for carrying such information from Uzhavar Udhaviyagam to fellow farmers exhibited a strong linkage.

Uzhavar Udhaviyagam had shown relatively stronger linkage with Pondicherry Agro Service Industrial Corporation (PASIC) (2.363), Krishi Vigyan Kendra, Madur and Sikkal (2.090) on extending the information providing function on agri-inputs availability. The stronger linkage was due to fact that the issue of subsidy on agri-inputs is vested with the Agriculture officer of the *Uzhavar Udhaviyagam* unit. Hence the agriculture officer concern often checks the availability of agri-inputs before issuing subsidy certificate to the farmers. Apart from this, he might have also provides information received from private dealers regarding the availability of inputs and such information passed to clientele whenever unforeseen shortage occurs from the PASIC. Pasumai Padai (2.363) had shown a strong linkage in this regard and thereby helps in spread the information on input availability to other farmers.

In connection with the market intelligence function, it established linkage with the institutions viz., regulated Market/ Market committee, Uzhavar sandhai, PAJANCOA &RI. However it is not justified with the full purview of market intelligence function. Because it provides, only the current market rate for the produce and not the forecasting price. The obvious reason being that the lack of availability of modern gadgets at the centre and as such these institutions are not empowered enough to utilize the market intelligence data to suit the farmers need.

The *Uzhavar Udhaviyagam* also shares the information on seed registration and certification. Since the Department of Agriculture operates both Pondicherry seed certification agency and *Uzhavar Udhaviyagam*, it naturally established a strong linkage in this regard. On the other hand the Pasumai padai also established a strong linkage with respect to information acquisition and dissemination functions to other farmers.

Hiring services of farm machinery & equipments to intended farmers through Pasumai padai is one the important function of the *Uzhavar Udhaviyagam.* To perform this function it had established necessarily a medium to strong linkage with Pasumai padai (2.363) and private agriculture machinery owners (1.818). It had comparatively lesser linkage with mechanic operators (1.363) and machinery suppliers (1.181). Though this particular function fully operated by the Pasumai padai,the Agriculture officer of the concern unit has to closely monitor and supervise for the smooth function. Thus it facilitates to have stronger linkage.

In order to provide technical information to farmers it depends on Agriculture Officer (Technical) Department of Agriculture (2.727), Scientists of PAJANCOA & RI (2.363), Subject matter specialist of KVK Madur and Sikkal (2.181), Agri portal (2.000), NGOs like CEE, Pradan (1.818) and for the information dissemination function it has mostly depends on Pasumai padai. On this line, it establishes medium to strong level of linkage with the allied institutions.

It has been observed that in all kinds of extension services of *Uzhavar Udhaviyagam,* in variably Pasumai padai has established a medium to strong linkage regarding the information dissemination function.

2. Mode of establishing and sustaining linkages

The officers in charge of the Uzhavar Udhaviyagam units have been adopted different extension methods for information sharing and dissemination to establish and sustain the linkage with the allied institutions and clientele. The relevant data on this subject was collected and given in the following Table 2. It is evident from the Table 2 that most of the officer concerns have used the, telephone/mobile communication (100.00%), establishing personal rapport (81.81%) personal/direct method (72.72%) for information sharing and dissemination. They have mostly used the occasion of participating in training /seminar/Workshop and ATMA meetings as a means to share the information.

Table 2 : Mode of establishing and sustaining linkages

n=11

SI.No.	Extension methods used	Number	Percentage
1	Personal /direct contact	8	72.72
2	Telephone/mobile communication	11	100.00
3	e-mailcommunication	0	-
4	Telephone/mobile communication	11	100.00
5	Organizing get-together	2	18.18
6	Establishing personal rapport	9	81.81
7	Forum	2	18.18
8	Participating training /Seminar/Workshop organized by PAJANCOA & RI, KVK	11	100.00
9	Participating in ATMA meeting	11	100.00

3. Constraints in establishing linkage with other organization

The constraints experienced in establishing linkage with the other organizations were collected and furnished in the Table 3. Delay in getting Government order to execute the scheme benefits, delaying in getting information from the hierarchical set up, lack of resources availability and lack of time due to multi targeted work were expressed as constraints by almost all of the officers of *Uzhavar Udhaviyagam. Beside that* lack of location specific appropriate technology to suit the tail end of cauvery (81.81%) was also expressed as a major constraint in building strong linkage with clientele.

Table 3 : Constraints in establishing linkage with the other organization

n=11

SI.No.	Constraints	Number	Percentage
1	Lack of location specific appropriate technology	9	81.81
2	Lack of resources availability	10	90.90
3	Lack of time due to multi targeted work	10	90.90
4	Poor transport facilities	7	63.63
5	Not sparing of staff from other organization for joint venture work	5	45.45
6	Delaying in getting information from the hierarchical set up	10	90.90
7	Delay in getting Government order to execute the scheme benefits	11	100.00

4. Suggestions to strengthening the linkage with the other organization

The various suggestions offered by the respondents were collected and presented in the Table 4. Farmer's participatory research and extension was suggested by majority of the respondents (81.81%) as one of the powerful intervention for technology generation, dissemination and refinement, which could provide solution for the lack of location specific appropriate technology. e-governance was suggested by another majority of the respondents (81.81%) to curb the delay in getting information or orders from the higher authority. Two-third of the respondents (63.63%) has suggested the constraints like lack of resources and infrastructure and it could be made available if public —private partnership mode established. The multi targeted work was one the most expressed constraints. It could be made easy by means of digitization of data. Majority of the respondents (81.81%) suggested that the organizing district level linkage meeting involving all the development departments would pave way for better linkage and mutual benefits.

Table 4 : Suggestions to strengthening the linkage with the other organization

n=11

SI.No.	Constraints	Number	Percentage
1	Farmer's Participatory research and extension	9	81.81
2	e-governance	9	81.81
3	Establishing private partnership for infrastructure development	7	63.63
4	District level linkage meeting	9	81.81

Conclusion

In the wake of increasing involvement of Farmers interest Group in participatory extension and meeting the multifarious demands of the farming community, partnership in various modes/forms can provide synergistic approach in the extension efforts. To conclude that, *Uzhavar Udhaviyagam* have *established* a medium to strong linkage with the Pasumai Padai for information dissemination function. It has inadequate linkage with organizations providing information regarding market intelligence and the input availability. The present level of linkage could be further increased by means of providing modern electronic gadgets to the units. The existing ATMA machinery at the district level could better utilized for the effective linkages on information sharing.

References

Van Den Ban. 1997. Changing Ideas on Agricultural Extension-A Global Perspective. *Journal of Extension Education,* Vol.8 No.2.

Policy frame work for Agriculture Extension. 2000. Department of Agriculture and Cooperation, Ministry of Agriculture, Govt. of India.

●●●

14

Status of Job Satisfaction of KVK Subject Matter Specialists

Nafees Ahmad[1] *and P. S. Slathia*[2]
[1-2] *Division of Agricultural Extension Education, FoA, SKUAST-Jammu*

Introduction

Beginning with the Green revolution organised training of farmers and extension personnel in the country has now come of age. Since, multiple training institutions government and non-govt have come up, playing a pivotal role in transferring the technical knowledge and know how to the farming community. The turning point and a giant leap in farmers' training came with the establishment of Farm Science Centre or Krishi Vigyan Kendras by ICAR during 1974. Having district as its jurisdiction area, KVKs have played crucial role in transfer of technology through a well qualified and trained work force at its disposal. As an organization, KVK has its own process for managing the trainers/ subject matter specialists in order to keep them motivated and dedicated to the service of poor farming community. Keeping in consideration some of the job related attributes/ factors that contributes towards effectiveness and efficiency of the SMSs, the present study has been conceived with the main objective- 'To study the job satisfaction of the KVK trainers'.

Methodology

The present research study entitled - 'Status of Job Satisfaction of KVK Subject Matter Specialists' was carried out in the state of Uttar Pradesh and Uttranchal as there are convergence of one of the largest number of KVKs in the country. For the study purpose, a total of five KVKs were selected, two each representing state agriculture universities and non-government organizations and the remaining one from ICAR research institutes. One of the KVKs under SAUs system represents the hill region of Uttranchal. One of the NGO KVKs belongs to Deen Dayal Upadhay research institute (DDU) and another is working under Kamla Nehru trust. While each KVK is a case by itself and focus of study, they also represent KVKs under different systems. The sample of respondents comprised of all the SMSs/trainers working at all of the KVKs selected under study. There were 49 such trainers working at these five KVKs i.e. 7 in ICAR KVK, 11 and 12 from NGO trust and NGO DDU KVKs respectively, and 9 and 10 from of SAU plain and hill KVKs respectively. The data were collected through personal interview technique with the help of interview schedule. The job satisfaction of trainers reflected the perceived degree or extent to which the trainers feel satisfied/dis-satisfied with a set of job related factors or attributes. These attributes are conceived in the present study as contributing forces towards job satisfaction level of trainers. They have been measured on 3 point rating scale. An attempt to assessment of job preferences and work environment of trainers as indicative of their job satisfaction has also been carried out in the present investigation. The result of the study as derived from the analyzed data have been presented in the following tables and discussed as under.

Findings and Discussion

Job satisfaction of the trainers had been assessed in terms of various job related factors such as- satisfaction with salary component, satisfaction with facilities/ amenities existing at the KVKs, recognition of achievement and utilization of potential. For the assessment of job preference of the trainers, related factors that has been considered under study include preference for nature of work, work area and work environment.

Table 1 : Attributes of Job satisfaction of KVK trainers

SI. No	Attributes of Job satisfaction	KVKs working under different systems					
		ICAR	SAU hill	SAU Plain	NGO DDU	NGO Trust	Total
1.	**Satisfaction with salary**						
	High	1 (14.28)	1 (10.0)	2 (22.22)	5 (41.67)	8 (72.72)	17 (34.69)
	Medium	5 (71.44)	3 (30.0)	5 (55.56)	6 (50.0)	3 (27.28)	22 (44.9)

	Low	1 (14.28)	6 (60.0)	2 (22.22)	1 (8.33)	-	10 (20.41)
2.	**Satisfaction with facilities/amenities**						
	Fully	1 (14.28)	-	-	1 (8.33)	9 (81.82)	11 (22.42)
	Partially	5 (71.44)	4 (40.0)	6 (66.67)	10 (83.34)	2 (18.18)	27 (55.16)
	Not at all	5 (71.44)	1 (10.0)	6 (66.67)	8 (66.67)	10 (90.9)	26 (53.05)
3.	**Recognition of achievement**						
	Proper	2 (28.72)	1 (10.0)	6 (66.67)	8 (66.67)	10 (90.9)	27 (55.16)
	Somewhat	5 (71.44)	5 (50.0)	2 (22.22)	4 (33.33)	1 (9.1)	17 (34.69)
	Not at all	-	4 (40.0)	1 (11.11)	-	-	5 (10.15)
4.	**Utilization of potential**						
	Fully	6 (85.72)	1 (10.0)	2 (22.22)	7 (58.34)	10 (90.9)	26 (53.05)
	Partially	1 (14.28)	9 (90.0)	7 (77.78)	4 (33.33)	1 (9.1)	22 (44.9)
	Not at all	-	-	-	1(8.33)	-	1(2.05)

(Figures in parentheses indicate percentages of corresponding value)

Satisfaction with salary

The study as per Table 1, has revealed that on an overall basis more number of trainers (44.9%) have medium level of satisfaction with salary component followed by high level (34.69%) and low level (20.41%) of satisfaction. 72.72 and 41.67 per cent of trainers respectively in NGO Trust KVK and NGO DDU KVK have high level of satisfaction. In case of SAU hill KVK, 60.00 per cent trainers had low level of satisfaction and only 10.00 per cent reported to have high level of satisfaction. It is interesting to note that proportionately more number of trainers at SAU hill KVK had low level of satisfaction as compared to other system of KVKs. It is obvious and understandable because trainers at hill KVK face a lot of hardship due to tough geographical conditions and have not found pay and perks as attractive. More number of trainers at NGOs KVKs had high level of satisfaction as they seem to be not too high on expectations.

Satisfaction with facilities and amenities

It appears from the Table 1 that majority of trainers (55.16%) are partially satisfied with facilities and amenities existing at their KVKs.

81.82 per cent of trainers at NGO trust KVK followed by 14.28 per cent at ICAR KVK reported full satisfaction with facilities. In case of SAU hill KVK, 60.00 per cent trainers were not at all satisfied with facilities. It is interesting to note that more number of trainers at NGO trust KVK had reported full satisfaction. The findings in this regard are consistent with the personal observation of the researcher. The KVK has well established infrastructure including training facility, demonstration units etc. which is one of the best in the country. In this case also, more number of trainers in SAU hill KVK had expressed their complete dissatisfaction. This may well be understood as the Kendra draws heavily from resources of hill campus of G B Pant University, Ranichauri where it is located.

Recognition of achievement

The Table 1 revealed that majority of respondents (55.16%) got proper recognition of their achievement by their respective organization. Most of the trainers at NGO Trust KVK (90.9%) followed by equal majority in NGO DDU KVK (66.67%) and SAU KVK (66.67%) expressed that their achievement was properly recognized. 40 per cent of the trainers at SAU hill KVK felt that their achievement was not at all recognized and only 10 per cent of them felt that their achievement was properly recognized. Once again more trainers at SAU hill KVK were of the opinion that their achievement was not at all recognized. This may be owing to the fact that over all dissatisfaction level of the trainers at hill KVK is high. It is noteworthy that more trainers in NGO trust KVK perceived that their achievement was properly recognized. The possible reason as already mentioned is that this KVK has been recognized for its excellence which certainly is not possible without proper mobilization of the workforce.

Utilization of potential

It is evident from the Table 1 that majority of the trainers (53.05%) had perception that their potential was fully utilized. Most of the trainers at NGO trust KVK (90.9%) followed by ICAR KVK (85.72%) and NGO DDU KVK (58.34%) were satisfied that their potential had been utilized fully by their respective KVKs. Exceptionally 8.33 per cent trainers at NGO DDU KVK felt that their potential was not at all utilized. It is obvious from the above findings that more number of trainers in NGO KVKs and ICAR KVKs were having full satisfaction with respect to potential utilization and more trainers in SAU KVKs have partial level of satisfaction in this regard. It may be because ICAR along with NGOs systems have a better system for managing and mobilizing their human resources.

Table 1 : Attributes of Job preferences of KVK trainers

SI. No	Attributes of Job satisfaction	KVKs working under different systems					
		ICAR	SAU hill	SAU Plain	NGO DDU	NGO Trust	Total
1.	**Nature/Type of, ob**						
	Training	3 (42.88)	4 (40.0)	2 (22.22)	6 (50.0)	10 (90.9)	25 (51.03)
	Teaching/ Research	4 (57.12)	6 (60.0)	7 (77.78)	6 (50.0)	1 (9.1)	24 (48.97)
2.	**Work area preference**						
	Rural	7 (100)	7 (70.0)	7 (77.78)	11 (91.67)	11 (100)	43 (87.76)
	urban	-	3 (30.0)	2 (22.22)	1 (83.34)	-	6 (12.24)
3.	**Working environment**						
	Fully Conducive	5 (71.44)	6 (60.0)	2 (22.22)	2 (16.66)	4 (36.36)	16 (32.66)
	Partially Conducive	2 (28.56)	6 (60.0)	2 (22.22)	2 (16.66)	4 (36.36)	16 (32.66)

(Figures in parentheses indicate percentages of corresponding value)

Nature of job

As evident from the Table 2, a little over half of the trainers (51.03%) preferred training job over teaching/ research (48.97%) A majority of the trainers at NGO trust KVK (90.9%) and NGO DDU KVK (50.0%) expressed their preference for training job. A majority of trainers at SAU plain KVK (77.78%) and SAU hill KVK (60.0%) indicated their liking for teaching/ research job. This finding is interesting in light of the fact that the KVK scientists have primarily training task to perform as per their mandate. The NGO KVKs seem to be better placed than SAU KVKs in this regard as their trainers have largely evinced interest and preference in training job which happens to be their primary business.

Work area preference

As evident from the Table 2, majority of the trainers (87.76%) have preferred rural area as the place of working. In this regard, cent percent trainers at NGO trust KVK and ICAR KVK followed by NGO DDU KVK (91.67%) outlined their preference to work in rural areas. 30 per cent trainers at SAU hill KVK and 22.22 per cent at SAU plain KVK reported their preference to work in urban areas. It is obvious from the findings and also encouraging to note that more number of trainers would like to serve in rural settings. One of the dominant reasons may be that an overwhelming large number of them came originally from rural background.

Work environment

The Table 2 reveals that majority of trainers (67.34%) had perceived that working environment in their respective organization was conducive. More number of trainers at NGO DDU KVK (83.34%) found the work environment fully conducive. It may be because this KVK belongs to a philanthropic organization which might have a harmonizing impact on overall working environment of the institution. More number of trainers at SAU hill KVK (60.0%) found the work environment partially conducive. It may be because of the difficult terrain in the hill region and also the fact that it is established in the hill campus of G B Pant University at Ranichauri, Tehri Garhwal with which it shares many resources.

Conclusion

It may be concluded that the trainers working across different systems were highly satisfied for one or the other job related attributes, reinforcing the fact that KVK under one system should emulate the positive aspects/ merits of the other. Though, the discourse has reflected that the trainers in NGO KVKs have a higher satisfaction level in respect of some of the job related attributes. It was also observed that KVKs under different systems lacked in sustaining proper satisfaction level of trainers which need to be worked out to keep the work force motivated.

●●●

15

Strategies for Technology Transfer through Adarsha Rythus in Andhra Pradesh

***N. Kishore Kumar*[1] *and A. Sailaja*[2]**

[1]Ph.D scholar and [2] Associate Professor, Acharya N.G Ranga Agricultural University, Hyderabad

Introduction

The share of Agriculture sector in GDP is declining over the years. This is because,of the a large gap in the latest known technology and technology being used by the farmers. Our human resources are not being utilized properly because our people in the villages are ignorant about many scientific innovations.

In Andhra Pradesh, more than 70.00 per cent of the population is dependent on agriculture but its contribution to the state GDP is only 30.00 per cent. It is felt that this low contribution to the GDP by the Agricultural sector is due to uneconomic returns

Indian agriculture is passing through a phase of reforms, owing to its shifting emphasis towards diversification, intensification, commerciali-zation and sustainability, which has posed newer challenges before the extension functionaries. Understanding and appreciating the information needs of different clients (farmers, farm women, labourers, youth, etc) and providing quality extension service is central to its success, which needs quality manpower at different levels. Among extension services training need resources is a key input for human recourses development and contributes substantially to face the challenges by all concerned. Farmer led extension approach gives farmers the opportunity to share their experience and practices through a method demo with fellow farmers in the area. It was noticed that farmers who are successful in their farming venture have established credibility among their peers. These very same farmers are just waiting to be tapped to act as extensionist.

In this context, The Government of Andhra Pradesh took an important decision to appoint about 50,000 “Adarsha Rythus” (Model Farmers) referred to as Farm Science Managers in Prof. Swaminathan’s Report and the National Policy for Farmers. The Adarsha Rythus were selected from various panchayats throughout the state at the rate of one for every 200 to 250 farmers in a phased manner. This back ground an objective.

1. To elicit the problems and suggest a suitable strategy for improvement of managerial abilities and job performance of adarsha rythus in Mahaboobnagar district of Andhrapradesh.

The results of the study would be of immense value to adarsha rythus, administrators and extension agencies in planning and conducting the programme successfully at the village level. Hence, the present study was undertaken with the following specific objectives.

Methodology

Ex-post facto research design was adopted for study. Mahaboobnagar district which ranks first with highest number of adarsha rythus in Andhra Pradesh was purposively selected for the study from three randomly selected mandals of the district and forty adarsha rythus were selected randomly from thus one hundred and twenty (120) respondents were selected for study.

The respondents were asked to express the problems and suggestions thereof and were ranked in the order of priority as presented in Table 1.

Findings and Discussion

The respondents were asked to express the problems and suggestions thereof and ranked in the order of priority as presented in Table 1.

Managerial abilities

It is evident from Tables that majority of respondents (72.50%) perceived ‘absence of need based and skill based training programmes’ as a major problem for which majority of respondents (71.66%) have given provision of need and skill based training programmes as a major suggestion.

It is evident from Table 1 that majority of adarsha rythus (55.50 %) perceived `absence of linkage with other line departments’ as a major problem for which majority of adarsha rythus (50.83 %) suggested ‘strong linkage with other line departments’. This clearly shows that the understanding gets enhanced through interaction with other functionaries.

It is evident from Table 1 that majority of respondents (40.50 %) perceived 'lack of proper guidelines' as next major problem in order for which majority of respondents (39.16 %) have given 'provision of work guidelines as a next major suggestion. This indicates that work guidelines serve as a major rider.

It is evident from Table 1 that majority of respondents (25.83 %) perceived 'unavailability of need based latest agricultural information materials' as next problem in order for which majority of respondents (25.50 %) have given availability of need based latest agricultural information material as suggestion. This helps him to keep abreast of latest technical knowledge.

It is evident from Table 1 that majority of respondents (23.33%) perceived 'unavailability of complete farm information on recent technologies' as next problem. Under for which majority of respondents (21.66%) have given 'availability of complete farm information on recent technologies' as another suggestion it helps him to update his technical knowledge in order.

Job Performance

It is evident from Table 1 that 'majority of respondents (85.33 %) perceived less honororium as a major problem for which majority of respondents (84.16 %) have given 'increase in honororium' as a major suggestion that government needs to fulfill this dire to increase their performance. Majority of respondents (83.33 %) perceived 'Irregular payment of honororium' as a major problem for which majority of respondents (81.66 %) have given regular payment of honorarium as a suggestion. This is important as it enables them to meet their day to day needs and perform better.

Table 1 also indicates that majority of respondents have perceived (65.83 %) 'non availability of inputs in time' as a major problem for which majority of respondents (63.33 %) have given availability of input in time as next suggestion in order as this helps them to win the credibility and trust worthiness of farmers of his village. Majority of adarsha rythus (50.80 %) perceived 'lack of good transport facilities' as next problem in order for which majority of adarsha rythus (47.50 %) have given provision of transport facilities as next suggestion. For attending monthly meetings and checking field along with extension staff monthly.

Majority of respondents (30.00 %) perceived 'no consideration of opinions of adarsha rythus while selecting farmer beneficiaries to different programmes' as a next problem for which majority of respondents (27.50%) have suggested consideration of opinions of adarsha rythus while selecting farmer beneficiaries to different programmes.

Table 1 : Results of Field Experiments at Kondayampatti

Field - 1, 2, 3
Variety: GRAND -9

Treatments	Finger Length (cm)	Finger Girth (cm)	Finger Weight (gms)	Volume of finger (cc)	Density Of fingers	No of fingers/ hand	Total Number of fingers/ bunch	No of hands/ bunch	Bunch Weight (Kg)	Yield t/ha	Yield increase (% over control)	TSS (^{0} Brix)
T_1	21.00	12.67	59.02	252.00	1.26	25.33	136.00	16.00	29.51	59.02	101.00	15.17
T_2	21.67	12.33	58.38	252.67	1.28	26.33	139.00	16.67	29.19	58.38	99.93	15.17
T_3	21.33	12.00	56.99	254.00	1.25	25.00	137.00	16.00	28.50	56.99	97.55	15.00
T_4	21.00	13.00	61.00	251.67	1.27	26.00	138.00	16.67	30.50	61.00	104.42	15.00
T_5	19.00	10.67	53.30	251.67	1.25	22.33	130.00	15.00	26.65	53.30	91.24	14.83
T_6	19.33	11.33	59.13	252.33	1.27	24.67	135.00	15.67	29.57	59.13	101.21	15.17
T_7	19.67	11.33	64.12	255.67	1.28	29.00	140.00	17.33	32.06	64.12	109.76	14.83
T_8	24.67	14.33	72.00	261.33	1.31	32.33	150.00	18.67	36.00	72.00	123.24	15.67
T_9	23.00	13.33	55.62	253.33	1.21	26.00	135.00	15.00	27.81	55.62	95.21	15.17
T_{10}	24.00	14.00	69.38	260.00	1.30	31.67	147.00	18.33	34.69	69.38	118.76	15.50
T_{11}	21.67	12.00	61.20	254.67	1.28	26.33	136.00	15.67	30.60	61.20	104.76	15.00
T_{12}	17.67	10.00	58.42	242.00	1.05	15.33	115.00	11.33	29.21	58.42	100.00	14.17

SEd: 0.890 0.515 2.545 10.649 0.052 1.087 5.723 0.671 1.272 2.545 0.632

CD (P=.05) 1.847 1.068 5.279 :22.084 0.108 2.255 11.87 1.39 2.64 5.279 1.312

Table 1 further indicates that majority of respondents (17.50%) perceived lack of appropriate performance appraisal procedures as next problem in order for which majority of respondents (16.66 %) suggested imposing of appropriate performance appraisal procedures as next suggestion in order to better work in the village. Majority of respondents (14.17%) perceived 'political interference in selecting adarsha rythus' for which majority of respondents (12.50%) suggested non political interference in selecting adarsha rythus as a next suggestion in order to have transparency of work in the village. Majority of respondents (5.00%) perceived 'lack of awareness among farmers to utilize the services of adarsha rythus for which majority of respondents (4.16%) suggested creation of awareness among farmer to utilize the services of adarsha rythus as next suggestion in order the farmers enlightened on the roles and duties of adarsha rythus they can approach in times of needs.

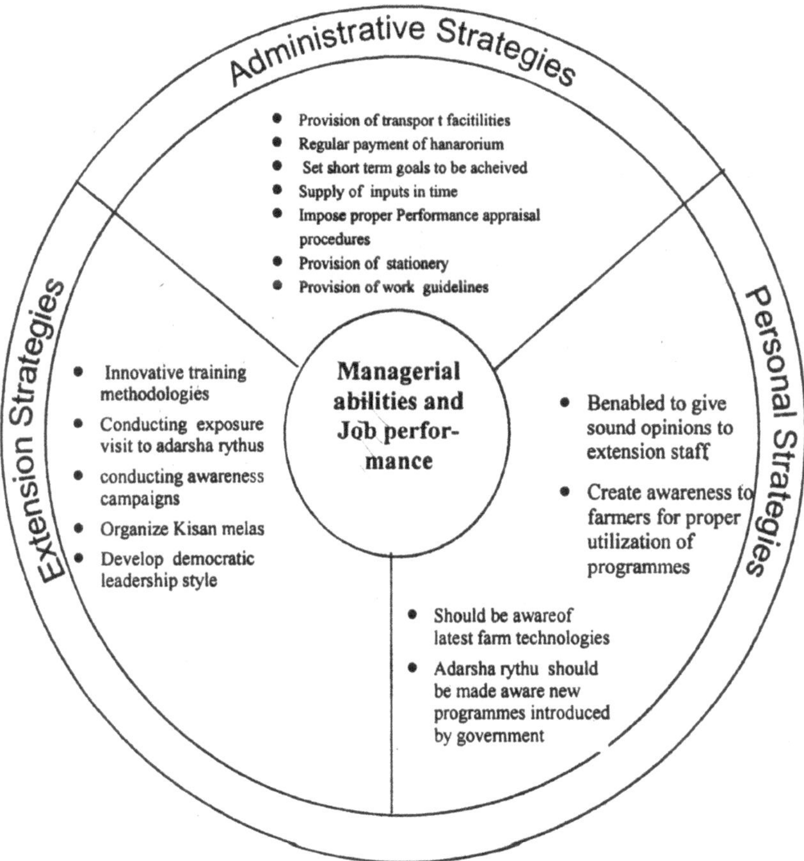

Fig. 1 : Strategies to improve Managerial Abilities and Job Performances

Conclusion

The present scenario of globalization calls for reorientation of extension management system so as to transfer knowledge based inputs at various levels. Extension management is more concerned with `dong things right' rather then`dong the right things. In this changing context, in addition to the technical skills, extension functionaries need to have a wide array of managerial skills to successfully complete the assigned jobs. Competence in managerial aspects will be useful in improving the performance of extension functionaries resulting in efficiency of various developmental programme.

References

Anil Kumar A, Joy M and Ram. achandram U. 2003 Job performance of agricultural officers in Kasaragod district of Kerala State. Indian Journal of Extension Education, 39 (3 & 4) : 167-171.

Chidananda M 2008 A study on entrepreneurial behaviour of dryland farmers in Karnataka State. M.Sc. (Ag.) Thesis, Acharya N. G. Ranga Agricultural University, Hyderabad.

Das 1991 A institution contigenencies limited job attitude- job performance relationship. Organisational Behaviour in Human performance Vol. (10), pp: 208-224.

Halakatti S V 1998 Job performance and job attitude of agricultural assistants and related factors. Karnataka Journal of Agricultural Sciences 11 (2) : 436-440.

Hemanthkumar B 2002 A study on attitude, knowledge and adoption of recommended practices by oriental tobacco farmers in Chittoor District of Andhra Pradesh. M.Sc. (Ag.) Thesis, Acharya N. G. Ranga Agricultural University, Hyderabad.

Khan N A 1990 A Study on organizational climate, job satisfaction and performance of managers and supervisors of Andhra Pradesh dairy development corporation federation. Ph.D. Thesis, Acharya N.G. Ranga Agricultural University, Hyderabad.

Kishorbabu B 2004 Marketing Behaviour of vegetable growers in Ranga Reddy district of Andhra Pradesh. M.Sc. (Ag) Thesis, Acharya N. G. Ranga Agricultural University, Hyderabad.

Krishna Prasad P 2005 Study on rural poverty and sustainable livelihoods in agrarian sector of Andhra Pradesh. Ph.D. Thesis, Acharya N.G. Ranga Agricultural University, Hyderabad.

•••

16

Problem Solving Capacity of Agricultural Expert System

***S. Helen*[1] *and F.M.H. Kaleel*[2]**

[1] *Associate Professor (Agricultural Extension), Communication Centre and*
[2] *Professor and Head, Department of Agricultural Extension, College of Horticulture (Kerala Agricultural University), Vellanikkara, P.O. Thrissur Dt.*
Kerala state, South India.

Introduction

Farming community is facing multiple problems to maximize crop productivity as well as to increase farm income. In spite of successful research on new agricultural practices concerning crop cultivation, majority of the farmers are not getting upper-bound yield and not earning profit due to several constraints. One of them is that expert advice regarding crop cultivation is not reaching the farming community in a timely manner. It is true that India possesses valuable agricultural knowledge and expertise. However, a wide information gap exists between research and actual adoption in the field. Farmers need timely expert advice to make farming more productive and competitive. Local information resource centers are gaining importance with computers carrying expert systems to help farmers in making decisions. It is known that many Agricultural Research Institutes are involved in the development of agricultural expert system to satisfy the information needs of farmers.

In this context, Kerala Agricultural University developed an Agricultural Expert System (AES) for diagnosing pests and diseases of nine major crops of Kerala called 'DIAGNOS-4' which had drawn tremendous attraction among extension personnel and farmers. The modified version of it was released for the benefit of all the stakeholders involved in agricultural development. It was a computer-assisted software for the identification of major and minor pests and diseases of nine identified crops such as rice, coconut, banana, pepper, cashew and vegetables like amaranthus, bhindi, cucurbits and brinjal. It also suggested management measures for combating pests and diseases. This package was aimed to support the agricultural extension workers and literate farmers for decision-making and aid them in suggesting suitable control measures of the major pests and diseases of important nine crops of Kerala (Ganesan *et al,* 2002).

As part of the project, the knowledge base of the system had been validated in consultation with the scientists of different research stations of Kerala Agricultural University and the officials of the Departrnent of Agriculture. No systematic users' level study had been conducted so far to assess how far the field level problems are tackled by using the software. Therefore, it was felt appropriate to conduct a study to assess the problem solving capacity of agricultural expert system before releasing the system. With this background, a study was conducted to analyze the problem solving capacity of Agricultural Expert System as assessed by extension personnel.

Methodology

'Diagnos-4' is the Agricultural Expert System, specially designed software for tackling the problems in transfer of technologies related to plant protection aspects of important crops of Kerala. Since extension personnel were expected to use such soft wares, the study was conducted among them in Palakkad district of Kerala, South India. The respondents were selected purposively who were mainly dealing with the cultivation of rice, coconut and banana as major crops. Between group-randomised design was considered in the study. This design could enable the researcher to select the groups randomly and separately for each treatment (Singh, 1986). Moreover, this was the classical experimental design to assess the performance of agricultural expert system at the users' level subjected to different treatments and facilitate the researcher for comparative analysis. The respondents were categorized into three groups. Different treatment groups were as follows:

T_1 - Groups discussed and prioritised their problems before exposure
T_2 - Groups exposed to human expertise alone
T_3 - Groups exposed to agricultural expert system alone
T_4 - Groups exposed to both human expertise and agricultural expert system

Stoner et al (1998) defined a problem as a situation that occurred when an actual state of affairs differed from a desired state of affairs. Problem solving process was referred to the methods of dealing with the treatments and opportunities in the environment. Vinayagam (1998) described problem solving as the tendency to solve problems rather than yield to the pressure of the problems. Srinivasa (2006) illustrated that problem solving involved taking 'yes' or `no' decision to a given problem. In the present study, problem solving capacity of Agricultural Expert System was operationalised as the ability of the `Diagnos-4' to provide solutions for the technical problems faced by the respondents in the field of plant protection aspects of rice, coconut and banana cultivation.

Technical problems faced by extension personnel were inherent to each region. In this study, technical problems were limited to plant protection aspects of rice, coconut and banana. The number of respondents in each treatment were thirty, grouped into three constituting ten respondents in each group. They were asked to discuss and prioritise the plant protection problems in rice, coconut and banana cultivation experienced by them. They were guided to prioritise the problems in a five- point continuum, '5' indicated 'most experienced' and '1' indicated `least experienced'. The scores in between in the decreasing order showed the degree of decrease in the level of the problems experienced.

After scoring the identified problems, the respondents were exposed to different treatments. They were advised to indicate in a five-point continuum '5' as the 'most sufficient' and '1' as 'least sufficient' against the prioritized field problems indicating the sufficiency of solutions obtained from different exposures. Total scores obtained by each problem were obtained by the summation of scores offered by the respondents. The total scores assigned to each problem and sufficiency of solutions derived by the different treatment groups during pre and post exposures were compared by working out percentage analysis.

Findings and Discussion

The problem solving capacity of the agricultural expert system assessed by extension personnel is furnished in the table 1. Careful examination of the data showed that stem borer (92.80 per cent) was the most important problem followed by gall midge and leaf folder (90.00 per cent each), bacterial leaf blight (77.14 per cent), brown plant hopper (73.40 per cent) and earhead bug (66.67 per cent) in rice cultivation. Human expert was able to offer solution to the control of stem borer at a tune of 74.14 per cent. Agricultural expert system could provide solution only to 65.30 per cent. The group which was exposed to agricultural expert system alone complained that biological control measures of stem borer was not explained in detail, which might be the reason for the lower percentage of solutions offered by agricultural expert system. At the same time the group that was exposed to both agricultural expert system and human expert recorded a gain of 91.16 per cent solutions. This group was able to get the information on biological control measures from human expertise and reported increase in percentage of solutions. In all other cases of plant protection problems in rice agricultural expert system could provide better solutions, but the combination of agricultural expert system and human expert served much more satisfactory solutions to the extension personnel. When the overall percentage of solutions offered by agricultural expert system was worked out, it was almost similar to the solutions given by human experts. However, it served better in combination with human experts.

Table 1 : Problem solving capacity of the agricultural expert system as assessed by extension personnel

SL No.	Experienced plant protection problems	Identified Problems T_1		Solutions received					
				Human expert alone T_2		Agricultural Expert System alone T^3		Human experts + Agricultural Expert System T_4	
		Mean scores	Percentage	Mean scores	Percentage	Mean scores	Percentage	Mean scores	Percentage
1	**Rice** Stern borer	4.64	92.80	3.44	74.14	3.03	65.30	4.23	91.16
2	Bacterial leaf blight	3.86	77.14	3.26	84.52	2.14	89.70	3.42	93.19
3	Brown plant hopper	3.67	73.40	3.46	58.31	2.14	55.48	3.56	97.00
4	Gall midge	4.50	90.00	3.32	73.78	3.41	75.78	3.91	86.89
5	Leaf folder	4.50	90.00	4.07	90.44	4.32	96.00	4.46	99.11
6	Ear head bug	3.33	66.67	3.12	93.60	3.20	96.00	3.28	98.40
	Mean	4.08	81.67	3.45	79.13	3.04	79.71	3.81	94.29
1	**Coconut** Stern bleeding	3.60	72.00	2.54	70.56	3.14	87.22	3.45	95.83
2	Eryophid mite	4.50	90.00	3.23	71.78	3.43	76.22	4.27	94.89
3	Termites	3.00	60.00	2.43	81.00	1.56	52.00	2.21	73.67
4	Bud rot	4.00	80.00	2.36	59.00	2.45	61.25	2.82	70.50
5	Root wilt	2.00	40.00	1.12	56.00	1.24	62.00	1.42	71.00
	Mean	3.42	68.40	2.34	67.67	2.36	67.74	2.83	81.18
1	**Banana** PseUdostem weevil	4.00	80.00	3.64	91.00	3.82	95.50	3.83	95.75
2	Bacterial wilt	3.50	70.00	3.00	85.71	3.32	94.86	3.45	98.57
3	Leaf spot	4.50	90.00	4.21	93.56	3.42	76.00	3.61	80.22
4	Bract mosaic virus	3.00	60.00	2.50	83.33	2.62	87.33	2.74	91.33
5	Bunchy top	4.67	93.33	3.52	75.43	4.34	93.00	4.46	95.57
	Mean	3.93	78.67	3.37	85.81	3.50	89.34	3.62	92.29

In coconut cultivation, eryophid mite, (90.00 per cent) was the most important problem followed by bud rot (80.00 per cent), stem bleeding (72.00 per cent), termites (60.00 per cent) and root wilt (40.00 per cent). For almost all the problems, agricultural expert system could offer better solutions similar to human expertise, except the termite problem for which users could not find satisfactory solutions. Similar to paddy cultivation, agricultural expert system could furnish details of control measures for the plant protection problems in coconut cultivation on par with human expert.

Bunchy top was (93.33 per cent) the common and most intensively reported problem followed by leaf spot (90.00 per cent), pseudostem weevil (80.00 per cent), bacterial wilt (70.00 per cent), and bract mosaic virus (60.00 per cent) in banana cultivation. Except leaf spot (76.00 per cent), all other problems could be served better solutions by agricultural expert system than human expertise. This is in agreement with the findings of CLAES (2006). Just like the other two crops, the combination of agricultural expert system and human expertise could render best solutions among all the treatment groups.

Conclusion

For almost all the technical problems identified in the plant protection of rice, coconut and banana cultivation, agricultural expert system could offer better solutions similar to human expertise, except in very few cases. Therefore it was concluded that agricultural expert system needed modifications with more of updated biological control measures. Extension personnel required complete orientation for using the software and the field problems could be solved by retrieving information from the agricultural expert system.

References

CLAES. [Central Laboratory for Agricultural Expert System]. 2006. Disease diagnosis evaluation graph Insects diagnosis evaluation graph Nutrition deficiency evaluation graph. [On-line]. Available: http://www.claes.sci. eg /expertsystem/ pestid.html [23.02.06]

Ganesan, V, Suma. N. V. S. and Kumaran, 2002. Expert System: The booming technology for the diagnosis of pests and diseases of crops. In: *Proceedings of 14th Kerala Science Congress.* 29-31 st Jan, 2002. Cochin. pp: 300-302.

Singh, A.K. 1986. Principles of test construction. *Tests, Measurements and Research Methods in Behavioural Sciences.* First reprint. 1993. Bharati Bhawan publishers, Patna. 563p.

Srinivasa, S. 2006. Connotations of problem solving. [On-line]. Available: Files/content IE5/G5EBGHIR/ Connotations of Problem Solving%5B1% 5D.ppt. [12.04.06]

Stoner, F. J, Edward F. A and Daniel A. Gilbert. J. R.1998. Decision making and Managing. *Organizational Change and Innovation Management.* (6thed). Prentice- Hall, Inc. USA. Pp: 240&420.

Vinayagam, S. S. 1998. Entrepreneurial behaviour of agri-business operators in Kerala. PhD Thesis. Kerala Agricultural University, Vellanikkara, Thrissur. 155p.

●●●

17

Technology Transfer through Tar Mode in Sugarcane Based Cropping System

T. Rajula Shanthy
Senior Scientist (Agri. Extension), Sugarcane Breeding Institute, Coimbatore

Introduction

Sugarcane based production system has to go a long way to turn hightech because of the existing technological gap in the irrigated agroecosystem. This has been realized more in the recent studies undertaken by social scientists. All the crops associated with the sugarcane based production need varietal introduction as well as the adoption of relevant crop management practices. Since the possibility of increasing area under sugarcane is remote in view of emerging low input high value crops, the projected target of 420 million tonnes by 2020 AD has to be achieved by enhancing the cane productivity and sugar recovery concurrently. The full exploitation of the existing improved sugarcane production technologies and development of a package of agro-techniques in the changing scenario of sugarcane based production system suiting to the socio-economic conditions of the farmers is as such the only master key to unlock the yield potential of sugarcane.

All the technologies available in the research stations cannot be used as such in all the situations. In the micro farming situations, the technology has to be refined to suit the sugarcane based production system. The new varieties identified in research system having enormous potential for production should be assessed and refined to fit in to the existing irrigated agro - ecosystem paving way for its adoption. It can be said that almost every crop cultivated is affected by at least one pest or disease and some crops are plagued by several pests and diseases. The yield reduction due to the pest or disease is enormous. Identifying suitable remedial measures which fit in the situations can be an apt solution for enhancing production.

It has been observed that 30-35% of the sugarcane production technology was adopted by the farmers and the rest remain unutilized (Saxena 1991, Verma *et. al.* 1986). The researchers in extension education had shown that there were several constraints with varying degree of seriousness in increasing agricultural production which were confronted not only by the farmers but also by the Scientists and extension agents (FAO 1986, Bat et. al., 1987, Anderson and Herdt 1980, Bylerlac and Tripp 1988, Singh and Sharma 1990, Kaimowitz 1990, Singh and Rajendra 1990 and Napier 1991).

New invention in farm machinery and agricultural implements can decrease the labour requirements for crop operation thereby reducing the cost of cultivation and increasing the production. Implements are also available for intercultural operation wherein the women folk are mostly engaged. It is imperative to reduce this drudgery by introducing such implements in agricultural operation.

Whatever be the benefit of new scientific practices, the attitude of the farmers towards scientific agriculture is a key factor in the success of any programme. 'Seeing is believing' & 'learning by doing'; this adage proves true even today in the era of cyber extension and globalization. The Technology Assessment and Refinement through Institute Village Linkage Programme under National Agricultural Technological Project is a unique approach in this direction.

In view of the above, the Institute Village Linkage Programme was implemented at Sugarcane Breeding Institute with the following specific objectives.

Refined objectives

- Identifying the problems in the sugarcane based irrigated agroeco-system and prioritizing them through farmers participation
- Locating technology alternatives available at ICAR/SAU institutes and deciding the alternatives for testing through discussion with target groups
- Designing the experiments for technology assessment and refinement through verification trails and on-farm testing

- Assessment of technologies in terms of agronomical, statistical, economical and farmers' perspective
- Evaluation of technologies and identifying the needs for refinement through discussion with target groups
- Evaluation of technologies for productivity and sustainability in the irrigated agro-ecosystem.

In the micro farming situations, the sugarcane technology has to be assessed / refined to suit the sugarcane based production system. The Institute Village Linkage Programme at Sugarcane Breeding Institute was organized to satisfy the technological need of the sugarcane based irrigated agro-eco system through technology assessment and refinement with participation of the farmers.

Methodology

Locale of study

The project was implemented in a cluster of three villages viz., Varappalayam, Nanjundapuram and Madathur in 22 Nanjundapuram Panchayat of Perianaickenpalayam Block in Coimbatore district during the year 2000 to 2005. These villages were selected due to the prevalence of traditional varietal management for sugarcane and the extent of adoption of recommended crop management practices was very low.

Participatory Rural Appraisal

Initial bench mark surveys conducted in the study area gave a clear picture about the socio-economic status of the villages. However, to get an in depth knowledge of the prevailing situation in these villages, participatory rural appraisal techniques was used. Participatory Rural Appraisal is a flexible tool which can be used in a lot of different situations to achieve very different objectives. It is an activity carried out by a group of people from different professional fields which aims to learn about a particular topic, area or situation. The study revealed that sugarcane, coconut, banana, rainfed sorghum and vegetables are the main crops grown there. The major constraints for low production in these crops were analyzed and prioritized. Accordingly suitable technological interventions were selected and introduced into the farming system for verification/ on farm trials.

Findings and Discussion

Identified and prioritized problems

The following were the important problems in the order of priority identified by the farmers in the three villages.

1. Low sugarcane yield
2. Low jaggery yield
3. Low banana yield
4. Low milk yield
5. Acute labour shortage
6. Low yield of vegetables
7. Inadequate rainfall
8. Low prices for jaggery
9. Low prices for banana
10. Very low cost for milk
11. Very low cost for vegetables
12. High cost of feed ration for cattle
13. Transportation problems of the produce
14. Non availability of tractor in time
15. High cost of fertilizers and chemicals in the nearby markets
16. Specific brand of fertilizers and chemicals are not available
17. Improper identification of pests and diseases
18. Frequent and prolonged power cut
19. Labour wages for agriculture is very high
20. Lack of knowledge about new varieties
21. Lack of knowledge about crop management practices
22. Lack of training facilities.

Problems that farmers were not aware of

1. Varietal deterioration
2. Productivity of new varieties
3. Fertility status of the soil
4. Recommended fertilizer application
5. Timely application of fertilizer
6. Infertility in cows.

Problem Cause Diagram

A problem cause diagram was prepared as in Fig.1.

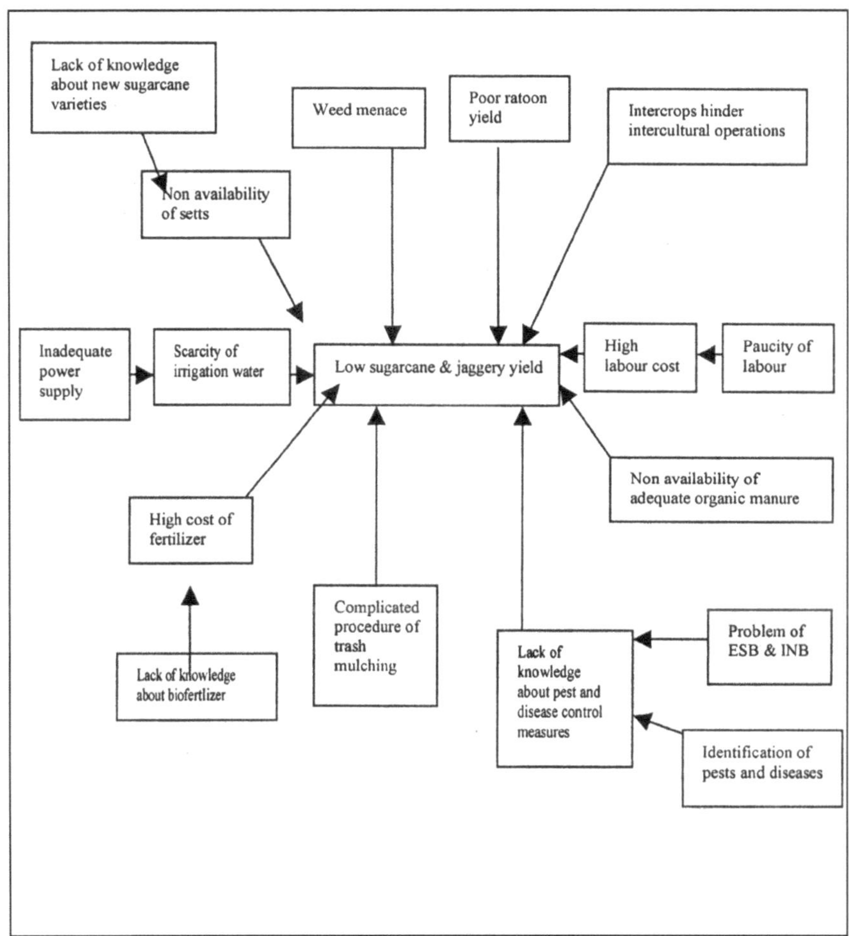

Fig. 1 : Problem cause diagram

Technological intervention

After prioritizing the problems, discussions were carried out with the farmers to find out suitable technological interventions to remedy the problems operating in their farms. Consensus was arrived with the farmers to conduct the following verification trials and on farm trials to increase sugarcane productivity.

I. Verification trials

In verification trials, two or three varieties were compared with the existing varieties by conducting experiments in farmer's fields under the three locations.

Verification trials were conducted to assess the performance of.

1. High sugar and high yielding sugarcane varieties for plant and ratoon crops

II. On farm trials

In on-farm trials, the recommended practices and farmers practices were compared by conducting experiments in the farmer's fields under the three locations. Ten on-farm trials were conducted to refine the following practices.

1. Closer and wider row spacing in sugarcane
2. Green manure intercrop in sugarcane
3. Weed control measures
4. Trash mulching
5. Inorganic fertilizer application
6. Bio-fertilizer application
7. Water management
8. Early shoot borer and Internode borer control measure
9. Sugarcane cutter planter
10. Ratoon management practices.

Farmers were facilitated to bring out the causes of the problems and then to prepare problem-cause diagram to arrive at the solution.

To improve the sugarcane based irrigated agro-eco system, 37 technological interventions were identified, of which 20 were related to sugarcane on the introduction of new varieties, correct and timely application of fertilizer, pest and disease management. These technological interventions were implemented in the farmers' field in the form of verification trails and on farm testing. They are:

A	**Sugarcane**
1	Assessing the performance of New Sugarcane varieties
2	Assessing the ratoon performance of New Sugarcane varieties
3	Refining the sett treatment in sugarcane
4	Refining the age of the seed materials
5	Assessing the performance of intercrops in sugarcane
6	Refining the fertilizer application in sugarcane
7	Refining the control of early shoot borer and internode borer in sugarcane
8	Refining the ratoon management practices in sugarcane
9	Refining water management in sugarcane
10	Refining trash mulching practices in sugarcane
11	Refining weed control in sugarcane
12	Refining the biofertilizer application practice in sugarcane
13	Assessing the performance of sugarcane with green manure crop
14	Assessing / Refining the wide row spacing package for sugarcane
15	Refining the sugarcane planting with sugarcane planter
16	Assessing the performance of multiratoonability of sugarcane
17	Refining the performance of sugarcane variety in wider and closer row spacing
18	Refining mechanical intercultural operations in sugarcane cultivation
19	Refining intercropping in wide row spaced sugarcane
20	Refining water management through skip furrow irrigation
B	**Brinjal**
21	Assessing the perforrriance of new brinjal varieties
22	Refining the fertilizer application in brinjal
23	Refining the control of fruit borer in brinjal
C	Tomato
24	Assessing the performance of new tomato varieties
25	Refining the fertilizer application in tomato
26	Refining the control of fruit borer and leaf curl in tomato
D	**Pulse crop**
27	Assessing the performance of short duration pulses intercrop in sugarcane plant crop
28	Assessing the performance of pulses intercrop in sugarcane ratoon
29	Assessing the performance of pure pulse crop
30	Assessing the performance of red gram (As intercrop in vegetables)
E	**Maize**
31	Assessing the performance of Maize crop (As intercrop in sugarcane)
F	**Vegetable cowpea**
32	Assessing the performance of vegetable cowpea (As intercrop in sugarcane)

	Animal based
33	Assessing the performance of sorghum crop (Fodder)
34	Assessing the performance of Napier grass
35	Improving the milk yield by assessing the performance of mineral mixture
36	Assessing the performance of fish farming in farm ponds
	Gender based
37	Removal of drudgery for farm women by introducing suitable farm implements

III. Training programmes

The following training programmes were given to the farmers of the three villages:

1. Sugarcane production technology
2. Cane agronomy for wide row spacing
3. Organic inputs for sugarcane agriculture
4. Ratoonabiliy of sugarcane
5. Micro-propagation in sugarcane
6. Use of pheromones to control sugarcane pests
7. Soil and water conservation

Results of on farm / verification trials

The results of the on farm and verification trials are summarized below:

Variety

The results of the verification trials with the varieties Co 62175, Co 85019 and Co 86032 indicated that the variety Co 86032 gave a jaggery yield of 230 bags (130 kg/bag) per ha and the local variety Co 62175 gave jaggery yield of 130 bags (130 kg/bag) per ha. The variety Co 86032 gave 76.92 per cent higher yield than the local variety. The income from Co 86032 was Rs. 2,31,500 whereas the income from Co 62175 was Rs.1,34,000. The variety Co 86032 showed a remarkable cost benefit ratio of 6.61, which had a great impact on the farmers in these villages. Nearly 90 per cent of the farmers in these villages are continuously cultivating the variety Co 86032.

In general, the variety Co 86032 performed well in the three adopted villages. On an average the variety Co 86032 gave 166.36 t/ha which is 32.05% higher than the variety Co 85019 (126 t/ha). The variety Co 86032 due to its performance in cane and jaggery yield has spread fast to the nearby villages, covering almost the entire area.

The results of the verification trials on ratoons indicated that the variety Co 86032 gave highest yield of 161.06 t/ha in ratoon and the reduction was only 7.43 per cent when compared with the variety Co 62175 which gave 123.09 t / ha in ratoon and the reduction in yield was 13.44 per cent.

Spacing

Results of the on farm testing indicated that the wider row spaced sugarcane gave highest yield of 167.75 t/ha than closely spaced sugarcane. The increase in yield was 4.03% and cost benefit ratio was 2.63. Wider row spacing, being a new technology is spreading fast in the study area as well as nearby villages.

Weed control

The results of the on farm testing indicated that the experimental field (Pre emergence application of atrazine and hand weeding at 45 th and 90th days after planting) gave highest yield of 157.43 t/ha and the farmers practice (Hand weeding at 30 th and 60th days after planting) gave the yield of 146.53 t/ha. The per cent increase in yield of experimental plot was 4.47 per cent with a cost benefit ratio of 2.43. Majority of the farmers are spraying atrazine to control weeds.

Trash mulching

The results of the on farm testing indicated the mulched plot gave the highest yield in all the three locations. The average yield of the treated plot was 161.96 t/ha and the control plot was 151.92 t/ha. The trash mulched plots gave 6.62 per cent increased yield than the control plots. The trash mulched plots gave cost benefit ratio of 2.53 and the control fields gave 2.31. The extent of adoption of trash mulching practice was 25% in these villages.

Inorganic fertilizer application

The results of the on farm testing indicated that the soil test based recommendation (310 kg N, 95 kg P_2O_5, and 80 kg K_2O) gave highest yield in all the three locations when compared with institute recommended practice and farmers practice. The per cent increase in yield for soil test based recommendation was 23.41 per cent than the farmers practice (305 kg N, 192.5 kg P_2O_5, and 187.5 kg K_2O) and showed 2.11 per cent increase than the institute recommendation (225 kg

N, 62.5 kg P_2O_5, and 112.5 kg K_2O). The cost benefit ratio was highest for the soil test based recommendation (2.74) followed by institute recommendation (2.66) and farmers practice (2.03). Farmers have now realized the importance of applying fertilizers based on soil test results.

Bio fertilizer application

To reduce the cost of sugarcane production it is essential to reduce the cost of fertilizers. Due to these reasons bio fertilizer experiments were carried out in the three villages with *Phosphobacteria* and *Azospirillum.* Experimental results have shown that bio fertilizers reduce 25% each of the phosphorus and nitrogen fertilizer requirement of sugarcane crop.

Application of biofertilizer (10 kg *Phosphobacteria* and 5 kg *Azospirillum* at 30th days after planting and 5 kg *Azospirillum at 60th* days *after planting)* gave a considerably higher cane yield (156.04 t/ha) than the fields where biofertilizers were not applied (144.83 t/ha). The per cent increase in yield of biofertilizer plot was 7.74 and the cost benefit ratio was 2.4.

Water management

The results indicated that the experimental field (32 irrigations) gave higher yield (153.17 t/ha) than the control (42 irrigations) plot (138.80 t/ha). The per cent increase in yield was 10.50 per cent and the cost benefit ratio was 2.37. The farmers in these villages are now aware of the economic use of water (36 irrigations) for getting more yield in sugarcane crop.

Plant protection

The on-farm results indicated that there was a reduction in yield in the control plot than the treatment plot for both early shoot borer (ESB) and internode borer (INB). The treatment plot for ESB (Application of 50 ml lindane mixed in 100 litre of water at 30 th days after planting) gave 16.22 per cent increased yield than the control plot. Likewise, the treatment plot for INB (Release of *Trichogramma* egg. parasite 2.5 cc/ hectare, six releases for every 15 days starting from 4th month onwards) gave 8.19 per cent increased yield than the control plot. The cost benefit ratio was high for treated plots than the control plots. Realizing the loss caused by these pests, farmers in these villages are following the recommended control measures.

Mechanical planting

The results of the experiments using cutter planter gave the highest cost benefit ratio of 2.52. This was due to saving of the cost in the operations such as cutting the setts, forniing ridges and furrows, planting the setts and covering of setts. This practice also gave higher yield of 162.5 t/ha than the farmers practice (152.5 t/ha). The increase in yield was 6.55% than the farmers practice. However, the percentage of adoption of sugarcane cutter planter was very low in the villages due to nonavailability of the equipment.

Livestock production

The livestock in the sugarcane based production system showed a slight improvement after the introduction of cross breed but needs adequate cattle management practices. The milk production can also be enhanced by way of wasteland and grassland management to supply fodder in the lean period besides improving the fodder sorghum in the rainfed agriculture through varietal introduction.

Thus the Technology Assessment and Refinement through verification trials and on-farm testing with farmers' participation could identify location specific technologies and thereby increase the productivity of sugarcane. However, the conviction of the people in these systems needs to be reoriented towards high tech agriculture through on-farm trials, raining, awareness campaign and by conducting field trips.

Impact of Assessment / Refinement

The farmer participatory appraisal mode of assessment and refinement of sugarcane technologies in the study villages resulted in instant widespread adoption of the technologies by the other farmers in the villages. The extent of adoption of assessed/ refined technology by the farmers of IVLP villages during the end of the project period is presented in Table 1.

Table 1 : Extent of adoption of assessed / refined technology by the farmers of IVLP villages

Sl. No.	Technologies	Extent of adoption by the farmers (%)
1	Variety Co 86032	98.0
2	Wide row spacing	22.0

Contd. ...

3	Intercrop (Coriander)	20.0
4	Intercrop (Black gram)	35.0
5	Green manure intercrop (Sunnhemp)	15.0
6	Fodder intercrop (Maize)	15.0
7	Herbicide application	80.0
8	Trash mulching	25.0
9	Soil test based fertilizer application	45.0
10	Bio-fertilizer application	25.0
11	Water management	55.0
12	Control of ESB & INB	35.0
13	Sugarcane cutter planter	4.0
14	Ratoon management practices	60.0
15	Intercrop in ratoons	35.0

The results indicated that the adoption percentage was high for the practices variety Co 86032, herbicide application, ratoon management practices, water management and soil test based fertilizer application. The adoption percentage was low for the practices viz. intercropping and use of cutter planter.

Team sprit in TAR mode

- The TAR/ IVLP projects core team members are drawn from different disciplines namely Agricultural Extension, Breeding, Agronomy, Agricultural Chemistry, Entomology and Plant pathology.
- The multi disciplinary team helped to diagnose the problems in sugarcane cultivation and to suggest remedial measures.
- The team was also helpful in giving training programmes concerning sugarcane production technology and offering latest information on sugarcane cultivation.

Conclusion

Sugarcane based production system is one of the important production systems operating in the irrigated agro-eco system. It comprises sugarcane, coconut and banana as the main component crops and brinjal, tomato and chillies as minor component crops. Being the main cash crop, sugarcane and banana are highly vulnerable to price fluctuation in the market. As an annual crop, it yields income once in a year. The investments on these crops for a year are locked without yielding any fruits. To compensate these losses the farmers are growing vegetables to get weekly income. The money obtained from these crops and from dairy enterprises is ploughed in the investments of cash crops. The intercrops and fodder crops such as fodder maize, sorghum, Napier grass and fodder cowpea are grown for the purpose of feeding the cattle.

In any farm, diversified industries are operating in an integrated manner. The diversification is mainly to sustain the entire farm. Output of one enterprise becomes the input for other enterprise. The enterprises are balanced in such a way to yield a higher annual turnover for the whole farm. Experiences indicate that a technology can be successfully introduced only if it fits in the integrated farming system as a whole. The package of technologies should be designed in a manner so that in the real farm situation it can have a complementary effect and utilize the available resources from other enterprises.

References

Anderson, J. R. and W.H. Herdt. (1988). Impact of new technology on food grains productivity to the next century. Hans, England, Dartmouth Publishing Company.

Bat R., B.V. Aron, B.V. and G.T. Karaska. (1987). Patterns of change in developing rural regions Boulder Colorado: West view press.

Byerlee, D and R. Tripp. (1988). Strengthening linkage in agricultural research through a farming systems perspective - The role of social scientists. Experimental Agriculture. p. 24-37.

FAQ (1986). The technology application gap: overcoming constraints to small farm development, Rome, Research and Technology paper-1.

Kaimowitz, D. (1990). Making the link: Agricultural researches and technology transfer in developing countries, Boulda, Colorado, West view press.

Napier, T.L., (1991). Factors affecting technology transfer. In Extension strategies for rainfed agriculture (Ed) C. Prasad and P.Das Indian Society of Extension Education, New Delhi.

Saxena A. P. (1991) Needed research in agricultural communication. Paper presented at All India training Seminar on communication Technology in augmenting agricultural production held at G. B. Pant Univ. of Agril.& Tech. Pantnagar.

Singh S.P. and Rajendra. (1990) Adoption of improved variety of sugarcane. Ind. J. Extn. Edu. 26 (1&2): 110-111.

Singh S. P. and R.K. Sharma. (1990). Commmunication behaviour of contract farmers under T&V system. Rural India 53(a): 42-46.

Verma R.P., J. Lal and M. Alam. (1986). Effective communication of sugarcane production technology (gains and constraints), Bharathiya Sugar 11(12) :19-20.

18

A Comparative Analysis of the Extension Approaches used by both Private and Public Extension Service Providers

K. Vengalamani [1], R. Arunachalam [2] and T. Masanaselvam [3]

[1 & 3] Post graduate scholars and [2] Associate Professor, Dept of Agricultural Extension,

Agrl. College and Research. Institute, Madurai

Introduction

Agriculture has always been. India's most important economic sector. To improve this agriculture betterment, it needs principally two important sectors, one is public sector and another one is private sector. In India, generally, Ministry of agriculture and co-operation, State department of agriculture, State agricultural universities, ICAR and its transfer of technology projects are the major public extension service providers

The contribution of public extension in attaining self-reliance in food production is very well recognized. But in this changing time of globalization, public extension alone is not sufficient to address the multi-faceted problems faced by farmers. Also the performance of public sector extension is under scrutiny. Rivera *et al* (2000) reported that the public extension system is now seen as outdated, top-down, paternalistic, inflexible, subject to bureaucratic inefficiencies and therefore unable to cope with the dynamic demands of modern agriculture.

Chandra (2001) expressed that Public extension alone can't meet the specific needs of various regions and different classes of farmers and it required a system to promote private and community driven extension to complement, supplement, work in partnership and even a suitable substitute for public extension. Experiences from earlier attempts and present situation of agriculture clearly' indicate that no single mode or model will give good results in all situations. For this purpose private sector involvement has to be encouraged in extension service.

In the Indian context, Private agri clinics, Agricultural consultants, Consultancy firms, Progressive farmers. Farmers organizations, Co-peratives, Non-governmental organizations, NGO run KVKs, Agribusiness companies, Input dealers, Newspapers, Agricultural magazines, Private television channels, Private sector banks, Internet and Donor agencies and Unemployed agricultural graduates are the major private extension service providers,

Research studies showed that, private extension service also has the problems like failure to provide affordable service to small farmers and many farmers were unable to pay the service of private sector. This reveals that the privatized extension will not embody the replacement of a monolithic public extension system.

Farmers in Tamil Nadu are availing the services of the public and private extension system depending upon their cropping needs. This study has been formulated with the basic objective to understand the extension approaches used by both public and private extension service providers in Madurai district. The results will be very much useful to revitalize the extension machinery in the effective transfer of technology

Methodology

The research was conducted in Madurai district of Tamil Nadu. Considering the maximum number of active private extension service providers and maximum area under crops, the taluks viz., Vadipatti and Melur were selected for this study purpose. By employing the same criteria, three blocks, viz., Vadipatti and Alanganallur from Vadipatti taluk and Melur from Melur taluk has been selected. In the same way from each block four villages were selected. At most care has been taken to select the farmers who have experienced both the public and private extension services. The farmers who have experience in both private and public extension services alone could judge the performance of the both the systems and express their perceptual status clearly. The list of beneficiaries from each village was obtained from the local agriclinics and it was decided to select one-third of total population for this study purpose. Accordingly, the sample size has been fixed as 120 respondents and they were selected from each village by employing proportionate random sampling method.

By having discussion with local extension workers, subject matter specialists of local Krishi Vigyan Kendra, crop scientists, staff of private agriclinics and local farmers 27 different extension approaches used by both public and private extension service providers were identified .These approaches were classified into 3 groups viz., Individual approach, Group approach and Mass approach. There were two response ategories namely 'used' and 'not used' with scores of 2 and 1 respectively. All the scores on sub-items were summed up to get the total score. Percentage analysis has also been done to get meaningful interpretation of the results.

Findings and Discussion

The relevant data are presented in Table.1 and the results are discussed in the following paragraphs.

Individual approaches used by both private and public extension service providers

With regard to the individual approaches it is found that, farm and home visit was the major approach adopted by the private agricultural extension service providers as experienced by majority of the respondents (63.30 per cent) followed by diagnostic field visit (52.50 per cent) and telephone calls (43.33 per cent). Around one fourth of the respondents (23.30 per cent) stated that they personally visited the office of the private extension service providers to get the required information.

No much variation was seen with regard to the usage of different individual approaches followed by public extension service providers. Around one third of the respondents (29.16 per cent) opined that they made personal visit to the office of public extension service providers to get the required information. Almost a similar percentage of the respondents (26.66 per cent) stated that the public extension service providers have adopted farm and home visit to transfer technological information. Only 16.66 per cent of the respondents made telephone calls to the public extension service providers.

Table 1 : Extension approaches used by private and public agricultural extension service providers

(n=120)

(Multiple response)

SI. No.	Categories	Private		Public	
		Number	Percent	Number	Percent
I	**Individual approach**				
	a) Farm and home visit	76	63.30	32	26.66
	b) Telephone calls	52	43.33	20	16.66
	c) Office calls	28	23.30	35	29.16
	d) Diagnostic field visit	63	52.50	27	22.50
II	**Group approach**				
	a) Method/Result Demonstration	58	48.33	27	22.50
	b) Training programmes	70	58.34	68	56.60
	c) Workshops	45	37.50	26	21.66
	d) Group discussion	-	-	15	12.50
	e) Trail plots	59	49.16	52	43.33
	f) Field trips	20	16.67	45	37.50
	g) Field days	5	4.16	24	20.00
III	**Mass approach**				
	a) Exhibitions	52	43.33	40	33.33
	b) Campaign	30	25.00	42	35.00
	c) Posters	60	50.00	20	16.67
	d) Farm broadcast	75	62.50	82	68.33
	e) Farm telecast	50	41.66	47	39.16
	f) Folders	55	45.83	40	33.33

From the above findings it could be inferred that among the individual approaches the private extension service providers widely adopted farm and home visit, diagnostic field visit and telephone advisory services. These three services are most effective services in the process of transfer of technology. As the private extension is a paid and committed service, the private service providers have concentrated widely on these three effective methods to render quality service and win the confidence of the local people. But the public extension service providers have almost least adopted these methods, for which they claim their poor staff strength, poor infrastructural and transport facilities as their main reasons.

Group approaches used by both private and public extension service providers

As regard the group approaches it is found that private extension service providers conducted various' training programmes for their clients as reported by majority of the respondents (58.34 per cent). Almost a similar percentage of the respondents stated that the private extension service providers used trial plots (49.16 per cent) and method/result demonstration (48.33 per cent). Around forty percentage of the respondents (37.50 per cent) informed that they have participated in the workshop organized by the private extension service providers. It is also found that they have organized field trips also as opined by 16.67 per cent of the respondents.

Similar to the private extension service providers the public extension service providers also conducted various training programmes to transfer the technological information as experienced by majority of the respondents (56.60 per cent). Trail plots (43.33 per cent) and field trips (37.50 per cent) were the other major group approaches followed by the public extension service providers. Around one fifth of the respondents felt that the method/result demonstration (22.50 per cent), workshops (21.66 per cent) and field days (20.00 per cent) were also adopted by the public extension service providers.

Among the group approaches the private extension service providers adopted training programmes, trail plots, method/result demonstration, whereas the public extension service providers mostly concentrated on the training programmes and trail plots. The other methods were also adopted in differential manner. The training programmes and method/result demonstration are the important group approaches where by the farmers are thought the skill involved in practice. Interestingly both the methods well adopted by private extension service providers than the public extension service providers.

Mass approaches used by both private and public extension service providers

Among the mass approaches it is found that the majority of the respondents (62.50 per cent) felt that they have experienced the farm broadcast of the private extension service providers. Posters (50.00 per cent) were also widely used by them. Folders (45.83 per cent), exhibition (43.33 per cent) and farm telecast (41.66 per cent) were the other mass approaches adopted by the private extension service providers. One fourth of the respondents stated that the private extension service providers conducted campaigns in the process of transfer of technology.

Almost the similar trend is observed in the mass approaches followed by the public extension service providers. Farm broadcast (68.33 per cent) was their major mass approaches in the process of transfer of technology, followed by farm telecast (39.16 per cent). Around one third of the respondents stated that the public extension service providers adopted the other mass approaches viz., campaign (35.00 per cent), exhibition (33.33 per cent) and folders (33.33 per cent). Posters were the least adopted approach (16.67 per cent).

Conclusion

With regard to the individual approaches, farm and home visit was the major approach adopted by the private extension service providers followed by diagnostic field visit and telephone calls. As regard the group approaches, they conducted various training programmes for their clients Demonstrations were also widely conducted by them for the dissemination of the information. Majority of the respondents have also experienced the farm broadcast of the private extension service providers. Use of posters, folders, exhibits and farm telecast were also their important mass approaches.

Regarding the individual approaches of the public extension service providers, around one third of the respondents opined that they made personal visit to the office of public extension service providers to get the required information. They have also adopted the farm and home visit to transfer the technological information. The study revealed that the public extension service providers conducted various group approaches. They organised training programmes to transfer the technological information Trail plots and field trips were their other major group approaches. Farm broadcast was their major mass approaches in the process of transfer of technology, followed by farm telecast Campaign, exhibitions and folders were also their widely used mass methods. The above findings revealed that the private extension service providers have widely adopted different extension approaches than the public extension service providers.

References

Chandra, R. Shiraj. 2001. Effectiveness of eco-friendly cultivation practices in paddy-analysis. Unpub. M.Sc. (Ag.) Thesis, AC&RI, TNAU, Madurai.

Rivera, W., Zijp, W. and A. Gary, 2000, Contracting for extension: Review of emerging practices. Agricultural Knowledge Information System (AKIS). Thematic Group, Washington.

●●●

19

Perception on Critical Technology Adopted by Vegetable Growers under Precision Farming –An Overview

K. Thangaraja*[1] *, R. Ganesan* [2]*, R. Sasikala* [3] *and R. Rajasekaran
[1]*Ph. D. Scholar (Agrl. Extension),* [2] *Director (Agri Business Development),*
[3-4] *Asst .Professor (Agri. Extension), TNAU, Coimbatore -3*

Introduction

India's population is projected to be around 1,350 million by 2020 AD which requires food grain production of 343 million tonnes. Considering the current food grain production levels, India needs to raise food grain growth by not less than 4 percent. On the other hand the available per capita cultivable land has declined from 0.48 ha to 0.14 ha and by 2050 it may further decline to 0.07 ha i.e. for every 14 persons only one ha land will be available to produce all essential items like food, fiber, fuel, fodder, fruits etc. (Harshal et al., 2006). About 75 percent of 105 million farm families belong to small and marginal farmers with less than 2 ha and 40 percent of the rural population lives below poverty line. We often talk about self-sufficiency in food supply but the fact is that population living below poverty line does not have the purchasing power.

In 1952, India had 0.33 ha of land available per capita, which is reduced to 0.15 ha in 2000. As the availability of land has decreased, application of fertilizers and pesticides become necessary to increase production. The major effect is that our agriculture has become chemicalized. In this situation, it is essential to develop eco-friendly technologies for maintaining crop productivity. Since long, it has been recognized that crops and soils are not uniform within a given field. The farmers have always responded to such variability to take actions, but such actions are inappropriate and less frequent.

In this context, Tamil Nadu State sponsored turnkey project of Tamil Nadu Precision Farming Project (TNPFP) implemented by Tamil Nadu Agricultural University (TNAU) at Dharmapuri and Krishnagiri districts in 400 ha to train the farmers, extension workers, agri-horti entrepreneurs and students for a period of three years (2004-2007) with a financial assistance of Rs.720 lakhs. TNAU was undertaking this project and implemented with the co-operation of the state departments of Horticulture, Agricultural Engineering, Agriculture, Agricultural Marketing and Agri-Business and the District Administration. High value crops like Tomato, Brinjal, Sugarcane, Banana, Gherkins, Hybrid Capsicum, Paprika, Baby Corn, and White Onion, Bhendi, Cabbage and Cauliflower have been proposed to be cultivated under this scheme.

Under this project, 100 percent subsidy was given to the farmers selected during first year. 10 percent of the cost of cultivation was collected from the farmers selected during second year and 20 percent of the cost of cultivation was collected from the farmers selected during third year. The project area lies in the districts of Dharmapuri and Krishnagiri of Tamil Nadu state. The area covered in these districts are 100 ha, 200 ha and 100 ha during the years 2004-05, 2005-06 and 2006-07 respectively. The experiences gained during the year 2004-07 are presented here. Keeping this in view, the research study was carried out with an an objective "To study the Perception on Critical Technology adopted by Vegetable Growers under Precision Farming"

Methodology

The study was conducted in Dharmapuri district of Tamil Nadu. Dharmapuri district was purposively selected for the study based on a pioneer attempt, maximum area covered by Precision Farming Project. Dharmapuri district comprises of five taluks namely Dharmapuri, Palacode, Pappirettipatti, Pennagaram and Harur. Among them three taluks viz., Dharmapuri, Palacode and Pappirettipatti were selected considering the criteria of maximum beneficiaries, area covered under Precision Farming Project. Sample of 90 Precision Farming farmers were selected for this study. The sample was proportionately selected from selected three taluks of Dharmapuri district. In each of the selected three taluks, two villages namely Jarugu, Nallanahalli in Dharmapuri taluk, Somanahalli, Agaram in Palacode taluk and Molayanur, Menasi in Pappirettipatti taluk were selected using simple random sampling technique. The data were collected with well-structured interview schedule. The percentage analysis was used for data analysis.

Findings and Discussion

Perception about the critical technology adopted by vegetable growers was considered as a prerequisite to get an overall understanding of Precision Farming project. Hence, nine dimensions of the technology adopted by the vegetable growers under precision farming project have been identified to get the opinion of the farmers. And each aspect was analyzed separately. The results are presented in the table 1.

Table 1 : Perception on critical technologies adopted by the vegetable growers (n=90)

Sl. No.	Critical Technologies	Perception	
1.	**Varieties / Hybrid**	**No.**	**%**
	Highly preferred	64	71.11
	Least preferred	17	18.89
	Not preferred	09	10.00
2.	**Raising seedlings in community nursery**		
	Highly preferred	46	51.11
	Least preferred	23	25.56
	Not preferred	21	23.33
3.	**Seed treatment**		
	Easy to practice	24	26.67
	Somewhat difficult to practice	39	43.33
	Difficult to practice	27	30.00
4.	**Planting method**		
	Highly suitable	45	50.00
	Moderately suitable	24	26.67
	Least suitable	21	23.33
5.	**Adoption of drip irrigation**		
	Highly preferable	64	71.11
	Moderately preferable	21	23.33
	Least preferable	05	5.56
6.	**Nutrigation**		
	Very simple	56	62.22
	Somewhat complicated	20	22.22
	Very complicated	14	15.56
7.	**Micro nutrient spray application**		
	Very simple	37	41.11
	Somewhat complicated	25	27.78

Contd. ...

8.	Very complicated	128	31.11
	Application of growth promoter and growth regulator		
	Very simple	38	42.22
	Somewhat complicated	29	32.22
	Very complicated	23	25.56
9.	**Use of supporting pole**		
	Most benefited	74	82.22
	Somewhat benefited	12	13.33
	Least benefited	04	4.44

Use of recommended varieties / hybrid

It reveals that nearly three – fourth (71.11%) of the respondents expressed that they 'highly prefer' to use recommended hybrids introduced under Precision Farming. About 18.89 percent of the farmers `preferred least' followed by 10.00 percent of the farmers opinioned that `not preferred' to use recommended variety or hybrid in tomato cultivation under Precision Farming.

The farmers expressed that introduction of hybrid US 618 tomato under precision farming was 'highly preferred' because of the reason that it gives more number of the harvesting, high quality, good demand in market, long preservability, high yield and resistance to transport damage.

Raising seedlings in community nursery

From the table it could be observed that, about half (51.11%) of the respondents expressed that they 'highly preferred' to raise seedlings in community nursery, one-fourth (25.56%) of them 'least preferred' followed by 23.33 percent of respondents expressed 'least preference' to practice raising seedling in community nursery.

From the opinion of farmers, it could be stated that practice of raising of seedling in community nursery was 'highly preferred' because the seedlings have uniform growth, seedlings mortality rate is less and less occurrence of pest and diseases.

Seed treatment with bio – fertilizers and fungicides

It is inferred from the table that 43.33% of the farmers opined that seed treatment with bio – fertilizers and fungicides adopted were 'somewhat difficult to practice' followed by more than one – third (30.00%) opinioned that 'difficult, to practices and 26.67 percent of them expressed as 'easy to practice'. The farmers expressed that the use of bio-fertilizers and fungicides are 'somewhat difficult to practice' due to lack of awareness about using bio –fertilizers and fungicides.

Planting method

Regarding planting method ,abouthalf (50.00%) of the respondents have expressed that paired row planting {(90x60x60 cm (Between pairs, rows and plants)} is 'highly suitable' for tomato cultivation under Precision Farming. This method of planting is highly preferable by the farmers because it helps to maintain optimum plant population, good aeration and absorption of sunlight leads to increased root growth and plant canopy and easy to intercultural operations. About 26.67 percent of the beneficiaries expressed that this method is 'moderately suitable' followed by 'least suitable'.

Adoption of drip irrigation

It was obvious from the table that 71.11 percent of the beneficiaries opined that the adoption of drip irrigation system was 'highly preferable' due to less water consumption, increased water-use efficiency; enabled exact quantity of water to reach the root zone, ensured uniformity in irrigation of individual plants and prevents run-off and soil erosion. About 23.33 percent of the farmers expressed 'moderately preferred' followed by 'least preferable'.

Nutrigation

It was observed from the table that majority (62.22%) of the respondents expressed that the adoption of nutrigation technique (200:62.5:250 NPK Kg /ha – 3 days interval) was 'very simple' and created much impact among the farmers . That is minimum wastage of fertilizers , labour cost, time saving, weed growth is prevented and immediate uptake of nutrient by the plant. Only 22.22% of the respondent opined that it is 'somewhat complicated' followed by 'complicated `(15.56%).

Micro nutrient spray application

It was obvious from the table that 41.11 percent of the farmers opined that application of micro nutrient (0.5% 30, 45, 60 DAP) was 'very simple' because there were good foliage, increased growth of the plant, increases colour and shining of the fruits. About 27.78 percent of the respondents expressed 'somewhat complicated' followed by 'complicated' (31.11%).

Application of growth promoter and growth regulator

From the table it was evident that 42.22 percent of the farmers perceived that the application of growth promoter and growth regulator i.e., use of Plano fix @ 0.25ml/ lit of water to control flower drops 60, 90, 120 DAP was 'very simple' under Precision Farming method. From the findings it could be inferred that the use of growth regulators leads to greenish plant growth, increased the flowering, controlled the flower dropping and increased size and shape of the fruit. About 32.22 percent of them opined that the application of it was 'some what complicated' followed by complicated.

Use of supporting poles

It is inferred from the table that majority (82.22%) of the beneficiaries expressed their opinion about the use of supporting poles as 'most benefited'; about 13.33 percent of them opined that 'some what benefited'. and meager (4.44%) percent of them expressed 'least benefited'. Majority of the farmers expressed that the use of supporting poles (Bamboo, Casuarina 30 DAP with 6 ft interval) helps to prevent rotting and cracking of fruits, duration of the plant is high, easy to ntercultural operation and harvesting.

Conclusion

The findings of the study clearly exposed that, the critical technology such as cultivation of recommended hybrid US 618 tomato, practice of community nursery and adoption of drip irrigation system as 'highly preferred'. Adoption of technologies like seed treatment with bio — fertilizers and fungicides were 'somewhat difficult to practice' and adoption of paired row planting method {(90x60x60 cm (Between pairs, rows and plants)} was 'highly suitable' for tomato cultivation under Precision Farming. Similarly the technologies such as adoption of nutrigation techniques, application of micro nutrient (0.5% 30, 45, 60 DAP), application of growth promoter and growth regulator i.e., use of Plano fix @ 0.25ml/ lit of water to control flower drops 60, 90, 120 DAP as 'very simple' and majority of the beneficiaries expressed their opinion as 'most benefited' for the use of supporting poles (Bamboo, Casuarina 30 DAP with 6 ft interval) during the cultivation of tomato under Precision Farming.

References

Harshal, E., Premod, Chaudhari, M., 2006, Precision agriculture: A new form of agriculture to maximize crop production and minimize the environmental damage, Agricultural update, 1 (2): 24-25.

Sanjay Arora. 2005. Precision Agriculture and Sustainable Development, Kurukshetra, 54 (2): 18-22.

Sudha, T. 2008. Prospects of Precision Farming in Dharmapuri District — A Multidimensional Analysis. Unpub. M.Sc.(Ag.)Thesis, AC&RI, TNAU, Madurai.

●●●

20

Performance Analysis of the Gram Panchayat Leaders in the Three Tier Panchayati Raj Institutions of Rajnandgaon District of Chhattisgarh

P. Shrivastava [1]***, K.K. Shrivastava*** [2] ***and J.D. Sarkar*** [3]

[1]*Ph.D. Scholar,* [2]*Professor and* [3]*Prof & Head, Dept. of Agricultural Extension, College of Agriculture, IGKV, Raipur.*

Introduction

Indian agriculture contributes to 25 per cent of the National Gross Domestic Product. As per Census of India, 2001, India's total population was 1,028.6 million and the percentage of decadal growth of population during 1991- 2001 was 21.34 per cent. Thus, the main challenge is to produce enough food for increasing population.

Many new agricultural technologies and improved practices are being developed in agricultural research stations. Effective communication of these agricultural innovations to millions of farmers in rural area is essential to bring about accelerated agricultural development. The impact of demographic pressure on reducing farm size further on the one hand and the need for longer or optimum sized farms to take more advantage of modern technology on the other hand would bring into sharper focus the growing contradiction in the farming sector.

Agricultural development programme however well planned and imaginative, cannot make an impact on rural life unless it is backed both by extension and research support. Research obviously cannot work in isolation, nor can extension, only together they form an effective team. A good deal of work in establishing closer linkage between research and extension has been done in the last few years. The result of painstaking efforts of research scientists have been carried to the farmers by an army of more than 70,000 village extension workers in the major states where professional agricultural extension system is functioning.

It is difficult for any country to provide enough number of extension workers to reach each and every family for its social welfare programme. With the awareness on the part of the people and the beginning of the process of modernisation, the need for intensive contacts become more and more prominent. This problem can be solved to some extent through the use of local leaders. A local leader who has adopted improved practices extends the same to others. The common man has much faith in the local leaders. A villager would like to hear and imitate his own neighbour as compared to accepting the advice of an outside change-agent.

Conceptually, panchayats are elected rural local bodies responsible for local government functions. But the Balwantrai Mehta Committee has given them a substantially development programme delivery orientation. While this has given them access to rural development programme funds, it has also simultaneously made them vulnerable and obsessed with these funds and their use.

In the post, 73rd amendment phase Indian states have responded with varying degree of enthusiasm. Chhattisgarh responded with innovativeness and remarkable commitment to making the system sustainable and successful. During the implementation of the .panchayati raj system Chhattisgarh has faced several opportunities and difficulties. The experience of the state is extremely rich and provides vital insights into the process of institutionalising panchayati raj.

In general panchayats should be able to facilitate member work effectiveness by influencing strategy choices more for complex jobs than for simple, straight forward or routine ones. The reason is that on simple jobs, strategy choices usually cannot make much of a difference in effectiveness; instead, how well one does is determined almost entirely by how hard one works. With the above points in reference a study was organised on the following objectives :-

1. To assess the performance of panchayat leaders with reference to rural development activities.
2. To analyse the relationship between the independent variables and the performance of panchayat leaders & their attitude towards Panchayati Raj Institutions.

3. To predict the variation in performance of panchayat leaders caused by independent variables.

Methodology

An ex-post-facto study was organised in Rajnandgaon district of Chhattisgarh in 2010 to assess the role performance of gram panchayat leaders. Two hundred leaders were randomly selected from 50 gram panchayat leaders of nine janpads of rajnandgaon district. They were personally interviewed with the help of a structured schedule developed on the basis of objectives of the study.

Performance here refers to the degree of accomplishment of the tasks that make-up the panchayat leaders assigned roles/functions, it indicates how well an individual is fulfilling the job demands. The performance dimensions are several. They include duties, responsibilities, behaviour and traits. For each of the relevant dimensions and sub dimensions there of standard will have to be fixed based on past performance or any other base. Relative weightages have to be assigned to each of these in turn. Performance assessment is an important component of the information and control system. Performance assessment is the systematic, periodic and an impartial rating of an individuals excellence in matters pertaining to his present job and to his potentialities for a better job. When properly conducted performance assessment should not only let the individual know how well he is performing but also influence the individuals future level of effort, activities, results and direction.

In all six major activities/roles of panchayat leaders were identified and 44 questions were framed to cover all the above roles. The respondents were asked to state the extent to which they performed the different assigned roles/activities. The responses of the panchayat leaders were obtained on five point Likert type scale viz. Always, most of the time, sometimes, rarely and never with the scores 5, 4, 3, 2 and 1 respectively. The scores of all the 44 questions were added to obtain the overall performance score of each respondent, which were used in statistical analysis and the panchayat leaders were classified into three categories.

Findings and Discussion

The study reveals that the majority of the gram panchayat leaders had medium role performance, followed by 23.00 and 21.50 per cent of them who had high and low role performance respectively. It can be conclusively said that majority of the gram panchayat leaders had medium role performance as regards rural development and other panchayat activities.

Table l : Ascertain of independent variables with role performance of gram panchayat leaders.

SI. No.	Independent Variables	Correlation Coefficient "r"
1.	Age	- 0.3354 **
2.	Gender	0.0787
3.	Caste	0.1534 *
4.	Education	0.7841 **
5.	Family size	0.1446 **
6.	Social participation	0.0067
7.	Occupation	0.2524 **
8.	Annual Income	0.2785 **
9.	Size of land holding	0.3647 **
10.	Material possession	0.4283 **
11.	Socio economic status	0.4537 **
12.	Achievement motivation	0.2023 **
13.	Cosmo politeness	0.2023 **
14.	Job satisfaction	0.1954 **
15.	Information sources	0.3199 **
16.	Extension participalion	0.8497 **
17.	Time allocation for panchayat activities	0.2909 **
18.	Experience	0.1101
19.	Training need	0.1015
20.	Political affiliation	0.1312
21.	Political ideology	0.1398 *
22.	Political efficacy	0.1012
23.	Attitude towards panchayati raj institutions	0.8717 **

* Significant at 0.05 level of probability,
** Significant at 0.01 level of probability

Correlation analysis was done to find out the influence of independent variables on the role performance of gram panchayat leaders. The results of correlation analysis depicted in table 2 reveals that the variables gender, social participation, experience, training need, political affiliation and political efficacy had non significant relation with role performance of gram panchayat leaders. In other words it can be said that per unit increase or decrease in the above stated variables would have non significant influence on the role performance of panchayat leaders.

However the variables caste, family size and political ideology had positively significant relationship with role performance of gram panchayat leaders at 0.05 level of probability. Whereas the variables education, occupation, annual income, size of land holding, material possession, socio economic status, achievement motivation, cosmopoliteness, job satisfaction, information sources, extension participation, time allocation for panchayat activities and attitude towards panchayati raj institutions had positive and significant influence on role performance of gram panchayat leaders at 0.01 level of probability. The positive and significant relationship indicates that increase/decrease in the above significantly related variables would correspondingly increase/decrease the role performance of gram panchayat leaders. Only, the variable age had negatively significant relationship with role performance of gram panchayat leaders at 0.01 level of probability.

The negatively significant relationship shows that if the value of variable age would increase there would be a decrease in role performance. In other words we may infer that younger gram panchayat leaders will have better role performance and older gram panchayat leaders are not likely to have better role performance.

Table 2 represents the results of multiple regression analysis of independent variables with role performance of gram panchayat leaders. Multiple regression analysis was carried out to ascertain the contribution of different independent variables under study on the dependent variable i.e. role performance of gram panchayat leaders. Comparisons were made between obtained "t" values and table values of "t" to know which independent variable made significant contribution towards role performance and which independent variable did not make any significant contribution towards role performance of gram panchayat leaders. The regression coefficient "b" showed the extent of contribution the independent variables had on the role performance of gram panchayat leaders.

It is observed that the variables gender, family size, size of land holding, material possession, socio economic status,. job satisfaction, experience, training need, political affiliation and political ideology had non significant contribution towards role performance of gram panchayat leaders. This indicates that any increase or decrease in the above independent variables would not contribute significantly in the increase or decrease of role performance of gram panchayat leaders. Only one variable i.e. age had negatively significant contribution on role performance of gram panchayat leaders. Negatively significant contribution means that one unit increase in age would have 0.775 unit decrease in role performance of gram panchayat leaders or one unit decrease in age would have 0.775 unit increase in role performance of gram panchayat leaders.

Table 2 : Multiple regression analysis of independent variables with role performance of gram panchayat leaders

SI. No.	Independent Variables	Correlation Coefficient "r"	Correlation Coefficient "r"
1.	Age	- 0.3354 **	2.428
2.	Gender	0.0787	0.629
3.	Caste	0.1534 *	1.990
4.	Education	0.7841 **	5.307
5.	Family size	0.1446 **	0.015
6.	Social participation	0.0067	2.498
7.	Occupation	0.2524 **	1.993
8.	Annual Income	0.2785 **	2.662
9.	Size of land holding	0.3647 **	0.245
10.	Material possession	0.4283 **	0.240
11.	Socio economic status	0.4537 **	0.249
12.	Achievement motivation	0.2023 **	2.223
13.	Cosmo politeness	0.2023 **	2.719
14.	Job satisfaction	0.1954 **	0.031
15.	Information sources	0.3199 **	2.537
16.	Extension participalion	0.8497 **	4.542
17.	Time allocation for panchayat activities	0.2909 **	1.995
18.	Experience	0.1101	0.286
19.	Training need	0.1015	0.556
20.	Political affiliation	0.1312	0.178
21.	Political ideology	0.1398 *	0.430
22.	Political efficacy	0.1012	1.977
23.	Attitude towards panchayati raj institutions	0.8717 **	5.309

* Significant at 0.05 level of probability $R^2 = 0.8610$

** Significant at 0.01 level of probability F value = 47.41

However the variables, caste, social participation, occupation, achievement motivation, information sources, time allocation for panchayat activities and political efficacy had positively significant contribution towards role performance at 0.05 level of probability. Whereas the remaining five variables viz. education, annual income, cosmopoliteness, extension participation and attitude towards panchayati raj institutions had positive and significant contribution towards role performance of gram panchayat leaders at 0.01 level of probability. From the positive and significant contribution we may infer that one unit increase in the independent variables caste, education, social participation, occupation, annual income, achievement motivation, cosmopoliteness, information sources, extensjon participation, time allocation for panchayat activities, political efficacy and attitude towards panchayati raj institutions would increase the role performance of gram panchayat leaders by 1.523, 9.787, 1.058, 0.632, 0.911, 0.938, 0.669, 0.328, 2.100, 0.498, 1.113 and 1.195 units respectively. The R-square value of 0.8610 also indicates that all the 23 variables jointly contributed towards role performance of gram panchayat leaders to the extent of 86.10 per cent.

Direct, total indirect and substantial indirect effects of independent variables on role performance of gram panchayat leaders

Direct Effect

The data in Table 3 reveals that the variable material possession exerted the highest positive direct effect on role performance of gram panchayat leaders as the path coefficient value was 0.3742, followed by attitude towards panchayati raj institutions (0.3222), education (0.2804), extension participation (0.2750), cosmopoliteness (0.0773), size of land holding (0.0627), information sources (0.0498), achievement motivation (0.0380), occupation (0.0354), political ideology (0.0132), political affiliation (0.0054) and family size (0.0009) whereas the variable socio economic status had maximum negative direct effect on role performance of gram panchayat leaders with path coefficient —0.2863, followed by political efficacy (-0.0650), social participation (-0.0468), age (-0.0467), caste (-0.0342), time allocation for panchayat activities (-0.0329), annual income (-0.0297), training need (-0.0202), gender (-0.0195), experience (-0.0091) and Job satisfaction (-0.0011).

Total indirect effect

It is obvious from the Table 3 that variable socio economic status had highest positive total indirect effect with path coefficient 0.7400 followed by cosmopoliteness (0.6227), extension participation (0.5747), attitude towards panchayati raj institutions (0.5495), education (0.5037), time allocation for panchayat activities (0.3239), annual income (0.3082), size of land holding (0.3043), information sources (0.2701), occupation (0.3170), job satisfaction (0.1965), caste (0.1676), political efficacy

(0.1662), achievement motivation (0.1643), family size (0.1437), political ideology (0.1266), political affiliation (0.1258), training need (0.1217), experience (0.1192), material possession (0.0541) and social participation (0.0535). While the variable age had highest negative total indirect effect (- 0.2887) followed by gender (- 0.0592).

Table 3 : Path coefficient showing direct, indirect and substantial indirect effect of independent variables on role performance of gram panchayat leaders

SI. No.	Vari-able	Independent Variables	Direct Effect	Total Indirect Effect	Substantial indirect effect through first
1.	X1	Age	- 0.0467	- 0.2887	- 0.1062 (X23)
2.	X2	Gender	- 0.0195	- 0.0592	- 0.0513 (X4)
3.	X3	Caste	0.0342	0.1876	0.0705 (X10)
4.	X4	Education	0.2804	0.5037	0.2804 (X4)
5.	X5	Family size	0.0009	0.1437	- 0.0899 (X11)
6.	X6	Social participation	-0.0468	0.0535	- 0.0468 (X6)
7.	X7	Occupation	0.0354	0.2170	0.0991 (X10)
8.	X8	Annual Income	- 0.0297	0.3082	0.2338 (X10)
9.	X9	Size of land holding	0.0627	0.3043	0.2704 (X10)
10.	X10	Material possession	0.3742	0.0541	0.3742 (X10)
11.	X11	Socio economic status	- 0.2863	0.7400	0.3682 (X10)
12.	X12	Achievement motivation	0.0380	0.1643	0.0559 (X23)
13.	X13	Cosmopoliteness	0.0773	0.6227	0.2104 (X23)
14.	X14	Job satisfaction	- 0.0011	0.1965	0.0536 (X23)
15.	X15	Information sources	0.0498	0.2701	0.0944 (X23)
16.	X16	Extension participation	0.2570	0.5747	0.2750 (X16)
17.	X17	Time allocation to panchayat activities	- 0.0329	0.3239	0.1163 (X10)
18.	X18	Experience	- 0.0091	0.1192	0.0435 (X16)
19.	X19	Training need	- 0.0202	0.1217	
20.	X20	0.0320 (X23) Political affiliation	0.0054	0.1258	0.0460 (X23)
21.	X21	Political ideology	0.0132	0.1266	0.0494 (X4)
22.	X22	Political efficacy	- 0.0650	0.1662	0.1128 (X4)
23.	X23	Attitude towards panchayati raj institutions	0.3222	0.5495	0.3222 (X23)

Substantial indirect effect

It is clear from the table 4 that the variable material possession had the highest positive substantial indirect effect (0.3742) on role performance through itself followed by socio economic status (0.3682) through material possession, attitude towards panchayati raj institutions (0.3222) through itself, education (0.2804) through itself, extension participation (0.2750) through itself, size of land holding (0.2704) through material possession, annual income (0.2338) through material possession, cosmopoliteness (0.2104) through attitude, time allocation for panchayat activities (0.1163) through material possession, political efficacy (0.1128) through education, occupation (0.0991) through material possession, information sources (0.0944) through attitude, caste (0.0705) through material possession, achievement motivation (0.0559) through attitude, job satisfaction (0.0536) through attitude, political ideology (0.0494) through education, political affiliation (0.0460) through attitude, experience (0.0435) through extension participation and training need (0.0320) through attitude towards panchayati raj institutions.

The variables which had negative substantial indirect effect on role performance of gram panchayat leaders were age (- 0.1062) through attitude towards panchayati raj institutions, family size (- 0.0899) through socio economic status, gender (- 0.0513) through education and social participation (- 0.0468) through social participation itself.

References

Kuraria, U.K. Khare, Y.R. and Swarnkar, V.K. (1997) Role perception of elected representatives in panchayati raj system, *Maha. J. Extn. Edn.,* 16 : 184-186.

Limbadri, R. (2007) Grass root democracy : The experience of Dalits in P.R.I.'s, Indian *Journal of Public Administration,* 53(4) : 788-796.

Mondal, S. and Ray, G.L. (1996) Socio-economic profile of gram panchayat pradhans, *Indian J Extn. Edn.,* 32 (1-4) : 77-81.

Reddy, S.V. (1971) Leadership in relation to interpersonal communication behaviour, Unpublished Ph.D. thesis, IARI, New Delhi.

Shrivastava, K.K. (1999) Role perception and role performance of formal leaders working
under panchayati raj system in Kheda district of Gujarat state, Unpublished Ph.D. Thesis, Gujarat Agricultural University, Anand Campus, Anand.

Shrivastava, P. (2003) Job performance of village panchayat leaders of Jabalpur block, district Jabalpur, M.P., Unpublished M.Sc. (Ag) thesis, JNKVV, Jabalpur.

Tripathi, P.C. and Shukla, A. (2005) A text book of Research Methodology in Social Sciences, Sultan Chand and Sons, Educational Publishers, New Delhi.

Trivedi, G. (1963) Movement and analysis of socio economic status of rural families, Unpublished Ph.D. Thesis, IARI, New Delhi.

Vohra, N.N. And Bhattacharya, S. (2002) Looking Back : India in the twentieth century, National Book Trust, India, New Delhi.

Vyas, K.C.R. (1984) A study of the job chart and job performance of agriculture development officers with reference to their training, working under IERP in Narsinghpur and Chhindwara district of Jabalpur Division M.P., Unpublished M.Sc. (Ag) thesis, JNKVV, Jabalpur.

Wilkening, E.A. (1952) Informal leaders and innovators in farm practices, Rural Sociology, 17: 272-275.

Yadav, G.C. (1988) A study on role performance of gram panchayat members in relation to their socio-personal attributes in Panagar block, Jabalpur district, M.P., Unpublished M.Sc. (Ag) thesis, JNKVV, Jabalpur.

●●●

21

Case Study of a Successful Rabbit Farmer

S. Vishwendar*

** Department of Veterinary and Animal Husbandry Extension, Madras Veterinary College, Vepery, Chennai- 7*

Introduction

Livestock and livestock products are an important source of food nutrition and livelihood security. Animal origin foods are an important source of high quality protein and micronutrients compared to cereals and plant-based foods. Production of animal- based foods has grown dramatically over the past two decades. Meat, Milk and Egg consumption has grown rapidly in east and Southeast Asia over past twenty years. The greatest increase in the production and consumption of animal-origin foods, as a result of rising incomes and dietary diversification away from staple cereals, has taken place in East and Southeast Asia. World meat markets reveal that the production and trade of bovine meat, poultry meat, pig meat and ovine meat are significant unlike rabbit meat. In spite of many advantages in rabbit farming and rabbit meat, the rabbit meat is less consumed when compared to other meat. So the objective was to study the farmer who involved in rabbit farming.

Table 1 : Meat production in world- 2008

Type of Meat	2008 (million tonnes)	2009 (million tonnes)
Beef	65.1	64.3
Chicken	91.8	91.9
Pork	104.6	106.5
Mutton	13.2	13.4

Source : www.faostat.fao.org

Table 2 : Rabbit meat production-2008

Rank	Country	Production (million tonnes)
1	China	660020
2	Venezuela	481000
3	Italy	239977
4	Korea	91000
5	Spain	74991
6	Egypt	69842
7	France	50863
8	Czech Republic	36744
9	Germany	33683
10	Ukraine	13099
11	Russian Federation	11297
12	Argentina	7260
13	Greece	7147
14	Algeria	7000
15	Hungary	6647
16	Mexico	4250
17	Slovakia	4086
18	Colombia	3900
19	Poland	3530
20	Peru	3240

Source : www.faostat.fao.org

Farmer's profile

A successful male rabbit farmer Mr. S. Kumar of Potheri village of Kancheepuram district in Tamil Nadu, India was studied through direct interview on 12th November, 2010 and presented here as a case report. Respondent was 29 years old, married, employed in private firm for a salary of 4000/- per month, living in a joint family consisting of 7 members. Moreover has a house in 3600 square feet plot. His educational qualification was ITI. In 2006 he had twenty Turkeys (15 females and five males purchased from Institute of poultry production and management, Nandhanam, Chennai with his own money) earned Rs. 2800/- per month by selling turkey meat at the rate of Rs. 180/- per kg. Unfortunately the farmer had to sell the birds because of frequent occurrence of diseases, and unable to tolerate the peculiar sound produced by the birds often.

KVK's intervention

Fortunately he came in contact with KVK (Krishi Vigyan Kendra, Kancheepuram district) there he was suggested to involve in rabbit farming. He underwent training on rabbit farming in KVK for two days at free of cost in August 2007. Consequently, he started a rabbit farm as a part time job in September 2007 with 6 rabbits (5 females and 1 male purchased at the rate of Rs. 500/- per rabbit from private farm in Thambaram, Chennai) with initial capital of Rs. 6000/- without assistance of any financial institutions. His motivational factors were the advantages of rabbit for farming (ex. breed and grows quickly, excreta used as fertilizer, easy handling of rabbits etc.).

He promoted his business by advertising the advantages of rabbit meat (ex. high in protein, low in fat etc...) in local papers, televisions & distributing leaflets and had good sales of rabbit meat in the nearby locality. From the 3rd month of starting the farm he started earning profit. He started earning more profit after 1.5 years. The feedback he received from eaters are different taste and softer than chicken meat.

Respondent sold the rabbits at 5th month of age for meat purpose. Each rabbit gained 2.5 kg (average) live weight at 5th month of age. On average respondent started selling 136 rabbits (340 kg) per month at the rate of Rs. 225/- per kg live weight to nearby hotels and chicken shops for meat purpose with the help of market linkage provided by KVK. Through this marketing the gross income of the respondent was Rs.76, 500/- per month. The approximate net income was Rs. 15,000/- per month. He also had his marketing in Kancheepuram district, Thiruvallur district and Vellore district.

New Zealand White, Soviet Chinchilla and California were the breeds maintained in his farm. Among the three breeds respondent noticed New Zealand White was good for breeding purpose but reduction of weight was noticed after giving birth to young one unlike Soviet Chinchilla. California breed was also liked by the customers. He was not comfortable with rearing Angora breed because not able to adapt to this climate and the marketing channel to sell the fur that was obtained from this breed was not good. He also had Giant breed but the farmer felt difficulties in investing to provide cage sealed with thermo coal for the purpose of thermoregulation and not suited for breeding. So the farmer stopped rearing these two breeds. Thatched house for rabbits have been maintained for past two years with length 22 feet, breadth 16 feet and height 18 feet.

He had regular personal contact with doctors in KVK. Subject matter specialists used to visit the farm and gave suitable guidance regarding treatment, techniques which are used periodically in KVK. He also underwent training in Namakkal and Coimbatore farm in 2009 arranged by KVK where he was taught advanced information regarding farming.

Constrains faced

The first and foremost problem he faced was delaying in the sanctioning of loan for rabbit farming. He invested his own money and started the rabbit farm.

Death of mother rabbits after 15 to 20 days of delivery was noticed by respondent. By consulting with KVK, came to know that the occurrence of bacterial infection because of weakness after giving birth to young rabbits. Respondent with the help of KVK overcame this problem by providing hot water or cold water added with one to two grams of Calcium hypochlorite (bleaching powder) in one litre for disinfection purpose along with dry foods.

At the beginning, farmer started giving cabbage and cauliflower to rabbits as the feeding cost of those was less, which increased the smell of urine as well as increased the number of times of urination and spoiled the farm environment. He solved this problem by avoiding cabbage and cauliflower to rabbits with his own knowledge.

Respondent noticed death of one rabbit which was about 3.5 to 5 months of age followed by death of seven rabbits belong to same batch (35 rabbits) continuously in starting stage of farming with symptoms of blue ears and abnormal appearance of liver (found through post-mortem). Finally through KVK respondent came to know that 35 rabbits while in pregnancy(gestation) their mother were treated with a drug called Ivermectin continuously in another farm. Respondent overcame this problem by advising them not to follow this procedure.

After finding solutions for the above problems respondent didn't face loss in any aspect.

Successful entrepreneur

He got six to ten young ones (kits) from each kindling (giving birth to young one) and eight yields per year from each rabbit. The mating was done in cage once in 45 days. Duration of mating was one minute for each mating. The mating between close relatives (Inbreeding) was prevented by bringing different batch of rabbits at appropriate time.

Spending time in a rabbit farm was very very less. He was employing a labourer for only two hours per day (morning one hour and evening one hour for feeding, mating and medicines) with a salary of Rs.1000/- per month. Feeding cost per day per rabbit was Rs. 1.6 /- maintenance and electric charge too very low. He was giving 100 grams of mixture of maize, wheat bran, ground nut husk in the morning that cost about Rs. 12/- per kg, 250 grams of green fodder in the evening that cost about Rs. 1.6/- per kg. Disease occurrence was rare because of preventive measures he followed.

One bag of droppings (excreta) was sold for Rs. 10/- since the droppings of rabbits were used in horticulture. The disinfection of the farm was done once in a month using spray that contains 1miprothrin 0 .07 percent & Cypermethrin 0.2 percent around the farm. Clearing of droppings was done once in a week followed by application of bleaching powder. Burial method was the disposal method followed for 10 rabbits that died in the past three years.

Deworming was not practiced since Deworming is not needed in caged rabbits. Preventive measures for disease occurrence is carried out by giving Tetracycline powder at a dose rate of two grams for each rabbit for the rabbits above 3 months of age and Sulpha tablet at a dose rate of lgm per rabbit for the rabbits above 3 months of age once in 15 days alternatively. Hitech solution (Ivermectin) was the drug used to remove ectoparasites at a dose rate of 2m1 /day for 3 days.

Mr. Kumar used to rear the rabbits in two types of wire mesh (14mm) cages (hutches). One type was 1.5 feet x 1.5 feet that can accommodate one adult rabbit or five to six young rabbits, another type was 2 feet x 2 feet that can accommodate two adult rabbits or eight to ten young rabbits along with mother. Separation of young rabbit from mother (weaning) is done by 1 month of age.

There was no record maintenance. Rabbits were not insured. Respondent has planned to start another rabbit farm in one acre which can accommodate thousand rabbits at a time.

Economic and social impact

He got good income which met out expenses in his family regularly. As a result of these he was satisfied his family economic status was increased, respected in a society.

Malligai Women Self Help Group purchased twelve rabbits (three males & nine females) eight months back from Mr. Kumar and started rearing with the help of knowledge shared by respondent. Similar to this group Mr. Magesh from Thambaram, Chennai purchased fifty rabbits (fifteen males & thirty five females) 2.5 years back started rearing. Now he is maintaining 70 rabbits as parents stock in his farm.

He is given a separate stall in all the exhibitions conducted by KVK.

Outstanding rabbit farmer

This case study reveals the factors that played a role in success of farmer. Those factors were Training in KVK, Regular contact with , KVK, Interest, Involvement, Advertisement, Prompt dealing to a customer, Sharing knowledge, Demonstrating procedures, Assisting others in starting a farm. These fetch him award as an outstanding rabbit fanner for his successful Adoption and Dissemination of technologies/ practices in scientific rabbit farming on the occasion of 20th extension education council meeting of Tamil Nadu Veterinary and Animal Sciences University held on 20-03-09 at Fisheries College & Research Institute, Thoothukudi.

On Average	2008	2009	2010
No. of Rabbits sold per month	50	85	136
Net income per month	Rs.5000/-	Rs.9000/-	Rs.15000/-

Conclusion

KVK played major role in outstanding performance of the responddnt. There was no taboo existing among people in terms of consumption'of rabbit meat. People liked this meat and had the habit of taking it regularly. Marketing channel was available. Advertising the advantages of rabbit meat could make people to start eating rabbit meat and create marketing channel.

The non availability of rabbit meat in the market when compared to other meat is the reason for less consumption. Owning and maintaining a rabbit farm is not a tough job as rabbit farming has many advantages. Rabbit farming gives unexpected profit to the farmer. What he has done, why don't the others can do?

22

Transfer of Technology Issues in Bt Cotton

S. Parthasarathi [1] ***and R. Ganesan*** [2]
[1] *Assistant Professor, PAJANCOA&RI, Karaikal and*
[2] *Dean, AC&RI, Killikulam, TNAU.*

Introduction

India is the third largest producer of cotton in the World. The cotton farmers of India are facing many problems. The stories of Vidharba are crucial lessons for the technocrats and policy makers. Most of the cotton growing areas are dry tract hence grown as rain fed crop. An amount of Rs. 12 billion worth of pesticides are used to control the serious pest i-e boll worm. In order to make the farmers to live debt free life and ensure the maximum reach out efforts of extension, it, needs a strategic TOT efforts.

In this juncture introduction of Bt cotton is a phenomenal change in the cotton industry and needs a careful practices to increase the adoption since it has some complex issues like growing refugee crops. TOT of Bt cotton is unique because of its newness attracts positive and negative comments. A study to assess the acceptance, non adoption and constraints related to Bt cotton was carried out and the findings are given below.

Methodology

Annur and Kinathukadavu blocks of Coimbatore district was purposively selected due to the fact that Bt cotton growers are predominantly more in this area. The study was conducted in 6 villages of Annur Block and Kinathukadavu Block.

Selection of Farmers

In order to study intensively 80 farmers were selected which comprises of 40 Bt cotton growers and 40 non Bt cotton growers by applying simple random sampling procedure from the selected villages.

Selection of other stake holders for constraint analysis and Suggestions

The other stakeholders includes 15 extension functionaries viz., extension personnel from SDA (2 ADOs , 4 AAOs) two from KVK, one scientist from CICR, one from Cotton Corporation of India , 5 from private companies (MAHYCO, Rasi seeds), 8 input dealers and officials of 10 cotton mills who associated with cotton trading.

Findings and Discussion

Issues responsible for the acceptance of Bt cotton by Bt farmers

The issues responsible for the acceptance of Bt cotton are presented in the Table 1.

Table 1 : Distribution of farmers according to issues responsible for the acceptance of Bt cotton

(n=40)*

SI. No.	Items	Number	Percentage	Rank
1.	Promotional efforts by the companies	35	87.50	I
2.	Reduction in the seed price	33	82.50	II
3.	Technological back up	29	72.50	III
4.	Increase in income	25	62.50	IV
5.	Marginal reduction in the cost of cultivation	22	55.00	VI
6.	Favorable attitude toward cotton farming	20	50.00	VI
7.	Peer group pressure	19	47.50	VII

* Multiple responses

It is observed from Table 1 that the major issues identified for acceptance of Bt cotton were promotional efforts by the companies (82.50%), reduction in seed cost (82.50%), assured technical back up (72.50%) and assured increase in income (62.50%). The spread of Bt cotton in Tam ilnadu is mainly due to the major seed companies like MAHYCO & Rasi seeds. The intensive efforts made by these agencies in the supply of Bt cotton seed along with technical back up, might be the probable reason for the acceptance of Bt cotton in the study area. Apart from this the government intervention in reduction of seed cost from Rs.1800/- to Rs.750/- for a pack of one acre was also identified as yet another issue for the Bt cotton spread. The assured income increase by adopting Bt cotton in trails and subsequent seasons induced the farmers to continue the Bt cotton, was also a probable reason for the acceptance of Bt cotton by the growers. Some of the other issues identified were marginal reduction in cost of cultivation (55.00%), favourable attitude towards cotton farming (50.00%) and group pressure (47.50%).

Issues Responsible for the non adoption of Bt cotton by non Bt farmers

The issues responsible for the non adoption of Bt cotton by non Bt farmers are given in the Table 2.

Table 2 : Distribution of farmers according to issues responsible for the Non adoption of *Bt* cotton

SI. No.	Items	Number	Percentage	Rank
1.	Rainfed cultivation	33	82.50	I
2.	Lack of technical support	20	75.00	II
3.	High seeds cost	25	62.50	III
4.	Small holdings	20	50.00	IV
5.	Increased cost of cultivation	19	47.50	VI
6.	Accommodation of more space for refuge crop	12	30.00	VI
7.	Performing as a regular hybrid	10	25.00	VII
8.	Occurrence of other sucking pest	5	12.50	VIII
9.	Environmental hazard	3	7.50	IX
10.	Causes damage to indigenous varieties	3	7.50	IX
11.	No reduction of pesticide consumption	2	5.00	X

* Multiple response

It is evident from the Table 2 that major issues expressed by the non Bt farmers were *Bt* hybrids are not suitable for raided cultivation (82.50%), lack of technical support (75.00%), high cost of seeds (62.50%) and small holdings (50.00%). The shortage and uneven rainfall were the major issues faced by the farmers for the past few years. Hence, they fear to go for hybrids which are sensitive to stress. The poor support from the input dealers, government agencies and private seed dealers were the major issues by the small and marginal farmers. The cost of the seed is high when compared to the varieties and the performance of non *Bt* varieties like LRA 5166, RAJAT, Varalakshmi was satisfactory as expressed by them. These major issues are before the non adopters and have to be focused by the promoters of the *Bt* cotton.

Constraints experienced by Farmers

The constraints experienced by *Bt* cotton growers and non Bt growers were gathered, analyzed and presented in the Table 3.

Table 3 : Distribution of respondents according to constraints experienced by farmers

Constraints	Bt Farmers (n=40)			Non Bt farmers (n=40)		
	No.	Percent	Rank	No.	Percent	Rank
Price fluctuation	30	75.00	I	40	100.00	I
Labour scarcity	30	75.00	I	40	100.00	I
No technical guidance	15	37.50	VI	35	87.50	II
Not suitable to Rainfed	18	45.00	V	29	72.50	V
Poor performance	--	--		20	50.00	VII
Occurrence of other pests	25	62.50	III	8	20.00	IX
Less profit	20	50.00	IV	23	57.50	VI
Performance of other crops	15	37.50	VI	30	75.00	IV
Uneven rainfall	28	70.00	II	32	80.00	III
Middleman	7	17.50	IX	12	30.00	VIII
Seed cost	10	25.00	VII	32	80.00	III
Less Govt. support	8	20.00	VIII	6	15.00	X

* Multiple responses

From the Table 3 it is observed that the majority (75.00%) of the Bt respondents and cent percent of the non *Bt* farmers revealed that fluctuation in price, scarcity of labour were their constraint. But the *Bt* farmers expressed that they some how manage this problem due to reduction in number of picking and increased yield. The need of government intervention in fixing the price and promoting machineries might be possible solutions for the said constraints.

Occurrence of other sucking pest was expressed by two-third (62.50%) of the Bt farmers. The minor pests like aphids were causing severe damage to the crop. These pests are minor in yester years and emerged as major. A comprehensive education is essentail to control occurred in the past, the reason may be that they were controlled by their frequent spray towards the control of bollworm. Hence, these minor pests are becoming a major one, a through education is needed to manage these problem is essential.

A less than fifty (45.00%) percent were expressed that *Bt* cotton is not suitable for rainfed cultivation. This might be due to the experience of farmers that Bt hybrids could not able to with stand drought as compared to the varieties.

Suggestions to create awareness increase, the adoption and increase the spread / the area under Bt cotton by the stakeholders of *Bt* cotton

The suggestions offered by various stake holders to create awareness, increase the adoption and increase the spread / the area under Bt cotton are furnished in the following Table 4

Table 4 : Distribution of respondents according to their suggestions to create awareness, increase the adoption and increase the spread / the area under Bt cotton

(n=115) *

SI. No.	Suggestions	Number	Percentage
I	**To create awareness**		
1.	Demonstrations by private and Government agencies	85	77.27
2.	Reward to successful farmers	78	70.90
3.	Combination of mass media to create awareness	62	56.36
II	**To increase the adoption**		
1.	Mechanization to over come the labour problem	90	81.81
2.	Price stabilization b ovemment	87	79.09
3.	Contract farming should be promoted	69	62.72
4.	Introduction of hybrids suitable to drought	35	31.81.
5.	Credit facilities	40	36.36
6.	Exclusive FFS for Bt cotton	22	20.00
III	**To increase the spread / the area under *Bt* cotton**		
1.	Reduction in seed cost	78	70.90
2.	Education to farmers	73	66.36

3.	Effective crop insurance	52	47.27
4.	Promotion of farmers' interest groups	50	43.47
5.	Market forecasting	40	36.36
6.	*Bt* varieties suitable to rainfed	35	31.81

* Multiple Response

It could be observed from the Table 4 that majority (77.27%) of the respondents revealed that more demonstrations have to be conducted by both private and government agencies to create awareness followed by reward to successful farmers (70.90%) and effective combination of mass media to create awareness and importance of *Bt* cotton (56.36%).

The integration of private and public extension system in the promotion of *Bt* cotton might increase the confidence among the farmers, gives field level exposure and easy supervision by the government officials, assessment of failure and effective implement of crop insurance scheme may be easy. Rewards to the successful farmers motivate others to share the experience of success and thus promotes the spread of Bt cotton. An effective combination of extension methods 'such as demonstrations, exposure visits, and publications will also increase the extent of awareness.

Further the Table reveals that mechanization should be promoted to over come the labour problem (81.81%), market price should be stabilized by government intervention (79.09%), contract farming should be promoted (62.72%), were suggested to increase the adoption of Bt cotton. To increase the adoption of *Bt* cotton mechanization have to be promoted for weeding and harvesting which requires more labour. The involvement of government agencies and cotton mills are essential to stabilize the price and to prevent the syndicate of merchants in the market while fixing the price. Contract farming system with effective credit facilities, buyback arrangement will also promote the adoption of Bt cotton. The results reveals that reduction of seed cost (70.90%), educating the farmers (66.36%), crop insurance scheme should be implemented effectively (47.27%) were the suggestions given by the respondents to increase the spread / the area under Bt cotton.

The spread of area under *Bt* cotton will increase, if the seed cost is reduced further. A difference of minimum rupees 300 exists between *Bt* and non *Bt* hybrid. This difference should be minimized further to increase the area under *Bt* cotton. A through education is essential to impart knowledge about the recent advances and self reliance for technology among the farming community. An effective crop insurance scheme should be implemented to safe guard the farmers from crop failure and thus economic loss, is also be a contributing suggestion by the stake holders.

Conclusion

Genetically modified crops are boon for the food consumption and farmers income in the developing countries like India. The research in food crop is also in the way. It is high time for the extension machinery to give importance in transferring technologies with new strategies. There should be best pet TOT packages for all sector of farming community with out any disparity by means of technology and cost of inputs. The involvement of all the stakeholders including the consumers will be essential for quality products and required quantity with assured income to the producers.

References

Anonymous 2007. Journal of South Indian cotton Association, Volume XXIX/17/2007.

Karthikeyan C. and P. Lavanya 2007. *Bt* cotton cultivation in India-A Feasibility Analysis, Intensive agriculture, January-June 2007.

●●●

23

Adoption of Pest Management Technologies by Tomato Growers in Mettur Taluk of Salem District

P. Suhirdha[1]*, G. Tamilselvi2 and J. Vasanthakumar*[3]
[1-3] *Department of Agrl. Extension, Annamalai University, Annamalainagar*

Introduction

Among the various vegetable crops, tomato *(Lycopersicurn esculentum)* is one of the important and popular vegetable and plays an important role in balanced nutrition and hence it is called as `poorman's orange'. Tomato fruits can be seen in the market round the year. In India, tomato occupies an area of 2,90,279 ha with the productivity of 15.85 t/ha and production of 46,03,446 tonnes (www.nhb.gov.in). Among the different districts of Tamil Nadu, Salem district has an area of about 6311 hectares under tomato with 17,359 tonnes of production. Pest management is one of the important dimensions in cultivating tomato crop that needs significant attention. The objective of pest management technologies are maximize pest control in terms of overall economical, social and environmental values. In order to restore ecological balance, it is necessary to integrate all the available pest management techniques in the most compatible manner with the least use of chemicals which is the principle of pest management. 'Pest management technology' utilizes all suitable techniques and methods in a compatible manner as possible and maintains the pest population below their economic injury levels given by Bikramjit et al. (2007).

In the recent past, efforts have been made to increase the tomato production by developing technologies like high yielding and pest resistant varieties / hybrids, which makes the plant more succulent and susceptible to pest for which farmers feel that there are no alternatives other than spraying chemical pesticides so as to minimize the crop loss. When these chemicals are used indiscriminately the problems of resurgences of pest, destruction of beneficial insects and imbalances in crop ecology emerge Lagnaouri *et al.,* (2004). Keeping these in view, the present study was taken to study the extent of knowledge, adoption behaviour and find out relationship between selected socio-economic and psychological characteristics of tomato growers and their adoption level of pest management technologies in tomato cultivation.

Methodology

Considering the maximum area and intensity of tomato crop in the state of Tamilnadu this study was conducted in Mettur taluk of Salem district, during 2008-09. In this taluk 31 villages were selected for the study based on maximum area under tomato cultivation, among these villages 120 farmers were selected proportionate random sampling for the study. The respondents were interviewed and data was collected through personal interview technique by using pre-tested schedule and were analyzed by using percentage analysis, cumulative frequency, zero order correlation and multiple regressions were used to categorize the independent variable and relationship of independent variables with dependent variables in the study. The dependent variables are knowledge level and adoption behaviour. The sixteen independent variables like age, educational status, occupational status. etc., along with the variables like constraints in adoption of pest management technologies and suggestions for greater adoption of those technologies were also studied.

Findings and Discussion

The findings of the study, in line with objectives are presented below.

1. Adoption behavior of tomato growers with respect to pest management technologies

a. *Adoption level of pest management technologies by tomato Growers*

The study reveals that majority of the respondents 51.67 per cent belong to low adopter group of pest management technologies in tomato cultivation. Whereas 34.17 per cent and 14.16 per cent of respondents belong to medium and high adopter group. This findings are accordance with the findings of Vimali (2001).

b. Practice wise extent of adoption of pest management technologies in tomato crop

Adoption level of the respondents was tested against twenty four items related to pest management technologies. Data in Table 1 projects the practice wise extent of adoption of various pest management technologies by tomato growers. In cultural methods, majority of the tomato growers adopted summer ploughing (80.00per cent), in mechanical method most of the tomato growers (70.83 per cent) adopted destruction of damaged fruits, in biological methods(45.83 per cent) respondents adopted the use of bio-fungicide and in chemical method, majority of farmer adopted chemical used to control sucking pest(70.83 per cent), out of twenty four items ,only few respondents (2.50 percent) respondents were adopted usage of virus in pest management technologies.

2. Relationship between selected socio-economic and psychological characteristics of respondents with their extent of adoption of pest management technologies

The data from Table 2 revealed that among the social, economical and psychological variables, the characters viz., age, area under tomato cultivation, experience in tomato cultivation, mass media exposure, innovativeness, risk orientation and scientific orientation were found to have positive and significant relationship at five per cent level of probability towards the adoption level of tomato growers. All other variables were found to be non-significant. The results are inline with the findings of prakasam (2005) and prabu(2004).

Conclusion

Finally study concludes that farmers need to be educated, convinced about the importance of IPM practices, adverse effects of indiscriminate use of insecticides on health hazards, pest resistance, eco-pollution through demonstration and follow up approach required facilities by the state department of agriculture is equally crucial. Immediate steps need to be taken by government developmental departments and agricultural university to train farmers in pest management practices by using mass media as a source of information.

Table 1 : Practice wise extent of adoption of pest management technologies by the respondents

(n=120)

SI.No.	Adoption items	Number of respondents	Per cent
1	**CULTURAL METHODS**		
i.	Summer ploughing	96	80.00
ii.	Trap crop for tomato cultivation	85	70.83
iii.	Crop rotation	85	70.83
iv.	Raised bed Nursery	32	26.67
2	**MECHANICAL METHODS**		
i.	Adoption of any one mechanical practice	25	20.83
ii.	Use of pheromone traps	10	8.33
iii.	Destruction of damaged fruits	85	70.83
3	**BIOLOGICAL METHODS**		
i.	Use of parasites	5	4.17
ii.	Quantity of parasites to control pest	6	5.00
iii.	Build up natural enemy fauna	10	8.33
iv.	Use of virus to control pest	5	4.17
v.	Dosage of virus	8	2.50
vi.	Use of bio-fungicide	55	45.83
vii.	Quantity of bio-fungicide	53	44.17
viii.	Use of neem based pesticide	62	51.67
ix.	Quantity of neem based pesticide	60	50.00
4	**CHEMICAL METHODS**		
i.	Chemical used to control bacterial wilt	55	45.83
ii.	Quantity of chemical used for control bacterial wilt	54	45.00
iii.	Chemical used to control sucking pests	85	70.83
iv.	Quantity of chemical used for control sucking pests	80	66.67
v.	Chemical used to control *Helicoverpa armigera*	85	70.83
vi.	Quantity of chemical used for control *Helicoverpa armigera*	82	68.33
vii.	Chemical used to control weeds	70	58.33
viii.	Quantity of chemical used for control weeds	65	54.17

Table 2 : Relationship between socio-economic and psychological characteristics of respondents and their extent of adoption of pest management technologies

Variable Number	Independent variables	Correlation co-efficient 'r' value
X_1	Age	0.201*
X_2	Educational status	0.073 NS
X_3	Occupational status	-0.112 NS
X_4	Annual income	0.127 NS
X_5	Farm size	0.076 NS
X_6	Area under tomato cultivation	0.215*
X_7	Experience in tomato cultivation	0.231 *
X_8	Extension agency contact	-0.068 NS
X_9	Mass media exposure	0.194*
X_{10}	Cosmopoliteness	-0.053 NS
X_{11}	Innovativeness	0.219*
X_{12}	Risk orientation	0.228*
X_{13}	Scientific orientation	0.266**
X_{14}	Economic motivation	-0.047 NS
X_{15}	Achievement motivation	0.073 NS
X_{16}	Market orientation	0.027 NS

References

Bikramjit, S. Randhir, S. and C. Dhrupad, 2007. Ecological Pest Management, LEISA News Letter, 9(4): 8-10.

Lagnaouri, A., Santi, E. and Santucci, F. 2004. Strategic Communication for Integrated Pest Management. First Paper Presentation at the Annual Conference of the International Association for Impact Assessment in Vancouver, Canada (IAIA 2004) under the Conference of Public Involvement and Risk Management.

Prabu, R. 2004. Adoption and Marketing Behaviour of Tomato Growers in Salem District, Unpublished M.Sc. (Ag.) Thesis, Annamalai University, Annamalai Nagar.

Prakasam, T. 2005. Knowledge Level and Extent of Adoption of Arecanut Growers of Salem District, Unpublished M.Sc. (Ag.) Thesis, Annamalai University, Annamalai Nagar.

Vehkatesh Gandhi. R., Hanchinal, S.N. Shivamurthy, M. and Shailaja Hittalmani. 2008. Adoption of Integrated Pest Management Practices among Tomato Growers, Karnataka Journal of Agricultural Sciences, 21(1): 17-19.

Vimali, D. 2001. A Critical Analysis of Awareness, Knowledge and Adoption of IPM Practices in Rice by the Farm Women, Unpublished M.Sc. (Ag.) Thesis, Annamalai University, Annamalai Nagar.

●●●

24

Impact Evaluation of DPAP-V Watersheds in Karur District

R. Sasikala [1] ***, A. Palaniswamy*** [2]***, K Thangaraja***[3] ***and M. Shanthasheel*** [4]

[1] *Assistant Professor,* [2] *Professor, Directorate of Extension Education,*
[3] *Ph. D Scholar and* [4] *Assistant Professor, Dept. of Agricultural Extension and Rural Sociology, TNAU, Coimbatore-3.*

Introduction

Land and water are important natural resources available for agriculture crop production. The resource is available either for agriculture or for human / animal consumption, from precipitation in the form of rain. According to National Commission of Agriculture, half of the available land in India will continue to depend on dry land even after our total surface and ground water potentials are fully exploited by 2025 A.D. So far efforts were made in improving the irrigation potential to step up crop production and very little attention has been paid to dry lands. Considerably improved but the farmers in dry land areas are still suffering.

The problems in dry regions are many. Uncertain climatic conditions, eroded soils, poor resources and unscientific management of these resources have added to the misery of the people in these areas. It becomes imperative now, that intensive efforts are essential to develop dry land for removing the imbalance between the irrigated and dry land farmers. The cultivable lands are declining either in extent or in productivity.

The problems of rain water is closely linked with various soil groups found in dry lands, water use efficiency is significantly agglutinated in dry lands on account of soil characteristics, namely swelling and shrinking, cracking of black soil, formation of hard-pan salinity and water logging in alluvial soils, poor moisture holding capacity, low fertility and compact sub- soil in red soil.

The concept of watershed development in wastelands/ dry lands has a special significance. The development program involves many of the watershed attributes like size, shape, orientation, topography, geology, climate, surface condition, land use, ground water and socio- economic aspects etc., and also includes components such as crop management, soil and water conservation, Water harvesting, alternate land use etc. Clearly these factors are inter- related and often an effect on the social environment in the present context is quite different because the central item is to stabilize productivity/ production in dry lands.

Drought Prone Area Programme (DPAP) is one of the major schemes launched and implemented to augment and conserve the land and water resources on the lines of watershed development approach. The Government of India and Government of Tamil Nadu have introduced this programme in all districts. In Karur district, the DPAP V is implemented in Aravakurchi and K.Paramathi block which covers an area of 5000 ha to reclaim the vast stretch of wasteland and bring them under productive use through afforstation and other water resource development activity such as construction of check dams, percolation ponds, farm ponds, cattle pond, renovation of existing structures, agro forestry plantation, horticulture plantation etc. This completed programme was evaluated and the report is presented.

Methodology

Impact evaluation of DPAP- Vth batch watersheds were conducted in all the ten watershed villages spreaded over in K. Paramathi and Aravakurichi block of Karur district in two phases; (i) Information were gathered from Project Implementing Agency (PIA), and discussions made with members of the Watershed Committee and Watershed Development Team (WDT) (ii) Field study was done and interaction made with the beneficiaries of the DPAP-V programme in order to study the impact of the programme. And also using simple random sampling technique, 100 beneficiaries were selected to study the community participation in implementing the watershed activities. The data were collected with well-structured interview schedule. Percentage analysis was used for data analysis.

Findings and Discussion

Like all other development programmes, watershed development also banks heavily on the participatory approach. Though, watershed development programme envisages an integrated and comprehensive plan of action for the rural areas, peoples' participation at all levels of its implementation is very important (K. Palanisami and D. Suresh Kumar 2009). The watershed projects are being implemented with the active support and involvement of the community. Theommunity participation is expected in all stages from planning to exectition of the programme. The action plan of the watersheds was prepared by the community during PRA exercise on need based and accordingly the activities were implemented. Participation of community in planning, decision making and site selection are presented in the following tables.

Table 1 : Participation of Village Community in Planning

SI. No.	Name of the Watersheds	Participation in percentage			Total
		V. Good	Good	Moderate	
1	Gudalure East- I	60.00	20.00	20.00	100
2	Gudalure East- II	71.40	28.60	-	100
3	Gudalur West -1	54.60	36.40	9.10	100
4	T. Venkata Puram	96.00	4.00	-	100
5	Chinnadharapuram	80.00	20.00	-	100
6	Lingamanaickanpatti	60.00	15.00	25.00	100
7	Rengappagoundanvalasu	82.00	18.00	-	100
8	Sellivalasu	85.00	15.00	-	100
9	E. Alamarathupatti	95.00	15.00	-	100
10	Kullamapatti	71.40	14.30	14.30	100
	Total	**75.50**	**17.60**	**6.80**	**100**

Table 1 interprets that with regard to very good participation by the village community in the planning the work, the maximum participation was found in T. Venkatapuram and E. Alamarathupatti i.e 96% and 95 % respectively and among the ten watersheds with regard to very good participation, minimum participation was found in Gudalur West -1 (54.6%). with regard to good participation in planning the work 36.40% was the maximum level of participation found in Gudalur West -1 followed by 28.6% participation in Gudalur East -II and 20% in Gudalure East-1. Regarding moderate participation 25% of the people from Lingamanaickanpatti, 20% from Gudalure East- 1, 14.3% from Kullamapatti, 9.1 % from Gudalur West -I shown their participation in planning the watershed activities. Maximum participation can be achieved in planning the programme if the village community is given more awareness programme through exposure visit to different districts where the watershed programme is implemented and interactive sessions can be arranged with the people who have already benefited by this programme.

Table 2 : Participation of Village Community in Decision Making

SI. No.	Name of the Watersheds	Participation in percentage			Total
		V.Good	Good	Moderate	
1	Gudalure East- I	80.00	20.00	-	100
2	Gudalure East- II	100	-	-	100
3	Gudalur West -1	75.00	25.00		100
4	T. Venkata Puram	55.60	44.40	-	100
5	Chinnadharapuram	85.70	14.30	-	100
6	Lingamanaickanpatti	45.40	36.40	18.20	100
7	Rengappagoundanvalasu	44.40	22.20	33.30	100
8	Sellivalasu	20.00	70.00	10.00	100
9	E. Alamarathupatti	62.50	25.00	12.50	100
10	Kullamapatti	42.90	28.60	28.60	100
	Total	**61.10**	**28.50**	**10.20**	**100**

It is observed from Table 2 that only in Gudalure East- II cent percent of the village community made their decision collectively in implementing the watershed activities. Less than 50% of the village community from Lingamanaickanpatti, Rengappagoundanvalasu, Sellivalasu and Kullamapatti involved in high level participation in decision making process. With regard to good participation 70% of the village community from Sellivalasu made their decision collectively in implementing the programme. 33.3% of the village community from Rengappa-goundanvalasu shown moderate participation in collective decision making process which follows 28.6 % in Kullampatti and 18.2 % in Lingamanaickanpatti. So it is inferred that most of the village community was aware of the watershed programme but only 60% of the people were involved in collective decision making regarding implementing the programme. So it shows that though the people were aware, they don't know that each individuals participation in decision making will result in good planning, better site selection and quick execution of the work.

Table 3 : Participation of Village Community in Site Selection

SI. No.	Name of the Watersheds	Participation in percentage			Total
		High Participation	Moderate Participation	Low Participation	
1	Gudalure East- I	80.00	20.00	-	100
2	Gudalure East- II	71.40	14.30	14.30.	100
3	Gudalur West -1	100	-	-	100
4	T. Venkata Puram	100	-	-	-
5	Chinnadharapuram	71.40	28.60	-	100
6	Lingamanaickanpatti	81.80	18.20	-	100

7	Rengappagoundanvalasu	77.80	11.10	11.10	100
8	Sellivalasu	50.00	20.00	30.00	100
9	E. Alamarathupatti	37.50	50.00	12.50	100
10	Kullamapatti	42.90	57.10	-	100
	Total	**71.20**	**21.90**	**6.70**	**100**

It is observed from the Table 3 that in T. Venkatapuram and Gudalur West —I cent percent participation of the village community was found in selecting the site for construction of the structures which fallows 81.8%, in Lingamanaickanpatti, 80% in Gudalure East-I, 77.8% in Rengappa-goundanvalasu, 71.4% in Gudalure East- II and Chinnadharapuram watershed. 57% from Kullamapatti and 50% of the village community from E. Alamarathupatti shown moderate participation in site selection process. Only in Sellivalasu 30% of the people were less interested to participate in site selection for the construction of watershed structures. Whereas in other watersheds minimum people were found in low participation category (i.e. below 15%). Except in E. Alamarathupatti and Kullamapatti, in other watersheds more than 70% of the people actively participated in site selection. Regarding implementation of the watershed activities site selection plays a major role. People should be actively involved in selecting the site for constructing the structures so that water can be harvested and which in turn will recharge the ground water level. So it is observed that the people were well aware of the importance of site selection.

From the field study, it was observed that Water, Soil and Moisture conservation works like Summer ploughing, Conture bunding, Land levelling and Retaining wall, construction of Check dam, Farm pond, Sunken pond, Percolation pond, Deepening of percolation pond, Desilting of irrigation well and channel etc., were made almost in all the watersheds. From the discussion made with the beneficiaries it was came to know that soil erosion was maximum controlled and the nutrient value of the land was maintained. Moreover water recharge was good in wells and even in summer the people made their sowing. Importance was also given for the development of Horticultural and Agro Forestry Plantation which paved way for additional income to the farming community.

Conclusion

On the whole it was observed that the people were well aware of the importance of the watershed programme and they were involved in the implementation of the programme. To achieve cent percent participation from the village community more awareness programmes such as exposure visits and interactive sessions can be arranged with the people who have already benefited by this programme. It was also noted that, there was a good increase of water level in wells with the construction of Check dam, Farm pond etc.,. And soil erosion was maximum controlled and the soil fertility status was maintained.

Agro forestry is another activity implemented under this programme which makes the watershed area ecologically balanced. Supply of horticultural plants like Sapota, Mango, Coconut sapling etc., provided an additional income which enhances the status of the community. It is concluded that, this Drought Prone Area programme implemented in Aravakurichi and K. Paramathi block of karur district is a great boon to the farming community.

Reference

K. Palatiisami and D. Suresh 2009. Economics Research Review, Vol.22, (conference number) page 387-396.

●●●

25

Participatory Technology Development in Banana (PTD)

M. Israel Thomas [1]***, T. Rathakrishnan*** [2] ***and S. R. Padma*** [3]

[1] *Associate Professor(Agricultural Extension), Kerala Agricultural University, Pattambi.* [2]*Director, Students Welfare and professor (Extension),* [3]*Ph.D., Scholar, TNAU, Coimbatore.*

Introdution

In developing countries, traditional methods have special advantages over modern agricultural techniques. The capital and technological skill requirements in the use of traditional technologies are generally low and their adoption often requires little restructure of the traditional societies. By adopting such indigenous knowledge, our ancestors did not face any problem of large scale pest out break or economic crisis unlike the today's farmers. The present study focused with this central theme of research.

Indigenous Technical Knowledge refers to the unique, traditional, local knowledge existing within and developed around the specific conditions of women and men indigenous to a particular geographic area (Grenier, 1998). Indigenous Knowledge is the accumulated knowledge, skills and technology of the local people. Indigenous Knowledge has two powerful advantages over scientific knowledge like it has little or no cost, and is readily available. Indigenous Knowledge can be defined as the knowledge built up by a group of people through generations of living in close contact with nature over the years; the farmer in India had been using cow dung and cow urine as manure in the fields.

Participatory Technology Development (PTD) is a creative process of joint experimentation and research by farmers and development agents in discovering ways of improving farmers' livelihoods. The growing number of documented examples in recent years reveal that PTD is now accepted as a research approach to agriculture and natural resource management (NRM). It has been recognized that research is effective in improving farmers' livelihoods if farmers play a vital role in the process.

Methodology

K. Dharmathupatti village of Reddiarchathiram Block, Dindigul District and Kondayampatti village of Alanganallur block, Madurai District is purposively selected for conducting the field experiments based on more area under banana cultivation. The banana belt areas were selected to conduct key informants workshop. In key informants workshop the most co - operative and responsive areas with progressive banana growers like K. Dharmathupatti and Kondayampatti villages were selected for field experiments. G-9, Robusta, Poovan and Red banana varieties of Banana crop were chosen for the experiments in the study. The experiment was laid out in a Randomized Block Design and replicated thrice. Randomized Block Design is the most widely used in on-farm experimentation for almost all research objectives.

Key Informants workshops were conducted with the progressive banana farmers and the scientists in the Banana belts of K. Dharmathupatti, Kondayampatti and Melakkal to finalize the treatments with farmer participatory approach. The 12 treatments were T_1 - House hold wood ash + Pregnant cow urine, T_2 - Brick chamber wood ash + Pregnant cow urine, T_3 - Panchakavia + Brick chamber wood ash, T_4 - Panchakfavia + Household wood ash, T_5 - Urea, T_6 - Urea + Sulphate of potash, T_7 - Fresh cow dung extract + Zinc sulphate + Borax, T_8 - Fresh cow dung extract + Zinc sulphate, T_9 - Fresh cow dung extract + Borax, T_{10} - Urea + Sulphate of potash + water + Fresh cow dung, T_{11} - Cow dung extract solution, T_{12}- Control.

The distal axis (male axis) was cut 15 cm away from the last hand fifteen days after the last hand was opened and after removal of the distal axis along with the navel bud (male bud) the aforesaid treatments ingredients were tied in a polythene bag. Biometrical observations on the yield parameters under each treatment replication wise and of bunch were done by following methods of Robin (1985). Data on the yield parameters were recorded at the stage of harvest. Bunch Characters were Yield t/ha, Bunch Weight, Number of Hands and Fingers, Average Weight of the Fruits (Fingers), Volume of the Fruit (Finger), Density of Fruit (Fingers), Average Length of the Fruits (Fingers), Girth of the Fruits (Fingers), Total Soluble Solids (T. S. S.)

Findings and Discussion

Table 1 : Results of Field Experiments at Kondayampatti

Field-1,2,3 **Variety: GRAND -9**

Treatments	Finger Length (cm)	Finger Girth (cm)	Finger Weight (gms)	Volume of finger (cc)	Density Of fingers	No of fingers/ hand	Total Number of fingers/ bunch	No of hands/ bunch	Bunch Weight (Kg)	Yield t/ha	Yield increase (% over control)	TSS (^{0} Brix)
T_1	21.00	12.67	59.02	252.00	1.26	25.33	136.00	16.00	29.51	59.02	101.00	15.17
T_2	21.67	12.33	58.38	252.67	1.28	26.33	139.00	16.67	29.19	58.38	99.93	15.17
T_3	21.33	12.00	56.99	254.00	1.25	25.00	137.00	16.00	28.50	56.99	97.55	15.00
T_4	21.00	13.00	61.00	251.67	1.27	26.00	138.00	16.67	30.50	61.00	104.42	15.00
T_5	19.00	10.67	53.30	251.67	1.25	22.33	130.00	15.00	26.65	53.30	91.24	14.83
T_6	19.33	11.33	59.13	252.33	1.27	24.67	135.00	15.67	29.57	59.13	101.21	15.17
T_7	19.67	11.33	64.12	255.67	1.28	29.00	140.00	17.33	32.06	64.12	109.76	14.83
T_8	24.67	14.33	72.00	261.33	1.31	32.33	150.00	18.67	36.00	72.00	123.24	15.67
T_9	23.00	13.33	55.62	253.33	1.21	26.00	135.00	15.00	27.81	55.62	95.21	15.17
T_{10}	24.00	14.00	69.38	260.00	1.30	31.67	147.00	18.33	34.69	69.38	118.76	15.50
T_{11}	21.67	12.00	61.20	254.67	1.28	26.33	136.00	15.67	30.60	61.20	104.76	15.00
T_{12}	17.67	10.00	58.42	242.00	1.05	15.33	115.00	11.33	29.21	58.42	100.00	14.17

SEd: 0.890 0.515 2.545 10.649 0.052 1.087 5.723 0.671 1.272 2.545 0.632

CD (P=.05) 1.847 1.068 5.279 :22.084 0.108 2.255 11.87 1.39 2.64 5.279 1.312

Table 2 : Results of Field Experiments at K. Dharmathupatti - Robusta

Field-1 **Variety: Robusta**

Treatments	Finger Length (cm)	Finger Girth (cm)	Finger Weight (gms)	Volume of finger (cc)	Density Of fingers	No of fingers/ hand	Total Number of fingers/ bunch	No of hands/ bunch	Bunch Weight (Kg)	Yield t/ha	Yield increase (% over control)	TSS (° Brix)
T_1	21.00	12.00	270.00	225.00	1.24	24.00	118.00	10.00	31.9	57.35	133.18	13.5
T_2	22.00	14.00	262.00	221.00	1.23	20.00	114.00	8.00	29.9	53.76	124.84	13.5
T_3	21.00	13.00	270.00	224.00	1.25	22.00	118.00	10.00	31.9	57.35	133.18	13.5
T_4	23.00	14.00	262.00	220.00	1.24	22.00	105.00	8.00	27.5	49.52	115.00	13.0
T_5	20.00	12.00	260.00	218.00	1.24	18.00	112.00	7.00	29.1	52.42	121.73	13.0
T_6	20.00	12.00	263.00	220.00	1.24	20.00	116.00	8.00	30.5	54.91	127.51	13.5
T_7	20.00	12.00	265.00	221.00	1.24	21.00	117.00	9.00	31.0	55.81	129.60	13.5
T_8	22.00	13.00	273.00	228.00	1.24	26.00	123.00	12.00	33.6	60.44	140.36	14.5
T_9	21.00	13.00	271.00	226.00	1.24	25.00	120.00	11.00	32.5	58.54	135.94	14.0
T_{10}	23.00	15.00	268.00	223.00	1.25	24.00	117.00	10.00	31.4	56.44	131.07	13.5
T_{11}	20.00	12.00	266.00	222.00	1.24	22.00	116.00	9.00	30.9	55.54	128.98	13.5
T_{12}	18.00	10.00	230.00	202.00	1.19	15.00	104.00	6.00	23.9	43.06	100.00	12.0

SEd: 0.877 0.532 11.06 9.281 0.052 0.912 4.841 0.384 1.278 2.30 0.564

CD (P=.05) 1.818 1.103 22.95 19.24 0.107 1.89 10.04 0.797 2.65 4.771 1.170

Table 3 : Results of Field Experiments at K. Dharmathupatti - Poovan

Field-2 **Variety: Robusta**

Treatments	Finger Length (cm)	Finger Girth (cm)	Finger Weight (gms)	Volume of finger (cc)	Density Of fingers	No of fingers/ hand	Total Number of fingers/ bunch	No of hands/ bunch	Bunch Weight (Kg)	Yield t/ha	Yield increase (% over control)	TSS (^{0} Brix)
T_1	13.00	12.00	120.00	76.00	1.45	15.00	158.00	10.00	19.0	35.55	147.38	14.5
T_2	12.00	11.00	116.00	77.00	1.38	14.00	145.00	9.00	16.8	31.54	130.76	14.5
T_3	14.00	12.00	118.00	76.00	1.42	16.00	156.00	11.00	18.4	34.52	143.11	14.5
T_4	14.00	13.00	105.00	75.00	1.27	12.00	140.00	9.00	14.7	27.56	114.26	14.5
T_5	13.00	12.00	109.00	74.00	1.34	11.00	146.00	8.00	15.9	29.84	123.71	14.5
T_6	12.00	11.00	108.00	76.00	1.29	15.00	158.00	10.00	17.1	32.00	132.66	14.5
T_7	12.00	11.00	100.00	76.00	1.18	12.00	163.00	8.00	16.3	30.56	126.69	13.5
T_8	15.00	13.00	120.00	75.00	1.47	16.00	160.00	10.00	19.2	36.00	149.25	14.5
T_9	13.00	12.00	130.00	75.00	1.60	15.00	160.00	10.00	20.8	39.00	161.69	14.5
T_{10}	14.00	13.00	110.00	77.00	1.30	13.00	150.00	9.00	16.5	30.94	128.27	14.0
T_{11}	12.00.	11.00	115.00	76.00	1.38	15.00	150.00	10.00	17.3	32.34	134.07	14.5
T_{12}	11.00	10.00	96.00	71.00	1.21	10.00	134.00	7.00	12.9	24.12	100.00	13.0

SEd: 0.543 0.493 4.746 3.165 0.057 0.583 6.393 0.393 0.725 1.360 0.599

CD (P=.05) 1.127 1.022 9.843 6.56 0.119 1.209 13.25 0.816 1.504 2.82 1.243

Table 4 : Results of Field Experiments at K. Dharmathupatti - Red Banana

Field-3

Variety: Red Banana

Treatments	Finger Length (cm)	Finger Girth (cm)	Finger Weight (gms)	Volume of finger (cc)	Density Of fingers	No of fingers/ hand	Total Number of fingers/ bunch	No of hands/ bunch	Bunch Weight (Kg)	Yield t/ ha	Yield increase (% over control)	TSS (^{0} Brix)
T_1	19.00	12.00	299.00	118.00	2.53	11.00	97.00	6.00	32.0	52.21	117.72	12.00
T_2	20.00	14.00	300.00	120.00	2.50	12.00	100.00	7.00	33.0	54.00	121.75	12.00
T_3	20.00	15.00	298.00	120.00	2.48	11.00	98.00	6.00	32.2	52.57	118.53	12.00
T_4	19.00	14.00	305.00	122.00	2.50	13.00	101.00	8.00	33.9	55.45	125.02	12.00
T_5	18.00	12.00	293.00	118.00	2.48	10.00	96.00	5.00	31.1	50.63	114.16	12.00
T_6	18.00	13.00	293.00	117.00	2.50	10.00	97.00	5.00	31.4	51.16	115.35	12.00
T_7	18.00	12.00	297.00	119.00	2.50	11.00	107.00	6.00	34.7	57.20	128.97	12.00
T_8	20.00	15.00	370.00	129.00	3.64	18.00	110.00	10.00	44.4	73.26	165.18	12.00
T_9	18.00	12.00	350.00	125.00	3.60	15.00	110.00	9.00	42.0	69.30	156.25	12.00
T_{10}	21.00	15.00	365.00	128.00	3.63	17.00	106.00	10.00	42.3	69.64	157.02	13.00
T_{11}	18.00	12.00	300.00	125.00	3.20	13.00	95.00	8.00	31.5	51.30	115.67	11.50
T_{12}	11.00	10.00	280.00	108.00	2.59	9.00	88.00	5.00	27.4	44.35	100.00	11.00

SEd: 0.772 0.548 13.16 5.069 0.121 0.532 4.22 0.303 1.467 2.405 0.514

CD (P=.05) 1.601 1.138 27.31 10.51 0.25 1.103 8.75 0.629 3.04 4.98 1.067

Results of the Experiments conducted at Kondayampatti Village with Grand — 9 Variety banana

The results of treatments conducted at Kondayamaptti village in Grand — 9 variety are depicted in Table 1. A perusal of the table reveals the selected yield parameters that effect the growth of Banana.

The average length of fingers was significantly increased by treatments as seen in Tablel. The application of fresh cow dung extracts and zinc sulphate in T8 (24.67) recorded maximum length of fruit followed by the application of urea, sulphate of potash, water and fresh cow dung in T10 (24.00) and were on par with each other. The control T12 recorded minimum length of fruit (17.67) which was on par with T7 (19.67) and T6 (19.33).The average girth of fingers was significantly increased by treatments. The application of fresh cow dung extracts and zinc sulphate in T8 (14.33) recorded maximum girth of fruit followed by application of urea, sulphate of potash, water and fresh cow dung in TI O (14.00) and were on par with each other. The control T12 recorded minimum girth of fruit (10.00) which was on par with T5 (10.67).

The size of the individual fruit of banana, as represented by the average weight of finger was significantly increased by treatments. The maximum fruit size was observed due to the application of fresh cow dung extracts and zinc sulphate in T8 (72.00) followed by the application of urea, sulphate of potash, water and fresh cow dung in T10 (69.38)and were on par with each other. The minimum size of the fruit (58.42) was recorded in the untreated plantain. The average volume of fingers was not significantly influenced by the treatments. However, the T8 recorded relatively higher volume (261.33) than the other treatments.

The average density of the fingers representing the filling capacity of the individual fruit, was found to be significantly higher in all the treatments than the control. Here the treatments from Tl to T11 were found to be on par and significantly higher than the control.The data on the number of finger's per hand revealed that the highest number of fingers per hand (32.33) was recorded in the treatment T8 followed by T10 (31.67) and were on par. These two treatments were found to be significantly superior to the other treatments. The untreated (Control) T12 (15.33) banana recorded the lowest number of fingers per hand.

The total number of fingers per bunch was significantly increased by treatments. The application of fresh cow dung extracts and zinc sulphate in T8 (150.00) recorded maximum number of fingers per bunch followed by the application of urea, sulphate of potash, water and fresh cow dung in T10 (147.00) and were on par with each other. The control T12 recorded minimum number of fingers perk, bunch (115.00).The most important yield component of the Banana, 'the hands per bunch was significantly altered by various treatments. The spraying of fresh cow dung extracts and zinc sulphate in T8 (18:6) recorded the highest number of hands per bunch followed by spraying of urea, sulphate of potash, water and fresh cow dung in T10 (18.33) And were on par with each other. The lowest (11.33) number of hands per bunch was observed in control.

The bunch weight was significantly influenced by treatments. The highest bunch weight (36.00) was observed in banana due to the application of fresh cow dung extracts and zinc sulphate in (T8) followed by the application of urea, sulphate of potash, water and fresh cow dung in T10 (34.69 kg/bunch) which were on par with each other. The untreated plantain (Control) recorded the lowest bunch weight (29.21 kg/bunch). The treatments had a significant influence on the yield of Banana and the highest yield was recorded in T8 (72.00 t/ha) followed by T10 (69.38 t/ha) which were on par and were significantly superior to the rest of the treatments. There was an increase of 23.24 per cent and 18.76 per cent over the control in treatments T8 and T10 respectively. The lowest yield (58.42 t/ha) was recorded in T12. The data on the TSS revealed that the highest TSS (15.67) was recorded in the treatment T8 followed by T10 (15.50) and were on par. These two treatments were found to be significantly superior to the other treatments. The untreated (Control) T12 (14.17) plantain recorded the lowest number of fingers per hand.

Results of the Experiments conducted at K. Dharmathupatti Village in Robusta Variety

The results of treatments conducted at K. Dharmathupatti village in Robusta variety are depicted in table 2. A scrutiny of the data reveals the yield parameters that effect the yield of Banana.

The average length of fingers was considerably increased by treatments. The application of *panchakavia* and household wood ash in T4 (23.00) and application of urea, sulphate of potash, water and fresh cow dung in T10 (23.00) recorded maximum length of fruit followed by application of brick chamber wood ash and pregnant cow urine in T2 (22.00) and application of fresh cow dung extracts and zinc sulphate in T8 (22.00) and were on par with each other. The control recorded minimum length of fruit (18.00) which was on par with T6 (20.00) T7

(20.00) and T11 (20.00).The average girth of fingers was significantly increased by various treatments. The application of urea, sulphate of potash, water and fresh cow dung in T10 (15.00) recorded maximum girth of fruit followed by application of brick chamber wood ash and pregnant cow urine in T2 (14.00) and application of *panchakavia* and household wood ash in T4 (14.00) and were on par with each other. The control T12 recorded minimum girth of fruit (10.00).

The size of the individual fruit of banana, as represented by the average weight of finger was not affected by the treatments. The maximum fruit size was observed due to the application of fresh cow dung extracts and zinc sulphate in T8 (273.00) followed by the application of fresh cow dung extracts and borax in T9 (271.00) and were on par with each other. The minimum size of the fruit (230.00) was recorded in the untreated plantain. The average volume of fingers was not affected by the treatments. However, the T8 recorded comparatively higher volume (228.00) than the other treatments. The average density of the fingers representing the filling capacity of the individual fruit, was not affected by the treatments. Here the treatments from T1 to T11 were found to be on par and considerably higher than the control.

The data on the number of finger's per hand revealed that the highest number of fingers per hand (26.00) was recorded in the treatment T8 followed by T9 (25.00) and were on par. These two treatments were found to be significantly superior to the other treatments. The untreated (Control) T12 (15.00) plantain recorded the lowest number of fingers per hand. The total number of fingers per bunch was significantly increased by treatments. The application of fresh cow dung extracts and zinc sulphate in T8 (123.00) recorded maximum number of fingers per bunch followed by application of fresh cow dung extracts and borax in T9 (120.00) and were on par with each other. The control T12 recorded minimum number of fingers per bunch (104.00) which was on par with T4 (105.00). The most important yield component of the Banana, the hands per bunch was significantly changed by various treatments. The spraying of fresh cow dung extracts and zinc sulphate in T8 (12.00) recorded the highest number of hands per bunch followed by spraying of fresh cow dung extracts and borax in T9 (11.00) and were on par with each other. The lowest (6.00) number of hands per bunch was seen in control.

The bunch weight was significantly influenced by treatments. The highest bunch weight (33.60) was observed in plantain due to the application of fresh cow dung extracts and zinc sulphate in T8 followed by the application of fresh cow dung extracts and borax in T9 (32.5 kg/bunch) which were on par with each other. The untreated plantain (Control) recorded the lowest bunch weight (23.90 kg/bunch). The treatments had a considerable influence on the yield of Banana. The highest yield was recorded in T8 (60.44 t/ha) followed by T9 (58.54 t/ha) which were on par and were significantly superior to the rest of the treatments. It was found to be an increase of 40.36 per cent and 35.94 per cent over the control in treatments T8 and T9 respectively. The lowest yield (43.06 t/ha) was recorded in T12. The data on the TSS revealed that the highest TSS (14.50) was recorded in the treatment T8 followed by T9 (14.00) and were on par. These two treatments were found to be significantly superior to the other treatments. The untreated (Control) T12 (12.00) plantain recorded the lowest TSS

Results of the Experiments conducted at K. Dharmathupatti Village in Poovan Variety

The results of treatments conducted at K. Dharmathupatti village in Poovan variety are specified in table 3. An analysis of the table reveals the following yield parameters that effect the yield of Banana.

The average length of fingers was significantly increased by treatments. The application of fresh cow dung extracts and zinc sulphate in T8 (15.00) recorded maximum length of fruit followed by application of urea, sulphate of potash, water and fresh cow dung in T10 (14.00), panchakavia and brick chamber wood ash in T3 (14.00) and application of panchakavia and household wood ash in T4 (14.00) and were on par with each other. The control recorded minimum length of fruit (11.00) which was on par with T2 (12.00), T6 (12.00), T7 (12.00) and T11 (12.00) respectively.The average girth of fingers was significantly increased by treatments. The application of panchakavia and household wood ash in T4 (13.00), application of fresh cow dung extracts and zinc sulphate in T8 (13.00) and the application of urea, sulphate of potash, water and fresh cow dung in T10 (13.00) recorded maximum girth of fruit and were on par with each other. The control T12 recorded minimum (10.00) girth of fruit.

The size of the individual fruit of banana, represented by the average weight of finger was significantly increased by various treatments. The maximum fruit size was gained due to the application of fresh cow dung extracts and borax in T9 (130.00) followed by the application household wood ash and pregnant cow urine in Tl (120.00) application of fresh cow dung extracts and zinc sulphate in T8 (120.00) and were on par with each other. The minimum size of the fruit (96.00) was traced in the untreated plantain. The average volume of fingers was not influenced by the treatments. However, the T2 and T10 witnessed relatively higher volume (77.00) than the other treatments. The average density of the fingers representing the filling capacity of the individual fruit, was significantly increased by treatments. The maximum density was obtained due to the application of fresh cow dung extracts and borax in T9 (1.60) and followed by application of fresh cow dung extracts and zinc sulphate in T8 (1.47) and were on par with each other. The control T12 recorded minimum density of fruit (1.21) which was on par with T24 (1.27) and T6 (1.29).

The data on the number of fingers per hand showed that the highest number of fingers per hand (16.00) was recorded in the treatment T8 and T3 followed by (15) in T1, T6, T9, and T11 and were on par. These treatments were found to be significantly superior to the other treatments. The untreated (Control) T12 (10.00) plantain confirmed the lowest number of fingers per hand. The total number of fingers per bunch was significantly increased by treatments. The application of fresh cow dung extracts, zinc sulphate and borax in T7 recorded maximum number of fingers per bunch (163.00) followed by application of fresh cow dung extracts and zinc sulphate in T8 (160.00), application of fresh cow dung extracts and borax in T9 (160.00) and were on par with each other. The control T12 recorded minimum number of fingers per bunch (134.00) which was on par with T4 (140.00)

The most important yield component of the Banana, the hands per bunch was significantly altered by differential treatments. The spraying of panchakavia and brick chamber wood ash in T3 (11.00) recorded the highest number of hands per bunch followed by T1, T6, T8, T9 and T11 (10.00) and were on par with each other. The lowest number of hands per bunch (7.00) was recorded in control. The bunch weight was significantly influenced by the treatments. The highest (20.80) bunch weight was observed in plantain due to the application of fresh cow dung extracts and borax in T9 followed by the application of fresh cow dung extracts and zinc sulphate in T8 (19.20 kg/bunch) which were on par with each other. The untreated plantain (Control) recorded the lowest (12.90 kg/bunch) bunch weight. The treatments had a significant influence on the yield of Banana. The highest yield was recorded in T9 (39.00 t/ha) followed by T8 (36.00 t/ha) which were on par and were significantly superior to the rest of the treatments. It was found to be an increase of 61.69 per cent and 49.25 per cent over the control in treatments T9 and T8 respectively. The lowest yield (24.12 t/ha) was recorded in T12. The data on the TSS revealed that the highest TSS (14.50) was recorded in all the treatment except T7 and were on par. These treatments were not influenced by the yield. The untreated (Control) T12 (13.00) plantain recorded the lowest TSS.

Results of the Experiments conducted at K. Dharmathupatti Village in Red Banana Variety

The results of treatments accomplished at K. Dharmathupatti village in Red Banana variety are illustrated in table 4. An inspection of the table reveals the following yield parameters that effect the yield of Banana.

The average length of fingers was significantly increased by treatments. The application of urea, sulphate of potash, water and fresh cow dung in T10 (21.00) recorded maximum length of fruit followed by application of brick chamber wood ash and pregnant cow urine in T2 (20.00), application panchakavia and brick chamber wood ash in T3 (20.00) and application of fresh cow dung extracts and zinc sulphate in T8 (20.00) and were on par with each other. The control recorded minimum (11.00) length of fruit. The average girth of fingers was significantly increased by treatments. The application of panchakavia and brick chamber wood ash in T3 (15.00), application of fresh cow dung extracts and zinc sulphate in T8 (15.00) and application of urea, sulphate of potash, water and fresh cow dung in T10 (15.00) recorded maximum girth of fruit and were on par with each other. The control T12 recorded minimum (10.00) girth of fruit.

The size of the individual fruit of banana, as represented by the average weight of finger had significantly increased by treatments. The maximum fruit size (370.00) was observed due to the application of fresh cow dung extracts and zinc sulphate in T8 followed by the application of urea, sulphate of potash, water and fresh cow dung in T10 (365.00) and were on par with each other. The minimum size of the fruit (280.00) was recorded in the untreated plantain in T12.The average volume of fingers had significantly increased by treatments. The maximum fingers volume was observed due to the application of fresh cow dung extracts and zinc sulphate in T8 (129.00) followed by the application of urea, sulphate of potash, water and fresh cow dung in T10 (128.00) and were on par with each other. The minimum size of the fruit (108.00) was recorded in the untreated plantain in T12.

The average density of the fingers representing the filling capacity of the individual fruit was found to be significantly higher in all the treatments than the control. The maximum average density of the fingers was observed due to the application of fresh cow dung extracts and zinc sulphate in T8 (3.64) followed by the application of urea, sulphate of potash, water and fresh cow dung in T10 (3.63) and were on par with each other. The minimum size of the fruit (2.59) was recorded in the untreated plantain in T12. The data on the number of fingers per hand revealed that the highest number of fingers per hand (18.00) was recorded in the treatment T8 followed by T10 (17.00) and were on par. These two treatments were found to be significantly better to the other treatments. The untreated (Control) in T12 (9.00) plantain recorded the lowest number of fingers per hand.

The total number of fingers per bunch was significantly increased by treatments. The application of fresh cow dung extracts and zinc sulphate in T8 (110.00), application of fresh cow dung extracts and borax in T9 (110.00) and were on par with each other. The control T12 recorded minimum number of fingers per bunch (88.00). The most important yield component of the Banana, the hands per bunch was significantly altered by various treatments. The spraying of application of fresh cow dung extracts and zinc sulphate in T8 (10.00) and the application of urea, sulphate of potash, water and fresh cow dung in T10 (10.00) recorded the highest number of hands per bunch and were on par with each other. The lowest number of hands per bunch was observed in control.

The bunch weight was significantly influenced by treatments. The highest bunch weight (44.4) was observed in plantain due to the application of fresh cow dung extracts and zinc sulphate in T8 followed ' by the application of urea, sulphate of potash, water and fresh cow dung in T10 (42.30 kg/bunch) which were on par with each other. The untreated plantain (Control) recorded the lowest bunch weight (27.40 kg/bunch). The treatments had a significant influence on the yield of Banana. The highest yield (73.26 t/ha) was recorded in T8 followed by T10 (69.64 t/ha) which were on par and were significantly superior to the rest of the treatments. It was found to be an increase of 65.18 per cent and 57.02 per cent over the control in treatments T8 and T10 respectively. The lowest yield (44.35 t/ha) was recorded in T12.The data on the TSS revealed that the highest TSS (13.50) was recorded in the treatment T8 followed by T 1 0 (13.00) and were on par. These two treatments were found to be significantly superior to the other treatments. The untreated (Control) T12 plantain recorded the lowest TSS (11.00).

The results obtained in the study were discussed with the support of relevant literature in this season. The yield of banana in all the four varieties namely Grand 9, Robusta, Poovan and Red banana were significantly increased by the treatments.

The application of fresh cow dung extract and zinc sulphate in treatment (T_8) significantly increased the yield of banana in all the varieties except the poovan variety where the highest yield was obtained in treatment (T_9) by application of fresh cow dung extract and borax. This was due to the significantly increased bunch weight. The bunch weight also found to follow the trend similar to that of the yield of Banana (t/ha). Yield in Banana is a function of bunch weight and number of plants per hectare. (Kumar & Kumar, 2008).

The higher yield of banana obtained in the elite treatment T8 might be due to more number of fingers/bunch, fingers/hand, hands/bunch, average finger weight, finger volume etc., Generally, the farmer practice is to cut the distal end of the inflorescence after the opening of the last hand. Due to the inadequate supply and flow of nutrients and hormones the distal hand in bunch may not be fully developed into fruit. In the present study in treatment T8 by application of fresh cow dung extract with zinc sulphate an increased the number of fingers/bunch. The presence of nutrients in the available form as well as plant growth hormones (Adegunloye, 2007) might have been responsible for the retention of the fingers in the last hands of the bunch.

Zinc is required for the synthesize of Auxin, a hormone responsible for increasing the fruit set. In the present study, the inclusion of zinc as one of the ingredients in T8 treatment also might have played an important role in increasing the number of fingers per bunch.

The number of fingers per hand, another important yield component closely follows the trend similar to that of total number of fingers per bunch. In fact the number of fingers per hand is reflected in total number of fingers per bunch. In the present study the T8 consistently registered more number of fingers per hand in all the four varieties.

In banana, the size of the finger attracts the consumer and fetches higher price. Any technology enhancing this important yield component is worth rewarding and will attract the farmers to adopt. In the present study the maximum weight of fingers had registered in T8 except poovan; In Poovan the highest weight was registered due to T_9 by application of fresh cow dung extract and borax .The weight of the bunch in poovan was also found to be high in T_9. The differential response of banana cultivars to treatments has been reported by Kotur and Keshava Murthy, 2007.

The maximum weight of fingers in T8 due to the higher mobilization of starch produced during photosynthesis from the leaves. (Kumar and Jeyakumar, 2001) and the above process was accelerated by the plant hormones present in the cow dung. This is also evident from the analytical report of cow dung, cow urine, panchakavia etc. The analytical data is furnished in the appendix - 4. Prasad and Power (1997) reported that the fresh cow dung is rich in plant growth hormones as the cow grazes the tip of the grasses. The tips of the grasses are usually rich in hormones viz., auxin, cytokinin and gibberlic acid.

The average finger weight was also reflected in the finger volume. As the size increases the volume also increased. The enhanced girth of fingers was also found to follow the trend similar to that of finger weight. The finger length, the other finger components differ widely due to treatment. Hence, the size of the fingers as measured by weight was due to increased girth rather than due to length. The hormones available in active form in cow dung as well as in inorganic zinc which is indispensable for the synthesize of IAA could have increased the cell size, which may be the probable reason for the increased girth of fruits.

TSS is the measurement of sugars which is responsible for the sweet taste in fruit, decides the quality of the fruit. In the present study, the highest TSS was registered in T8 in all the four varieties. This is in close conformity of the findings of Kumar and Jeyakumar (2001). This might be probably due to the highest accumulation of starch and subsequently disintegrated into starch on ripening. (Prevel, 1989). The correlation co — efficient between yield and yield components irrespective of varieties and treatments revealed that the bunch weight was significantly and positively associated with number of fingers per bunch, volume of fingers and fingers weight.

Conclusion

The study concluded that the application of fresh cow dung extract and zinc sulphate (T8) to the cut distal end of inflorescence (male axis). The distal end of inflorescence leaving 15 cm was cut, 15 days after opening of the last hand enhances the yield of banana in cultivars Grand 9, Robusta, Poovan and Red banana. The enhanced yield was due to increased number of fingers and weight of fingers. The application of fresh cow dung extract and borax (T9) resulted in higher yield of banana in Poovan.

References

Adegunloye, D. V., Adetuyi, F.C., Akinyosoye, F.A. and M. 0. Doyeni. 2007. Microbial Analysis of Compost Using Cow dung as Booster, Pakistan Journal of Nutrition 6 (5): 506-510.

Grenier, L. 1998. Working with Indigenous Knowledge – A Guide for Researchers. Internal Development Research Centre, Ottawa, p.317.

Kothur, S.C., and Keshavamurthy. 2007. Enhancement of Bunch Size of Robusta and Ney Poovan Banana, ICAR News, A Science and Technology News letter, Vol 13(2):1-2 p.

Kumar, N. and P. Jeyakumar. 2001."Influence of Micro Nutrients on Growth and Yield of Banana (Musa sp.) cv. Robusta (AAA). W. J. Horst et al. (Eds.), Plant nutrition – Food security and Sustainability of Agro–ecosystems. 354-355, Kluwer Academic Publishers, Nether lands.

Kumar, A.R. and N. Kumar. 2008. Studies on the Efficacy of Sulphate of Potash (SOP) on the Physiological, Yield and Quality Parameters of Banana cv. Robusta (Cavendish – AAA), Eur Asian Journal of Bio Sciences, 2, 12, 102 -109.

Prasad, R. and Power, J.F., 1997. Soil Fertility Management for Sustainable Agriculture, Lewis Publishers, New York, 356. p.

Prevel, M.P.J. 1989. Physiological Processes Related to Handling and Storage Quality of Crops. In: Methods of K Research in Plants. Proceedings of the 21 st Colloquium of the International Potash Institute held at Louvain – la- Neuve, Belgium, 19-21 June 1989. 1P1, Bern, Switzerland. Pp. 219-248.

●●●

Section III

Natural Resources and Livelihood Management

26

Environmental Issues and Management of Natural Resources by Farmers for Sustainable Agriculture

R. Arunachalam*

**Associate Professor, Dept. of Agrl. Extension and Rural Sociology, Agricultural College &Research Institute, TNAU, Madurai - 625 104*

Introduction

In India the present population of over 1000 million, accounting for about 18 percent of the world's population, is estimated to become 1.4 billion by 2025 and 1.7 billion by 2050 AD, needing annually about 380 Mt and 480 Mt food grains respectively (Yadhav 2008). This scenario along with the increasing industrialisation and urbanisation will place tremendous pressure on the shrinking natural resources. However, we still have the capacity to produce enough food to meet the basic requirements of the burgeoning population, but the moot question is whether, we can do so while maintaining our natural resources and preserving the bio diversity at the same time. Land are the basic natural resources which are essential for agriculture.

Basic natural resources in agriculture

Land

The dwindling per capita availability of land that decreased from 0.5 ha in 1950-51 to 0.15 ha in 1999-2000 because of population escalation. It is likely to reduce further to 0.08 ha in 2020AD. Continued and excess application of chemical inputs and erosion due to wind and water have made our soils unproductive and nearly unfit for cultivation of crops. In India at present 3.90 million hectares of the land is used for non agricultural purposes and 8.53 million hectares of land is found in water logged condition. Alkali soils accounts to 3.58 million hectares and salinity in 4.04 million hectares of land.

Water

It would be of interest to know that the irrigated area in India has increased from about less than a million hectares in 1800 to about 5 m hectares in 1900 and now it is 88 million hectares. In India, rivers, tanks and underground water forms the basic source for the irrigation. Every year nearly 1.75 lakhs of tube wells are added and the water table is declining steeply everywhere. Tanks are the most popular mode of cheap in situ and consumer friendly way of water storage system in ancient India. Now, tanks got silted up and converted in to housing plots or sewage tanks (Leena 2006). However, tanks irrigate 10 per cent of the cropped area. The country now faces shortage of water a key input for agricultural production despite an annual average precipitation of 400 mha m. Erratic monsoons and faulty water management practices have contributed to the present state of affairs. A substantial portion of ground water in arid and semi arid regions and in some humid coastal areas is of poor quality. The growing competitive demands for fresh water from other sectors such as industry, power and households are further curtailing agricultural share. There has been excessive withdrawal of ground water.

Vegetation

India is still far away from the goal of bringing 33 per cent of its geographical area under trees as stipulated in the vision statement of National Forest Policy. At present the forest cover in our country is 637,293 sq kms which is 19.39 per cent of the geographical area of the country. The declining trend due to deforestation is - 0.01 per cent / annum. There is steady increase in the deforestation rate (Arunachalam 2007).

Animal

Livestock is an another major natural resource in Indian farming. In the absence of favourable condition for agriculture, livestock rearing is the only alternative source of livelihood for majority of the rural population. During 1951 the livestock population was 292.2 million and now it is

652.7 million. India has got about 9.5 per cent of the world's cattle population, while the total geographical area is only 2 per cent of the worlds geographical area. Variations in the number of cattle from state to state seen due to climate, grazing facilities and the area under fodder crops. The cattle dung annually available amounts to about 1200 million tonnes of which 400 million tonnes are used as fuel and 215 million tonnes as manure and the balance being wasted (Pandey, 2007).

Labour

In India there are four common labour types available namely., family labour, hired labour, skilled labour and contract labour and among which family labour use pattern is most common in rural India. It may be noted that the proportion of agricultural labourers to the total labour force is at present 25.70 per cent (Singh 2008). Even though there is a steady increase in agricultural labour force, the demand for labour in agriculture is highly seasonal and uneven due to the seasonal nature of agricultural operations. At present, agricultural production in the country faces severe labour related constraints such as, non availability of labourers during peak seasons and high cost of labour. The agricultural labourers also migrate to neighbouring towns and cities for white collar jobs because, the employment in agriculture is seasonal and uncertain. More over, the payment is also comparatively low than such non agricultural works.

Environmental Issues and Natural Resources

A perusal of different literatures on natural resource management clearly indicates the following threats to the sustainable agricultural production.

- Unabated land degradation
- Depletion of soil organic matter
- Global warming and shrinking of water bodies and ground water
- Depletion of grazing resources and soil erosion
- Lack of adequate efforts for ground water recharge, recycling and water disposal
- Soil erosion and run off
- Soil salinisation
- Rise in water table and water logging
- Lowering of water table in intensive cropping zones.
- Nitrate pollution of ground water
- Excessive use of chemical inputs which spoils clean atmosphere, weather system and soil productivity

- Monocropping leading to depletion of soil fertility, pest out break and loss of natural bio diversity
- Spurious inputs and over or under doing of pest management practices
- Increased fragmentation of lands and absentee landlords
- Decreasing crop and livestock bio diversity
- Rapid industrialisation, poor waste / effluent disposal and recycling facilities
- Inadequacy of location specific research
- Subsistence farming and
- High cost of labour.

A thorough understanding of the awareness and perception about environmental issues and natural resource management practices will help in designing suitable extension strategies to make the farmers aware and perceive better about the issues concerned and adopt remedial practices. Keeping this in mind the present study has been carried out with the following objectives

1. To identify and document the environmental issues prevailing in the study area
2. To study the awareness and perception of the respondents towards environmental issues
3. To study their utilization pattern of the natural resources

Methodology

Kanyakumari district of the state of Tamil Nadu was selected for this study as this district falls under high rainfall zone, and it is most benefited by both South west monsoon and North east monsoon. Various types of crops are cultivated in this district and hence, we can expect various types of resource use pattern in this district. A sample size of 200 farmers were selected by employing proportionate random sampling method

An exhaustive list of environmental issues related with agriculture was prepared based on the knowledge gained in the pilot study and also in consultation with the local extension workers and environmentalists. Sixteen important environmental issues alone were considered for further analysis. On the identified issues farmers' awareness and perception were studies using well structured interview schedule. Based on the experience gained in the pilot study and also in consultation with the local extension workers and environmentalists an exhaustive list of land, water, vegetation, farm animal and labour management practices was prepared. Using this list the natural resource management utilization behaviour was studied by employing appropriate statistical techniques.

Findings and Discussion

Awareness about environmental issues

A thorough perusal of the Table 1, revealed that almost all the respondents were aware about the environmental issues namely, increased soil erosion, reduced ground water potential, uncertainty in the onset of seasonal rainfall, disappearance of traditional crops and varieties, loss of real taste in grains of field crops, fruits and vegetables, intensive weed growth, health hazards during cropping, increased tmospheric temperature, conversion of cropped area for residential purposes and consequently, the loss of green cover. An overwhelming majority of the respondents, were aware about the environmental issues namely, development of resistance to pest and diseases (95.50 %) and more incidence of pests and diseases (93.50 %). Further, it is noted that 90.00 per cent of the respondents were aware that, the rain fall pattern has been disturbed to a greater extent, ground water became more salty and increased level of soil salinity. Nearly 90.00 per cent of the respondents were aware about the loss of soil inherent fertility. Around three fourth of the respondents (74.50 %) were aware about more water logging and poor drainage.

Perception about environmental issues

The findings in the Table I, shown that cent per cent of the respondents had agreed that, due to lack of vegetative cover and deforestation, heavy wind and rainfall displace the soil particles and consequently, it lead to soil erosion. All the respondents had also opined that due to the introduction of synthetic rubber, the area under rubber plantations was converted in to residential plots, which lead to the loss of green cover. Further, they have also agreed that the deforestation is the main cause for the increased atmospheric temperature. Almost an equal percentage of the respondents (99.00 %) had right perception that, excessive withdrawal of ground water and inadequate application of water harvesting techniques to recharge the same, lead to the reduction in ground water potential.

Almost all the respondents have agreed that the introduction of hybrids and high yielding varieties lead to the disappearance of the traditional crops like `Jambaka', 'Bread fruit', Valuthulanga brinjal', 'Kanthari chillies', 'Red rice' and 'Nei poovan banana'.

An overwhelming majority of the respondents (98.50%) had perceived rightly that improper handling of chemical inputs was the real cause for the increased health hazards during cropping. Almost an equal percentage of the respondents (98.00%) have admitted that deforestation is one of the major causes for the prevailing uncertainty in the onset of seasonal rainfall. They further stated that the use of high yielding varieties and application of more chemical inputs lead to the loss of real taste in the grains of field crops, fruits and vegetables. Further, 93.50 per cent of the respondents viewed that the intensive weed growth was due to poor tillage and poor crop management practices.

Most of the respondents (87.50%) were of the view that deforestation is again one of the major causes for the present disturbances in the pattern of rainfall. But 84.50 per cent of the respondents had agreed that the repeated and excess application of the same chemicals over a period of time against pests lead to the development of resistance in pests and diseases.

About three fourth of the respondents (74.50%) possessed right perception that the gradual increase in soil salinity is due to the application of more chemical inputs. The same percentage of the respondents agreed that the incidence of particular pests and diseases were found more in the monocropped areas.

Table 1 : Distribution of respondents according to their awareness and perception about Environmental issues

(n = 200)

SI. No.	**Environmental issues and perception statements**	**Awareness**				**Perception**			
		Aware		**Not aware**		**Agree**		**Disagree**	
		No	**%**	**No**	**%**	**No**	**%**	**No**	**%**
1.	E.I. Soil salinity - gradually increasing trend	180	90.00	20	10.00				
	Per: Application of more chemical inputs makes the soil saline					149	74.50	51	25.50
2.	E.I. Incidence of soil erosion more	200	100	--	--				
	Per : Lack of vegetative cover, deforestation, heavy wind lead to loss of soil thro' wind and rain water						200	100	--
3.	E.I. Soil inherent fertility lost	177	88.50	23	11.50				

	Per: More application of chemical inputs lead to loss of soil inherent fertility					136	68.00	64	32.00
4.	E.I. Ground water potential reduced	200	100	--	--				
	Per: Excessive withdrawal and absence of mechanism to recharge it lead to the issue					198	99.00	02	01.00
5.	E.I. Ground water became more salty now a days	180	90.00	20	10.00				
	Per : Excessive withdrawal of ground water over a long period of time leads to this issue					101	50.50	99	49.50
6.	E.1. More waterlogging and poor drainage	149	74.50	51	25.50				
	Per: Poor water management practices are the causes for this issue					119	59.50	81	40.50
7.	E.I. Onset of seasonal rainfall uncertain now a days	198	99.00	02	01.00				
	Per : Deforestation is one of the major causes for this issue					196	98	04	02.00
8.	E.I. Rainfall pattern disturbed to a greater extend	180	90.00	20	10.00				
	Per : Deforestation is one of the major cause for the issue					175	87.50	25	12.50
9.	E.I. Pest and diseases developed resistance	191	95.50	09	04.50				
	Per : Repeated and excess application of same chemicals over a period time against a pest leads to this issue					169	84.50	31	15.50
10.	E.I. Traditional crops like 'Jambaka', 'Bread fruit' Valuthulanka Brinja;', 'Kanthari chillies', 'red rice' 'Neipoovan Banana' have disappeared	198	99.00	02	01.00				
	Per : Introduction of hybrids and high yielding varieties lead to the disappearance above crops					198	99.00	02	01.00
11.	E.I. Grain of field crops, fruits and vegetables lost real taste	199	99.50	01	00.50				

	Per : use of high yielding varieties, hybrids and application of more chemical inputs make this issue					194	97.00	06	03.00
12.	E.I. Intensive weed growth	198	99.00	02	01.00				
	Per : Poor tillage operations and poor crop management practices lead to this issue					187	93.50	13	06.50
13.	E.I. Incidence of particular pest and diseases are more	187	93.50	13	06.50				
	Per : In monocropped fields such incidence are greater					149	74.50	51	25.50
14.	E.I. Respondents suffer due to health hazards during cropping.	200	100	--	--				
	Per : Improper handling of chemical inputs lead to this issue					197	98.50	03	01.50
15.	E.I. Area under plantation crops are being converted for residential purposes and loss of green cover	200	100	--	--				
	Per : Introduction of low cost synthetic rubber leads to this issue					200	100	--	--
16.	E.I. Atmospheric temperature has increased now a days	200	100	--	--				
	Per : Deforestation is one of the main reasons for this gradual increase in atmospheric temperature.					200	100	--	--

E.I : Environmental issues **Per :** Perception statements

Land management practices

There were totally twelve land management practices selected for this study (Table 2) Out of the twelve land management practices, the adoption behaviour of the respondents on the following practices were found conducive for achieving sustainability in agriculture.

- Selection of right crop for the soil (98.50 per cent)
- Use of traditional implements (86.50 per cent)
- Use of mechanised implements (56.00 per cent)
- Ploughing across the slope for soil and water conservation (58.50 per cent)
- Practicing mono crops (03.00 per cent)

Use of bio pesticides to soil - Recommended dose (48.50 per cent) and Excess application (50.00 per cent)

Almost all the respondents have selected right type of crops for their soil. Traditional implements were optimally utilised by most of the respondents and optimum use of mechanised implements were seen with majority of them. Majority of the respondents followed the intercultivation operation of ploughing across the slope for soil and water conservation.

Most of the respondents followed mono cropping practices sometimes and adequate application of soil bio pesticide was seen with nearly half of the respondents. Hence, it could be concluded that the above management behaviour of the respondents could contribute to sustainable agriculture. In the case of adoption of maintaining optimal cropping intensity, use of organic manure, use of inorganic manure, crop rotation and chemical application to soil (like herbicides) the respondents were found to be on the wrong footing.

It was found that most of the respondents were distributed between high to medium level of cropping intensity. This clearly indicates that majority of the respondents were over exploiting the soil by cultivating no of crops. If this trend is continued, the soil will loose its fertility and production capacity. Majority of the respondents inadequately applied organic manure and excess application of inorganic manure was seen with 43.00 per cent of the respondents. The above two conditions will definitely affect the soil fertility and also its structure. Nearly three fourth of the respondents have sometimes followed crop rotation practices and excess chemical application was seen with 40.00 per cent of the respondents. When crop rotation is not followed, the farmers could not effectively utilise soil nutrients available at various depths. It is a well established fact that excess application of chemicals in to the soil will definitely spoil its structure and ultimately affect its productivity. The above findings clearly indicate the need for formulating extension strategies to educate the farmers and also to make them use practices favourable to achieve sustainable agriculture.

Water management practices

Out of the nine selected practices (Table 3), the management behaviour of the respondents on the following practices confirmed to the prescribed norms of achieving sustainability in agriculture.

- Rain water harvesting (61.50 per cent)
- Optimal irrigation (90.00 per cent)
- Optimum tillage to increase soil water holding capacity (90.50 per cent)
- Organic management practices like use of mulches (61.50 per cent)

Majority of the respondents have adopted rain water harvesting methods and optimal irrigation to the crops was done by 90.00 per cent of the respondents. Majority of the respondents (61.50 per cent) used mulches to improve the soil water holding capacity.

Hence, it could be concluded that the management behaviour of the respondents on the above practices could contribute to achieve sustainable agriculture. But in the case of construction of farm pond, regular cleaning of irrigation channels, lining of channels, disposal of excess water and use of drip system, the respondents were found adopting them in a way that may hinder sustainable agriculture. The above mentioned practices can be considered as very important as they relate to water harvesting, avoiding wastage of available water and the effective utilisation of available water for increased production. Though the department of agricultural engineering has programmes to help farmers on the above aspects, the present condition is highly disturbing. They should take adequate effective measures to convince the farmers, to adopt water conservation and use technologies for which the government is providing liberal subsides too.

Vegetation management practices

Out of the nine vegetation management practices (Table 4), the management behaviour of the respondents on the following eight practices were found to be in line with achieving sustainable agriculture.

- Farm forestry (73.00 per cent)
- Agro forestry (92.00 per cent)
- Growing trees in field bunds (88.00 per cent)
- Growing wind brakes (76.00 per cent)
- Growing vegetation for natural fence (90.50 per cent)
- Burning vegetation for soil sterilization (76.00 per cent)
- Burning crop wastes for non agricultural purposes (68.50 per cent)
- Use of bio pesticides against crop pests and diseases (57.00 per cent)

It is observed that a vast majority of the respondents have adopted the above practices, which are essential for achieving sustainability in agriculture. Regularly following the above practices will aid in the conservation of this resource for future use.

Further, it is seen that only in the use of green and green leaf manure , their management behaviour was not in confirmity with the standard norms of achieving sustainability in agriculture. Only 8.00 per cent of the respondents were found adopting this practice.

In the study area, the extent of land use is very high compared to other districts and this has reduced the availability of source for green leaf manure. Under such circumstances, it is essential to make farmers in the study area to grow green manure crops in order to enrich their soil and to reap the benefits continuously in the years to come.

Table 2 : Distribution of respondents according to their adoption pattern of land management practices

(n = 200)

SI. No	Management of the land resource	Adoption	
		No	%
1.	**a) Selection of right crops for the soil**		
	i). Yes	197	98.50
	ii). No	003	01.50
2.	**Crop intensity**		
	i). High	35	17.50
	ii). Medium	119	59.50
	iii). Low	46	23.00
3.	**Use of organic manures**		
	i). Adequate	53	26.50
	ii). Inadequate	132	66.00
	iii). Not applied	15	07.50
4.	**Use of inorganic manures**		
	i). Minimum	14	07.00
	ii). Recommended	100	50.00
	iii).Excess	86	43.00
5.	**Tillage**		
A).	**Use of traditional implements**		
	j). More dependence	9	04.50
	ii). Optimum	173	86.50
	iii).Less dependence	18	09.00
B).	**Use of mechanised implements**		
	i). More dependence	40	20.00
	ii). Optimum	112	56.00
	iii). Less dependence	48	24.00
6.	**Intercultivation operation**		
A).	a). Ploughing across the slope in uplands prevents soil erosion		
	i). Done	117	58.50

	ii). Not done	83	41.50
B).	a). Ploughing across the slope in upland helps in water conservation		
	i). Done	117	58.50
	ii). Not done	83	41.50
7.	**Types of crops**		
A.	**Monocrops**		
	i). Always	6	03.00
	ii). Sometimes	141	70.50
	iii). Never	53	26.50
B.	**Crop rotation**		
	i). Always	15	07.50
	ii). Sometimes	146	73.00
	iii). Never	39	19.50
8.	**Plant protection in soil**		
A).	**Chemical application (Like herbicides) to soil**		
	i). Minimal	16	08.00
	ii). Recommended	104	52.00
	iii). Excess	80	40.00
B).	**Use of biopecticides to soil**		
	i). Minimal	3	01.50
	ii). Recommended	97	48.50
	iii). Excess	100	50.00

Table 3 : Distribution of respondents according to their adoption pattern of water management practices

(n = 200)

SI. No	**Management of the land resource**	**Adoption**	
		No	**%**
1.	**Water sources: Rain water conservation**		
	i). Done	123	61.50
	ii). Not done	77	38.50
2.	**Irrigating crops: Optimal irrigation to crops to conserve irrigation water**		
	i). Optimum	180	90.00
	ii). Excess	20	10.00
3.	**Tillage practices : Optimum tillage to increase soil water holding capacity**		
	i). Done	181	90.50
	ii). Not done	19	09.50
4.	**Farm pond : Construction of farm pond in the low lying area for rainwater harvesting**		
	i). Done	50	25.00
	ii). Not done	150	75.00

5.	**Organic management practices: Use of mulches to improve the soil water holding capacity**		
	i). Done	123	61.50
	ii). Not done	77	38.50
6.	**Cleaning of channels : Regular cleaning of irrigation channels to arrest loss of water**		
	i). Regularly	50	25.00
	ii). Sometimes	149	74.50
	iii). Never	01	00.50
7.	**Lining of channels : with concrete slaps / polythene sheets to avoid seepage loss**		
	1). Done	48	24.00
	ii). Not done	152	76.00
8.	**Disposal of excess water: Diverting excess water to farm pond as a ground water recharging method**		
	i). To farm pond	50	25.00
	ii). Left to runoff	150	75.00
9.	**Water saving devices: like drip and sprinklers in conserving irrigation water**		
	i). Done	56	28.00
	ii). Not done	144	72.00

Table 4 : Distribution of respondents according to their adoption pattern of vegetation management practices

(n = 200)

SI. No	Management of the land resource	Adoption	
		No	%
1.	**Farm forestry**		
	i). Done	146	73.00
	ii). Not done	54	27.00
2.	**Agroforestry**		
	i). Done	184	92.00
	ii). Not done	16	08.00
3.	**Growing trees in the field bunds for fire wood/timber for farm implements**		
	i). Done	176	88.00
	ii). Not done	24	12.00
4.	**Leaf manures (Use of green and green leaf manure)**		
	i). Always	16	08.00
	ii). Sometimes	182	91.00
	iii).Never	02	01.00
5.	**Wind brakes: Growing of trees as wind belts to protect crops from wind damages**		
	i). Done	153	76.50

	ii). Not done	47	23.50
6.	**Vegetation for natural fences**		
	i). Done	181	90.50
	ii). Not done	19	09.50
7.	**Burning vegetation** **Burning of tree leaves, weeds and tree wastes for sterilising soil (agrl. purposes)**		
	i). Done	152	76.00
	ii). Not done	48	24.00
	b). Burning of crop wastes for non agricultural purposes		
	i). Done	63	31.50
	ii). Not done	137	68.50
8.	**Biopesticides: Plant parts and products used as biopesticides against crop pests**		
	i). Done	114	57.00
	ii). Not done	86	43.00

Table 5 : Distribution of respondents according to their adoption pattern of animal management practices

(n = 200)

SI. No	**Management of the land resource**	**Adoption**	
		No	**%**
1.	**Keeping farm animals and birds**		
	i). Done	193	96.50
	ii). Not done	07	03.50
2.	**Carrying out farm operations with animals**		
	i). Fully	05	02.50
	ii). Partially	187	93.50
	iii).Never	08	04.00
3.	**Balanced feed for animals**		
	i). Provided	118	59.00
	ii). Not provided	82	41.00
4.	**Periodic health check up for animals and birds**		
	1). Done	85	42.50
	ii). Not done	115	57.50
5.	**Maintaining cleanliness in the animal sheds**		
	i). Done	183	91.50
	ii). Not done	17	08.50

Table 6 : Distribution of respondents according to their adoption pattern of labour management practices

(n = 200)

SI. No	Management of the land resource	Adoption	
		No	**%**
1.	**Types of labour**		
	a).Use of family labourers		
	i). Always	26	13.00
	ii). Sometimes	97	48.50
	iii). Never	77	38.50
	b). **Use of external labourers (over and above the available family labourers)**		
	i). Always	25	12.50
	ii). Sometimes	97	48.50
	iii). Never	78	39.00
	c). **Working along with hired labourers**		
	i). Always	50	25.00
	ii). Sometimes	120	60.00
	iii). Never	30	15.00
2.	**Effective management (supervision) of hired labourers.**		
	i). Done	196	98.00
	ii). Not done	04	02.00
3.	**Labour system : Operation specific contract**		
	i). Fully	60	30.00
	ii). Partially	107	53.50
	iii). Never	33	16.50
4.	**Optimal engagement of labourers to reduce cost of cultivation**		
	i). Done	179	89.50
	ii). Not done	21	10.50
5.	**Use of skilled labourers**		
	i). Always	64	32.00
	ii). Sometimes	114	57.00
	iii). Never	22	11.00

Animal management practices

Out of the five selected practices (Table 5), the management behaviour of the respondents on the following practices confirm that they have taken adequate measures to achieve sustainability in agriculture.

- Keeping farm animals and birds (96.50 per cent)
- Carrying out farm operations with animals — Partially done (93.80 per cent)
- Providing balanced feed for the animal (59.00 per cent)

Maintaining cleanliness in the animal shed (91.50 per cent)

Compared to olden days, the farmers have slowly reduced the no of animals maintained in their household and some of them have even stopped the practice due to acute shortage of labour and their changed attitude. But it is heartening to note that majority of the farmers are maintaining few animals and birds in their houses, which not only helps in performing various farm operations, but also supply considerable quantity of organic manure. Thus, the findings of this study by and large have clearly indicated that most of the respondents were managing their available resources very effectively in line with achieving sustainability in agriculture. It may be due to the fact that, it was Kanyakumari district which has cent per cent literacy level in the state of Tamil Nadu.

Labour management practices

There were seven labour management practices (Table. 6) selected for this study. The adoption behaviour of the respondents on the following practices confirmed to the standard norms in achieving sustainable agriculture.

- Use of external labour – engaged sometimes (48.50 per cent)
- Effective supervision of hired labourers (98.00 per cent)
- Optimal engagement of labour (89.50 per cent)

Majority of the respondents used external labourers sometimes. Most of the respondents have effectively supervised the labourers and have engaged optimal labourers in their farms. Labourers were engaged to perform various farm operations, most of which involved the handling of various resources like water, manure and chemicals. Effective supervision will ensure the optimum use of available resources and also correctly performing various tillage operations. So the findings of this study in this regard showed that the respondents were consciously following labour management practices which can be believed to lead to sustainable agriculture.

In the case of use of family labourers (13 .00 per cent), working along with hired labourers (25.00 per cent), operation specific contract system (30.00 per cent) and use of skilled labourers (32.00 per cent), the respondents were found lacking to the expected level. Effective extension education efforts should be taken in this regard to improve their management behaviour.

Conclusion

The study indicated that vast majority of the respondents were aware about various environmental issues and they have rightly perceived the cause of the issues. However the facts remain that the people themselves are responsible for such a state of affairs. Therefore, it is necessary to educate them not to do anything that may adversely affect their environment. Extensive use of mass media and personal approaches are suggested to achieve this.

The findings of this study indicated though the respondents were aware about various natural resource management practices, they were found lacking in the adoption of the natural management practices like, use of green and greenleaf manure, maintaining optimum cropping intensity, water conservation and optimal use of animal power, Intensive use of extension efforts to convince the farmers is suggested. They further expressed that lack of proper guidance and motivation on various natural resource management practices were their main constraints in adopting them. Formulating and standardizing appropriate natural resource management practices for achieving sustainable agriculture by the agencies concerned is essential in this regard.

References

Arunachalam, R. 2007. "Impact of Integrated Wasteland Development Programme on Farmer Beneficiaries", Compendium of Research Papers, All India Conference of KAAS. Nagercoil, Tamil Nadu, India Pp 27 - 31.

Leena, S.N. 2001. Modernisation of Indian Agriculture, Akshaya Vikas - Sustainable Development, Chennai, India.

Pandey.S.N. 2007. "Resource Management in India : Status and Concern, Indian Agriculture, India. Vol 32 (1): 10-15.

Singh.K.K, 2008. "Environmental Protection and Sustainable Agriculture". Agriculture Today, Hyderabad, India. Vol 54 (1-2) : 12 —16.

Yaday. J.S.P 2008. Natural Resource Management for Agricultural Production in India International Conference on Managing Natural Resources for Sustainable Agricultural Production in the 21st Century. Feb 14 - 18, 2000, New Delhi. India. Pp 22-28.

●●●

27

Eco-Friendly Agricultural Practices in Paddy — An Impact Analysis in Tamil Nadu

A. Janaki Rani. H.Philip and P.P.Murugan

[1] Asst. Professor (Agri. Extension), Dept. of Trade and Intellectual Property, TNAU, Coimbatore, [2] Professor (Agri. Extension), DOEE, TNAU, Coimbatore

and [3] Associate Professor (Agri. Extension), ATIC, TNAU, Coimbatore

Introduction

During the last four decade, spectacular progress has been achieved in agricultural production in the country. Compared with 51 million tones of food grain production in 1950-51, India achieved food grain production of 211.17 million tones during 2001-02, making it not only self sufficient in food production but also in a buffer stock of over 30 million tones. This achievement was because of high intensity cropping and cultivation of high yielding varieties that were highly input responsive. Farmers used chemicals and synthetic pesticides, fertilizers and growth promoters above the level of recommended dose with a motive to get higher yields. In this process the soil, water and environment got polluted besides degradation of natural resources. Paddy is the principle food crop and ranks second in the consumption of pesticides in Tamil Nadu. In order to protect our environment, the state department of agriculture had started advocating eco-friendly agricultural practices in paddy cultivation. In addition to the government efforts, number of non-governmental organizations (NGO's) also taking intensive efforts to make the farmers to adopt eco friendly agricultural practices. Among the several NGO's in Tamil Nadu, Agricultural Man Ecology Foundation (AMEF) is a Netherland based NGO found to have taken strenuous efforts to advocate and popularize eco -friendly agricultural practices. It has network with other NGO's situated all over Tamil Nadu and rendered technical and financial support to them for the promotion of eco-friendly agricultural practices. Keeping this in view, a study was conducted to assess the impact of eco-friendly agricultural practices among the paddy farmers who have trained by the AMEF network NGO's.

Methodology

Pudukottai District in Tamil Nadu state was selected purposively since more number of Agriculture Man Ecology Foundation network NGOs were placed and involved in the promotion of ecofriendly agricultural practices. Among the NGOs, two NGOs namely Biosphere and Manushi were selected randomly. A total of 120 trained paddy farmers selected randomly under those NGOs, constituted the respondents. For this study, expost facto research design was followed. A well-structured interview schedule was used to collect data on impact of ecofriendly agricultural practices. To make the study exhaustive the impact was studied under two heads namely direct and indirect impact. Under direct impact yield, income and cost of cultivation were studied. Under indirect impact the sub items like personal impact, social impact, economic impact and environmental impact were studied. Thus a set of 16 statements were selected and a score of three, two and one was given for the responses of increase, decrease and no change respectively. Percentage analysis was done to find out the number of respondents who were coming under the above categories.

Findings and Discussion

The results of percentage analysis on impact of eco friendly agricultural practices are presented in Table 1 and 2.

Direct impact

The results of percentage analysis on direct impact of eco-friendly agricultural practices presented in Table 1.

Table 1 : Distribution of trained farmers according to their direct impact of eco-friendly agricultural practices (n=120)

SI. No.	Impact	Increased		Decreased		No change	
		No	%	No	%	No	%
I	**Direct impact**						
1.	Yield	39	32.50	31	25.83	50	41.67
2.	Income	35	29.16	32	26.67	53	44.17
3.	Cost of cultivation	17	14.17	80	66.66	23	19.17

It was quite interesting to note that, two fifth (41.67 percent) of the trained paddy farmers reported that due to the adoption of eco-friendly agricultural practices no change in yield level at the initial stage. The farmers in the study area were in the transition stage from chemical agriculture to eco-friendly agriculture. In this process, they observed initial stagnation in yield before it got stabilized might be the reason for no change in yield.

With regard to income level, more than two-fifth (44.17 per cent) of the farmers expressed that no change in the income level. In chemical agriculture cost of cultivation is higher due to internal inputs like chemicals, fertilizers and pesticides. Consequently results high in yield, where as in eco-friendly agriculture due to the application of locally available inputs the cost of cultivation is low. The yield is also comparatively less until it sustained. So, in the initial stage there might be no much variations in the income level.

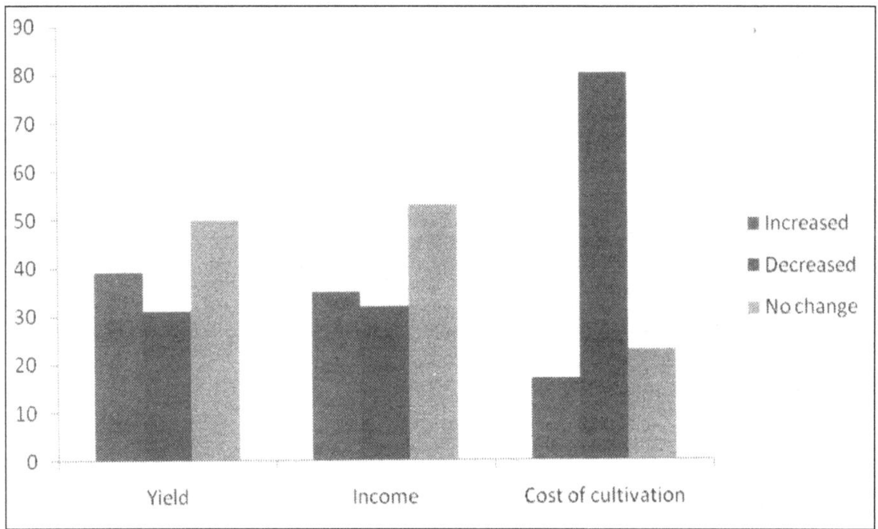

Fig. 1 : Direct impact of Eco-friendly Agricultural practices in paddy

Decreased cost of cultivation was expressed by two-third (66.66 per cent) of the trained farmers because of the practices involved were cost effective and making use of locally available farmyard manure, green manure, green leaf manure / compost, biofertilizers and herbal repellents. This would have reduced the cost of cultivation to some extent.

Indirect impact

The results of percentage analysis on indirect impact of eco-friendly agricultural practices presented in Table 2.

Table 2 : Distribution of trained farmers according to their indirect impact of eco-friendly agricultural practices (n=120)

Sl. No.	Indirect Impact	Increased		Decreased		No change	
		No.	Per cent	No.	Per cent	No.	Per cent
A.	**Personal impact**						
	Confidence in eco friendly cultivation	82	68.33	14	11.67	24	20.00
	Opportunity to know about development activities	76	63.33	5	4.17	39	32.50
	Exposure to media sources	68	56.67	17	14.17	35	29.16
	Consultation by fellow farmers	63	52.50	20	16.67	37	30.83
	Health condition	52	48.33	24	20.00	44	36.67
	Decision making power	76	63.33	12	10.00	32	26.67
B.	**Social impact**						
	Participation in social organizations	93	77.50	9	7.50	18	15.00
C.	**Economic impact**						
	Investment in savings	47	39.17	13	10.83	60	50.00
	Purchase of agricultural implements	36	30.00	17	14.17	67	55.83
	Purchase of milch animals	25	20.83	0	0	95	79.17
D.	**Environmental impact**						
	Soil fertility	109	90.83	0	0	11	9.17
	Improvement in environmental condition	104	86.67	0	0	16	13.33
	Beneficial insects	112	93.33	0	0	8	6.67

Increased confidence was noticed with more than two-third (68.33 per cent) of the paddy farmers. The slow but visible impacts of using eco-friendly agricultural practices in paddy cultivation probably are the reason.

Two-third (63.33 per cent) of the paddy farmers expressed that it had increased the opportunity to know the development activities. The NGOs in the study area formed farmers associations in the name of Vikas Vahini Voluntary Club (VVV) for promoting eco-friendly agriculture. The club facilitated the farmers to share information like developmental activities with each other. This might be the reason to know the developmental activities.

The NGOs in the study area screened film shows and video lessons on eco-friendly practices might have been the reason for 56.67 per cent of the farmers to have increased exposure to media sources.

Participation in trainings, FFS, demonstration etc., by the trained paddy farmers made other farmers in the village as resource persons and seeking advisory services. This might be the reason for increased level of consultation by fellow farmers.

About half (48.33 per cent) of the trained paddy farmers expressed that, quality rice, chemical free food and keeping quality are the advantages in eco-friendly agricultural practices made the farmers to have increased health condition. Increase in self confidence might be the reason for increasing in decision making power by 63.33 per cent of the paddy farmers.

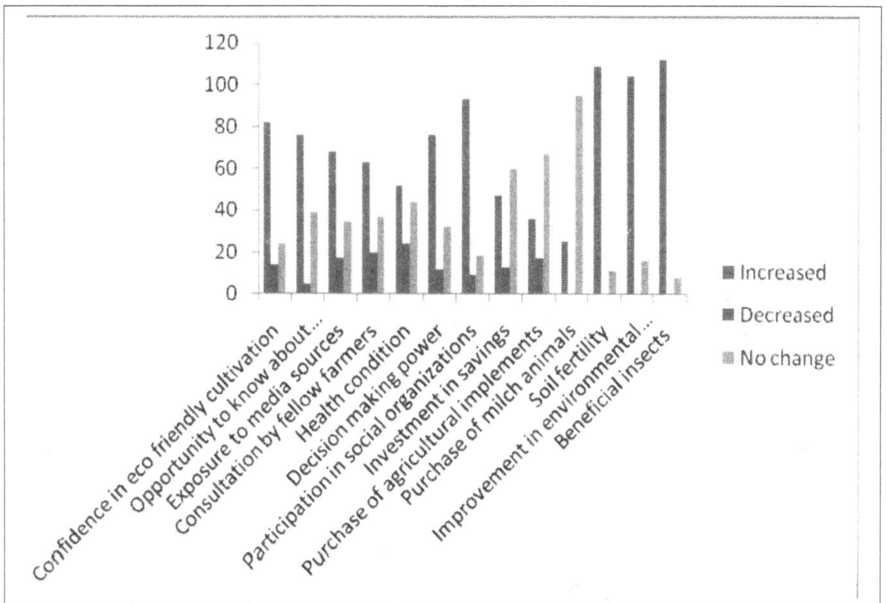

Fig. 2 : Indirect impact of Eco-friendly Agricultural practices in paddy

Regarding the social impact, involvement in farmers association and self help groups organized by NGOs paved way for increased participation (77.50 per cent) in social organizations.

With respect to economic impact, they were at the initial stage, less impact was observed regarding their investment in savings, purchase of agricultural implements and purchase of milch animals.

Regarding the environmental impact due to the adoption of eco-friendly agricultural practices the soil fertility got increased was reported by more than 90 per cent of the farmers. Based on their own observation seed on quality parameters of soil, farmers agreed with this condition.

The farmers in their own field experienced the effect of eco-friendly practices by seeing the loosening of soil, increased quality of water, existence of earth worms and beneficial insects. This resulted in increased positive response from 86.67 per cent of the trained paddy farmers about the improvement in environmental condition.

As a result of eco-friendly agricultural practices like summer ploughing, FYM/compost, Vermi composting, biopesticdes, biocontrol agents, number of beneficial insects got increased. Hence majority of the trained paddy farmers (93.33 per cent) accepted the increased level of beneficial insects.

Conclusion

The overall observation on the impact of eco-friendly agriculture practices among trained paddy farmers showed that there was no much change in yield and income at the initial stage and reduced the cost of cultivation. Increased the confidence, increased opportunity to know development activities, exposure to media sources and consultation by fellow farmers, health condition, decision making power and participation in social organizations. Also it has created significant impact on increased soil fertility, improvement in environmental condition and beneficial insects.

Since the farmers in the study area have been practicing eco-friendly practices for the past three to four years only by the effort of NGOs through farmers field school, Self Help Groups, V V V club etc., so, that these results were seen. But in future the yield will be sustained if there eco-friendly practices followed continuously and also by the involvement and effort of government, private and NG0s.

References

Anusuya, A. 1997. Impact of farmer's field school on farming community. Un pub .M.Sc. (Ag) Thesis AC & RI, TNAU, Madurai.

Sunitha.V.1998. Knowledge and adoption of eco-friendly farm technologies in paddy.

Shivaraj R.2001. Effectiveness of eco-friendly agricultural practices in paddy An — analysis, Unpub M.Sc.(Ag) Thesis, AC&RI, TNAU, Madurai.

Snehalatha, E.S. 1991. Spread and Acceptance of Farm Technologies and their impact in production and productivity and socio economic conditions of farmers. Unpub M.Sc.(AG) Thesis, AC&RI, TNAU, Madurai.

Thurstone,l.l (1946). The measurement of attitude. American Journal of sociology, Chicago university press:39-50.

●●●

28

Intensive Agriculture and Environment

D. Jebapreetha [1] ***and Rexlin Selvin*** [2]
[1] *Assistant Professor, Roever Agricultural College, Perambalur and*
[2] *Professor (Agri. Extension), Agrl. College & Research Institute, Madurai*

Introduction

Soil, water and vegetation are three basic natural resources. The survival of God's creation depends upon them and nature has provided them as assets to human beings. The management of natural resources to meet people's requirements has been practised since the pre-Vedic era. Farmers were ranked high in the social system and village management was in their hands. Over-exploitation of natural resources by growing population resulted in various severe problems. Destruction of vegetation has resulted in land degradation, denudation, soil erosion, landslides, floods, drought and unbalanced ecosystems. A balanced ecosystem is an urgent need.

There is an increasing realization throughout the world, about the state of our environment, caused by over-exploitation of natural resources for economic development. Resource exploitation in the developing world has largely been geared to meet the very basic needs of food, fodder, fuelwood and shelter of a large section of deprived societies. On the other hand, developed parts of the world have largely concerned themselves in trying to maintain and accelerate the already very high levels of resource consumption that they have achieved for a much smaller segment of the world's population, based on an early initiative taken and advantage gained through industrialization. Natural resource base has often viewed as limitless by the industrialized world either due to a myopic view of the future or because of their immense faith in technology being able to substitute for the natural resource capital. This faith in technological quick-fixes for world's problems linked with over-consumption is still prevalent.

Pariyar (2006) observed that due to mismanagement of water resources, around six million hectare of area is under water logged condition in India. With four per cent water resources plus 16 per cent of the world's population, the situation will soon worsen. Farmers are exploiting ground water by means of modern technologies at an alarming rate without considering it as a precious natural resource.

Murugesan *et al.* (2005) revealed that qualitative changes occur in ground water due to pollution, indiscriminate disposal of industrial, human and agricultural wastes, which pose great threat to the quality of ground water in many regions. Water quality deterioration had made the potable water resources a scarce one and endangered for living systems.

Thus it was decided to document the environmental issues and its causes that affect the natural resource in intensive cropping area with following objective

1. To document the soil related and water related issues
2. To enlist the causes for documented environmental issues.

Methodology

The study was conducted in Theni and Dindigal district of Tamil Nadu. The study area was selected purposively since intensive cropping is practiced vigorously and cash crops like grapes were cultivated intensively in a larger area and for longer years (25-30 yrs). Proportionate allocation was employed to draw the 150 respondents from the three blocks of the two districts.

Findings and Discussion

A. Documentation of environmental issues

The stakeholders like grapes growers, extension personnel, marketing agents and input dealers were consulted to document the issues related to soil, water and produce residues besides collecting data from secondary sources. The information gathered were grouped and presented as below.

I. Soil Related Issues

1. Worn out and weak soil

Due to the favourable weather prevailed in the study area farmers have gone for intensive cultivation of grapes crop which yielded three crops in a year and five crops in two years. Being a perennial crop, it exhausts nutrients from the soil. Intensive cultivation of any crop with inorganic fertilizers depletes the nutrient status of soil in general and micronutrients in particular. The recommended dose of Farm Yard Manure, rich in natural plant nutrients were not applied by the farmers. Continuous depletion of nutrients without enrichment made the soil weak.

2. Loss of inherent fertility of the soil

Cumbam valley and the foothills of Sirumalai were known for their rich fertility status of the soil about 20 to 30 years ago. But now, because of the intensive cultivation, the fertility status has come down, to the extent it needed 17,723.4 m. tonnes of fertilizers in 2007-08 to replenish the soil. Even in 1992-93 the total fertilizer consumption was only 57 m. tonnes in the study area.

Excessive use of chemicals had degraded the soil by killing beneficial microorganisms and earthworms which are the biological indicators of soil health. Indiscriminate application of fertilizers and inorganic inputs by the grapes growers had degraded the inherent soil fertility. Grapes growers preferred chemical fertilizers, as they were easily available and easily transportable to the fields. Moreover the results were immediate and visible. They dumped the chemical fertilizers for gain, which inturn affected texture and the natural capacity of the soil to nourish the crops. The soil in the study area lost it's inherent fertility and soil breeding was done by grapes growers with imported soil from river bed.

3. Soil compaction

Soil compaction is a form of physical degradation (formation of a panlayer) resulting in distortion of the soil, where biological activity, porosity and permeability are reduced and soil structure partly desil'oyed. Compaction can reduce water infiltration capacity of the soil. Because of the application of herbicides, both hand weeding and mechanical weeding were neglected which resulted in compaction of soil. The high wage rate and non-availability of skilled labours forced the farmers to rely more on weedicides which saved their time and money. As the result gradually, the soil got compacted diminishing it's capacity to absorb water and nutrients. The farmers were more concerned about the present than the future.

4. Human activities and mismanagement

Intensive cultivation, indiscriminate, unscrupulous and excessive use of chemicals had turned the fertile land in to barren patches of land. In the study area, it was reported by the grapes growers that about 25 years ago they had used lesser quantity of chemical inputs, than the recommended doses. But now, they apply more than the recommended doses. Moreover there is a shift in the variety cultivated. Earlier they cultivated the variety panneer (100 %) without much competition for yield among the farmers. But, now as they have shifted to Thompson seedless variety (75 %) which is a highly export oriented variety, requires high amount of chemical inputs. Such indiscriminate and excessive use of chemical inputs had turned the fertile land to barren patches.

II. Water Related Issues

1. Depleted aquifers and lowered water table

Intensive agriculture necessitated heavy extraction of groundwater for irrigation which depleted the aquifers and lowered the water table. Being perennial crop it stands for 15 years in the field and requires continuous water supply. More over the area under grapes is also increased which led to heavy and continuous extraction of ground water by the grapes growers. Groundwater was extracted by means of deep bore well in the study area. It was told by the grapes growers that within a span of 20 years the ground water table level had gone down to 1000 feet from 30 feet indicating the heavy extraction of groundwater by the grapes growers.

III. Loss of Biodiversity

Bio-diversity refers to the variety of species that exist in an ecosystem. Greater the bio-diversity, healthier the ecosystem is. Pesticides kill not only the pests but also the beneficial organisms. Extension personnel and the input dealers stated that earlier less amount of known pesticides were sprayed with hand operated sprayers and the number of sprays was also less and sprayed as and when required. But now, many unknown pesticides have been sprayed with power operated sprayers as routine affair once in three days. Thus pesticide usage has increased in number and quantity over years. Insecticides kill not only the targeted pests but also the many beneficial natural predators. Indiscriminate use of insecticides for immediate knock down effect has led to the destruction of natural enemies. The grapes growers were not concerned about the long-term effect of these poisons they spray on the crop and soils. On enquiry, grapes growers reported that they were not able to see natural predators like spiders, lady bird beetle, grasshoppers, dragonfly, preying mantis and pollinators like honey bees in their field. The plant biodiversity was also very much affected. Earlier they cultivated food crops like cereals and millets, pulses, oilseeds, sugarcane, cotton and vegetables. Now they concentrate only on cash crops- the grapes. If they get very low yield from the crop due to low fertility of the soil, they cultivate the vegetable lablab for a year. Similarly, farm animal biodiversity is also affected in the study area. Earlier they had farm animals like cattle, buffaloes, sheep and goats in large number which gave a balanced farming environment. But, now they have only limited farm animals due to many reasons. This is an obvious evident for the loss of biodiversity in the study area.

B. Causes for the documented environmental issues

The actual practice will decide the extent of environmental issues, hence the extent of environmental issues was measured by studying the adoption of recommended grape growing technologies especially the agronomic practices and plant protection measures disseminated by the State department of Agriculture and Horticulture.

Table 1 : Distribution of respondents according to their practice-wise adoption of recommended grapes growing technologies

(n=150)

SI.No.	Items	Adopted	%	Non-adopted	%
I	**Agronomic practices**				
1.	Manures	3	2.00	147	98.00
2.	Fertilizers	5	3.30	145	96.70
3.	Time of application	150	100.00	-	-
4.	Irrigation	150	100.00	-	-
5.	Training	150	100.00	-	-
6.	Pruning	150	100.00	-	-
7.	Level of pruning	150	100.00	-	-
8.	Pruning season	150	100.00	-	-
9.	Special practices	150	100.00	-	-
10.	Growth regulator	150	100.00	-	-

II	**Plant protection measures**				
11.	Disease management	2	1.30	148	98.70
12.	Pest management	2	1.30	148	98.70
13.	Weed management	3	2.00	148	98.70

Agronomic practices

The grapes growers adopted all the agronomic practices like irrigation, training, pruning, level of pruning, pruning season, special practices as per recommendations. But, they deviated in application of manures and fertilizers and applied growth regulators along with chemicals.

1. Manures

Regarding the application of organic manure, an overwhelming majority (98.00%) of the respondents had not applied the recommended dose of manures. Organic manure is essential to maintain the fertility status of the soil. Grapes being a perennial crop, it removes the nutrients continuously from the soil. Hence, soil needs to be enriched with farm yard manure, compost and green manures. But organic manure was applied less than the recommended dose by the grapes growers which made the soil weak.

2. Fertilizers

With respect to fertilizer application, an overwhelming majority (96.70%) of the respondents had not applied the recommended dose of fertilizers. Fertilizers were applied more than the recommended doses. High dose of fertilizers not only destroy the flora and fauna, which maintains the soil fertility but also the texture and structure of the soil. As a result soil has lost it's inherent fertility. Further an enquiry was made with the respondents of the study area about the adoption of recommended dose of manures and fertilizers and the findings are presented in Table 1.

Table 2 : Application of manures and fertilizers by the respondents

Sl. No.	Manures and fertilizers	Recommended dose kg/vine	Adopted dose kg/vine	Deviation
1.	**Panneer**			
	Farm Yard Manure	50 kg/vine	15 kg/vine	- 35 kg/vine
2.	N	0.20 kg/vine	0.45 kg/vine	+ 0.25 kg/vine
3.	P	0.16 kg/vine	0.30 kg/vine	+ 0.14 kg/vine

4.		0.40 kg/vine	0.07 kg/vine	- 0.33 kg/vine
1.	**Thompson**			
	Farm Yard Manure	50 kg/vine	20 kg/vine	- 30 kg/vine
2.	N	0.30 kg/vine	0.56 kg/vine	+ 0.26 kg/vine
3.	P	0.16 kg/vine	0.36 kg/vine	+ 0.20 kg/vine
4.	K	0.80 kg/vine	0.24 kg/vine	- 0.56 kg/vine

From the Table 2 it is clear that the grape growers applied manures (FYM) in less quantity against the recommended dose and applied nitrogenous fertilizers and phosphatic fertilizers more than the recommended doses. Potash is applied in lesser quantity than the recommended dose.

During investigation it was known that the grapes growers were not able to meet out even the minimum requirement of FYM to fill the planting pit at the planting stage due to non-availability. When the farmers were not in a position to fill the pit with FYM during planting stage, which is needed only in a small quantity, then the question of applying FYM, during the crop stand which is required in a large quantities does not arise. They depended only on the inorganic fertilizers which are readily available in required quantity. As a result soil is worn out and weak without enrichment.

Nitrogenous fertilizers were applied in more than the required quantities to have thick foliage and to maintain the crop vigour, as three crops were harvested in a year. Thick foliage attracted more number of pests and diseases, which needed the administering of more number of pesticide sprays. Excess of nitrogenous fertilizers especially Ammonium Phosphate turns the soil acidic and excess nitrogenous fertilizers in soils are subjected to leaching. Consequently the nearby water bodies get polluted.

Overuse of fertilizers leads to imbalanced nutrition in the soil, resulting imbalanced nutrition for the crops. Besides, this the excess fertilizers not only reduce the microbial population in the soil, which diminish the soil fertility but also affect the physical nature of the soil like texture, structure, porosity etc., Excess of phosphatic fertilizers in soil have an antagonistic effect on other nutrients especially micronutrients, leading to low fertility status. Thus, extraction of all available nutrients will lead slowly to barren patches of land. Further, excess application of chemical fertilizers also destroy the beneficial microorganisms and earthworms in the soil.

Plant protection measures

1. Disease management

From Table 1 it is vivid that majority of the respondents (98.70%) had applied dosages higher than the recommended and only 1.30 per cent of the respondents applied as per recommendation and even lesser than recommendation. Downy mildew and powdery mildew were the most devastating diseases that attack inflorescence, early stage fruit and led to poor production of berries in terms of quality and quantity, ultimately affected the marketing price of grapes.

To control the diseases and to get more income they used all the new fungicides available in the market as per input dealers' recommendations. Moreover they applied fungicides based on their perception and experience, without considering the health of soil, water, animals and even human beings. Lack of awareness about healthy environment, lack of knowledge on the mode of action of various fuzgicides, pesticides and anxiety over the crop yield would have motivated them to use new fungicides. Only two respondents shifted to the organic manure panchakavia and used less (or) no fungicides in their field and obtained good yield. On enquiry it was found that they were self motivated to adopt the organic manure.

2. Pest management

Majority of the respondents (98.70%) had not adopted the recommended dose of insecticides to manage pests. Unscrupulous, indiscriminate and excessive use of insecticides were witnessed in the study area. Excessive application of insecticides led to pest resistance, resurgence and emergence of new pests. Moreover new pests emerge which make the situation even worst. It was reported by the grapes growers that, mite problem was severe in Theni district. They were unable to control the mites, inspite of their use of new insecticides.

Excessive and indiscriminate use of pesticides had not only killed the targeted pests but also the natural predators and pollinators that exist in the environment, resulting in loss of bio-diversity. And also excess pesticides polluted the soil and the nearby water bodies through leaching. Besides, these pesticides polluted the environment. The quality and yield of berries are very much affected by the type of pests and diseases which affected the crop. The occurrence of pests and diseases in the study area was very much high as they cultivated three crops in a year. Moreover, enquiry with the respondents revealed that only 45.00 per cent of them adopted the recommended spacing depending on the variety. Closer spacing with thick foliage (due to application of more of nitrogenous fertilizers) also contributed for frequent occurrence of pests and diseases. As a result they adopted plant protection measures voluntarily more than the required amount with their own knowledge and experience, and also in consultation with the input dealers. Hence, an attempt was made to know the amount of various pesticides applied by the respondents. The details are given in Table 3.

Table 3 : Application of pesticides by the respondents

SI. No.	Chemicals	Recommended dose	Adopted dose	Deviation
	Downy mildew			
1.	Mancozeb 75 WP, 35 % SC (Indofil M-45, Dithane M-45) NS	1.5-2.0 g/1 (24g/121it)	50 g in 121it	+26 g
2.	Cymoxanil + Mancozeb S%+64 WP (Curzate M-8) S+NS	2 g/ I (3 00 g /150 lit)	600 g in 150 lit	+ 300 g
	Powdery mildew			
3.	Flusilazole** 40 EC (Nustar) S	25 ml / 2001 (2.5 ml in 12 lit)	5 ml in 12 lit	+ 2.5 ml
	Anthracnose			
4.	Carbendazim 50 WP, 46.27 SC (Bavistin) S	1 g/L, I ml/1 (120 g in 120 lit)	250 g in 120 lit	+ 130 g
	Mealy bug			
5.	Methomyl 40 SP (Lannate) NS	1.0 g/I (12 g in 12 lit)	25 g in 12 lit	+ 13 g
6.	Dichlorvos 76 EC (Nuvan) NS	1.0 ml (12 ml in 12 lit)	15 ml in 12 lit	+ 3 ml

Over use of pesticides get in to the produce affecting it's keeping quality, taste and persist in the produce as residues affecting the health of human beings. These pesticides were sprayed neglecting the safety measures to be followed while handling and spraying, resulting in health hazards of the spray men and other farm labourers.

Conclusion

Environmental problems pose a serious threat to the world's natural resources and to all of us who depend on them for food and shelter. Agriculture is one of the major contributors of the environmental problems. The cropping pattern in India is changing fast, particularly in favour of export-oriented commercial crops. Thus intensive agricultural practices affect natural resource base (soil, water and air). Hence, it is need of the hour to educate the farmers about the environment, its role in survival of mankind and motivate them to follow eco-friendly agriculture by adopting recommended scientific practices in order to protect the diminishing natural resources.

References

Murugesan, A., Ramu .A and N. Kannan. 2005. Characterization of Ground Water Quality in Madurai Region. Indian Journal of Environmental Protection, 25 (10): 885: 892.

Pariyar, Chitramal. 2006. Agricultural and its impact on Ecological balance. Agricultural Today, 9(10) ; 48-49.

●●●

29

Sustainability of Traditional Institution in Tank Irrigation Management: An Example from Madurai District of Tamil Nadu, India

M. Jegadeesan[1] *and Koichi Fujita*[2]

[1] *Asst. Professor, Dept. of Home Science Extension, Home Science College & Res Inst. Madurai- 625104 and*

[2] *Professor, Center for Southeast Asian Studies, Kyoto, University, Kyoto, Japan- 606 8501*

Introduction

The prosperity of Tamil Nadu depends on the development of rural areas. As per the 2001 Census, Tamil Nadu's rural population was 36.2 million, amounting to 58% of the total population. Of these 90 percent are earning their livelihood through agriculture and allied activities. The agricultural situation in Tamil Nadu largely depends on the quantum of rainfall received during seasonal rainfall of South West and North East monsoon. North East monsoon which occurs around September to November is more crucial as South West monsoon which will give enormous rainfall to other Indian states is blocked by Western Ghats. So Tamil Nadu is receiving low rainfall and it also comes in three to five heavy showers during October and November. This limits the wet period (good condition for cultivation) to two to five weeks for the whole season. The wet period is further shortened by fast wind and soil type in some area. Thus, the success of the crop is largely determined by one's capacity to exploit the short wet period. Even though the Tamil Nadu is relatively better position in tapping available ground water resources (92 percent of potential has been tapped), 82 percent of the well is owned by medium and large farmers who is only 10 per cent of total cultivators. About 74.3 percent of marginal farmers and small farmers have very limited access to water. Hence they are fully depended on public funded water resources like tank. Thus, the tank irrigation system has been developed since ancient time to conserve available water for agricultural production and for all water related need for the villagers. The small earthen bunds are formed across the small streams and rivulets wherever feasible to form tanks in which water is collected and stored during monsoon period and let down for irrigating small command area controlled through sluices and distributaries. Mosse (2003) described that the tank irrigation system is one of the vast network of thousands of water bodies that constitute a distinctive landscape which is medieval in origin but still the basis of livelihood in the dry southern plains. It is true that tank irrigation significantly contributes to agricultural production in India in general and in particular Andhra Pradesh, Tamil Nadu, Karnataka and parts of West Bengal, Uttar Pradesh and Rajasthan as these states account for 63 percent of the total area under tank irrigation. Over the years more modern form of irrigation such as canal irrigation and energized well irrigation have pushed back the tanks from their place of prominence. While the large farmers supplemented their water need through energized well, the hardest hits are the poor and marginal farmers who depend on common property resources like tanks for their livelihood. During Tenth Plan Period (2002-2006), the state is aimed an annual growth rate of 4% and hoped for sustainable agricultural development through assured water supply, employment generation and poverty eradication. As failed in achieving most of the Tenth Plan goals in agriculture, the union planning commission is planning to achieve these goals in Eleventh Plan (2007-2011) (Swaminathan, 2006). Besides the government's policy intervention to make agricultural sector still economically viable, the present situation in the rural Tamil Nadu is not adorable. The increasing trend of fallow lands which was 2.3 million ha during 1990-91 has increased to 3 million ha in 2003-04 (Policy Note, 2006). The cropping intensity also reduced from the average level of 120 percent in the state to 112.4 percent. In Southern dry district the intensity of cropping is still low and hovering around less than 100 percent. This situation displaced million of farmers who were solely dependent on farming forced to think about alternatives and feared to lose their habitual livelihood opportunities. Apart from the farmers, the landless labourers and farm women still in high numbers have lost their employment. Studies showed that reduced cropped area coupled with farm mechanization have already reduced 50 mandays available to the rural poor. While opportunity for work in agriculture is seasonal and it is highly unstable due to various factors, the farmers and landless labourers are searching employment within 10 to 20 km radius from the villages. These changes have potential to create unprecedented eco systemic and socio- economic imbalances among rural society in Tamil Nadu.

Thus, the present study is designed to capture current scenario of agriculture in Madurai district of Tamil Nadu. Three villages namely Kadaneri, Koovalapuram and K.Meenachipuram in the southwest part of Madurai district were studied. This study is an attempt to identify how well the tank has the ability to sustain by itself since in the study villages the irrigation tank is considered to be single most important productive asset, the farmers who have little land under tank command and highly insufficient water and other input resources faced continuous crop failure and their hope in agriculture in future is bleak. Most of the farmers and labourers start migrating from villages as industrialization has been running at uneven but increasing pace during last 15 years. Our field survey villages directly or indirectly affected since it lies close to the Sivakasi — Rajapalayam industry belt which major center for match box, fireworks and cotton industry. Thus the prospect of agriculture is shrinking, the opportunities for off farm work tend to be limited in this area and households strategies that one person from family migrate out from the village and remaining member seek employment within and close to the village to earn additional income is increasing. Acknowledging this, the government of India and Tamil Nadu are implementing National Rural Employment Guarantee Programme (NREGP) which envisaged providing 100 days employment per year for one person from every willing household. Apart from this Govt. of Tamil Nadu also provides food and other groceries through Public Distribution System (PDS) with highly subsidized prices.

Methodology

Madurai district of Tamil Nadu has been purposively selected for this study as it is the most urbanized, less industrialized and resource poor dry district in the state. Average rainy days in the state are 56 days and the highest is 102 days in Nilgiris and the lowest is 48 days in Western Madurai district and South and Central Ramanathapuram. Madurai district is increasingly urbanizing and the conditions for practicing agriculture is becoming miserable and three village located western part of this district has been selected. Three villages has fully depended on tank for irrigation and located on same administrative unit (Taluk and Block) which facilitate us to collect secondary data relatively easier from the government offices. After indentifying the study villages using secondary data the farmers list has been prepared based on the land holding in the tank ayacut (command) area. The fifty percent of the farmers having land in tank ayacut is considered as sample and additionally to represent landless labour, 10 percent of landless labour from every village is included in the sample. Thus total sample has been constituted about 195 households (Kadaneri 85: Koovalapuram 50: Meenachipuram 60) . The data has been collected through pre-tested, semi structured interview schedule, paying personal visit to the villages. The data were collected through personal interview; focus group interaction and discussion with opinion leaders and village informal heads. This study has been conducted during Sep. 2007 to Feb. 2008.

Sustainability of Informal Tank Institution

Tank management in Tamil Nadu is confronting a series of challenges.-It is largely a result of economic development and industrialization, which offered better income earning opportunities to the rural poor. Since agriculture has become increasingly less lucrative, farmers have few incentives to stay in agriculture or to invest in agriculture-related activities such as maintaining common resources. It is a potential threat to the sustainability of tank irrigation system. With understanding of this background, we attempted to analyze the sustainability of traditional tank system with help of design principles offered by the Ostrom (1990: 90). The data pertaining to the sustainability of the traditional institution is presented in Table 1.

Table 1 : Evaluation of Sustainability of Informal Tank Institutions

Design principles proposed by Ostrom (1990) for sustainability of institution in CPR	**Characteristics of informal tank institution**	
	Past	**Present**
Clearly defined boundaries	Command area and individual land rights are clearly defined by the state	No change
Congruence between appropriation and provision of rules and local condition	A set of well developed informal rules were in force at both level	Farmers are aware about the rules, which are not followed fully as in the past

Collective choice arrangement	Widespread arrangement for collective action during different period of tank season	Still existed but intensity was reduced
Monitoring	A set of irrigational functionaries was in place to do monitor and guard	Still existed but intensity was reduced
Gradual sanction	An established sanction system was advocated	Still existed but intensity was reduced
Conflict resolution mechanism	Through negotiation	No change
The right of users to devise the own institution are not challenged by the external authorities	Frequently challenged	Challenged but less intensified since international agencies favor local informal institutions
Nested enterprise	Informally existed and mediated by the *Zamindar* or other community leaders	Possible through Water User Association

Source : Ostrom (1990:90) and Our Field Survey in June 2008.

Regarding the first principle, the tank institution has clearly defined boundaries and beneficiaries with identifiable land rights in the *ayacut* which is confirmed by the state through issuing *patta* (land right document) to individual farmers. Secondly, the informal institution had well developed rules for both appropriation and allocation of resources among the farmers. The reasons for the decline of implementation of informal rules were explained in detail in the previous section of this paper. Thirdly, the collective choice management is also widely practiced in the tank villages in various forms such as contribution of common labor for supply channel cleaning, repairing tank bund during breaching and contribution of money to the common fund. Fourthly, monitoring is practiced by employing professionals in the name of *Kaval or Kankani* or *Madai Kaval* until the recent past. The government takes revenues from the tank resources, but is not spending for its maintenance and improvement. A study of tank irrigation system in Tamil Nadu found that revenue from the tank may go to the forestry department, fisheries department, and village *panchayat* and even to the department of mines. None is directly paid to the agency which is responsible for tank maintenance (Palanisami and Meinzen-Dick, 1999). Hence the informal system faces difficulties to pay for the guards and other irrigation functionaries. Still the monitoring of the resources is practiced by the *Neerkatti* in some villages. Hence, when the government is ready to accept complete turnover of the system to the informal institution, monitoring will be more effectively carried out by the villagers. Fifthly, the sanction system existed in the villages is functioning reasonably, which was discussed by Table 5. Sixthly, conflict resolution is cleverly managed by the informal village leaders through established negotiation skills. The role of traditional institution in the conflict resolution is reported by many researchers in India (Ananthpur and Moore, 2007; Ananthapur 2007). Seventhly, the impact of external influence that challenges the sustainability of institution is considerable in the recent past. But now the intensity of external influence is diffused since international donor agencies favor the growth of local institutions.

Lastly, ability of institution to nested with other institution to reap higher benefit for the development of the resource base through institution. Ability to nested with another institution is largely depends on favorable environment existed in the outside of the purview of informal institutions. Earlier the villages in Tamil Nadu has arrangement such as eight villages together form one cluster and three such cluster form an unit called *"18 patti panchayat"* literally means 18 villages association. This type of arrangement has deteriorated after the abolishment of *Zamindar* rule. But even today we can see for example among the seven villages under study five villages were earlier ruled by the *Zamindar* in Peraiyur. Even if the *Zamindar* in Peraiyur has no power now, the villagers used to approach him whenever the conflict goes beyond their control. But presently, the government passed the law to form water user association in each tank and it will be nested with other government organization at block, *taluk* and district level and thereby reach state bureaucracy at last. Many problems cannot be solved at local levels or with just local resources. Most cases local community has made serious efforts to solve priority problems by themselves with available resources. Then, they are in stronger position to get assistance from higher levels which is nested with other institution for the problems that cannot be redressed locally (Uphoff, 1992). Some NGOs such as Dhan Foundation in the study area are also organizing Tank Farmers Associations (TFAs) called *Vayalagams* with the farmers. Tank Cascade Associations *(Cascade Vayalagams)* are based on how the tanks are linked and are formed with the relevant TFAs. Tank Farmers Federations *(Vayalagam Federations)* are then formed at the block level where all the Tank Farmers Associations and Tank Cascade Associations become members. The Federation is registered under the Societies Act. There exists a nested relationship among the various Tank Farmers Associations, Tank

Cascade Associations and Tank Farmers Federations. This institution protects and develops the tanks often seeking assistance from the state government (Shanmugham and Gurunathan, 2007). These arguments supported us to derive some important conclusion that the informal tank institution has strong and established norms and customs to govern the common resources in the village. These customs were deteriorated over the period of time due to antagonistic approach of state bureaucracy towards tank irrigation and it was accelerated by the recent service sector-oriented economic development. More importantly, the government claim authority over the tank resources and not considers the villager's association and their control over the resources. Hence, farmers begin to develop mindset that the government is responsible for tank maintenance and that the farmers need only asking for help. From the farmer's perspective, irrigation management has been shifted from domains of local community to the domains of the state (Lam, 2001). The local farmers keep themselves off from participation and it eroded their informal norms and rules significantly.

Farmers in the study area were aware about the deterioration of collective action and unraveling of informal tank institution. Part of the reason why the deterioration is rather conspicuous is that tank institution in this area had long tradition of effective operation; that deterioration occurred in the relatively shorter period of time and was sharp contrast with its past history. More importantly, given that water is scarce and intensive irrigation management practices is much needed for equal distribution of water, degeneration of informal institution can have significant effect on irrigation performance (ibid, 1583). However, our study provided some points to reconsider the conditions of informal institution. The gap between past and present institutions is not so large and the farmers' participation in tank management is diminished but not entirely disappeared.

The informal tank institution assessment provides some insights. It appears that informal tank institution rate highly on most design feature criteria yet all available evidence suggest that informal institutions are in decline (Ananthpur, 2007). In these circumstances, it is worth to notice Sengupta's (1985) argument over existence of informal tank institution in India. He argues that from time to time government raised concern over the conditions of tanks. PWD and other agency responsible for tank could not offer anything more than occasional repair. How could it be that the great majority of the tanks survived with those occasional repairs did by them? Besides, the government never took up other works such as water distribution. How was it arranged without much conflict? All these questions suggest that the informal tank irrigation is alive, though in a deteriorating state of health. It is important to consider that the earliest inscription dated back to 200 B.C. which describes about the construction of tanks by the community itself. Tanks which were numerously constructed during the third *Sangam* period (300 B.0 to 200 A.D) are still functioning and inscriptional evidence provides sufficient information of institution and their function (Gomathinayagam and Ratnavel, 2007: 183-196). The tanks which have passed on the test of time are only available here and prove they are survival of the fittest. The differences have seen in past and present is explained in terms of caste based *status quo* and uneven development of groundwater and opportunity developed in the non-farm sector. An important point to note here is that the informal tank institution functioned under a totally different socio-political set up dominated by feudal or semi-feudal production relation (Reddy, 1988). There is also evidence that elements of social and economic heterogeneity and mutual trust heavily influence the emergence and sustenance of institution (Kahkonen, 1999).

Hence, the importance of external or contextual factors to the performance of water institution cannot be overlooked (Ananthpur 2007). Often government programs such as tank modernization pushed ahead of frantic face after experiencing hard realities in tank irrigation is not provided the momentum to reestablish the institution for tank water management. The way to approach to this problems are perhaps most incisive but it contribute enormously to rural poor in terms of providing livelihood opportunities. It is here the opportunity for the state to rebuild frayed relationship between villager and government since both government and farmers in the tank command are realizing the need to strengthen the bond of co operation after experiencing successive droughts in this area in recent past. Since the local irrigational institution are well localized, exhibited their ability to manage tank water and built with all the traits needed for sustainability could possibly succeed in the better management of tank water resources provided the felt need for cooperation among the farmers induced constantly.

Caste discrimination and heterogeneous nature of users is said to be the major reason for dismantling of informal institutions. But some researchers found that heterogeneity is not strong predictor of collective activity; rather heterogeneity is the challenge that can be overcome by good institutional design when the interest of those controlling collective choice mechanism are benefitted by investing time and effort to craft better rules as argued by Varughese and Ostrom (2001: 747). It will give us some important implications such as villagers are able to recognize the importance of Neerkatti in the tank water management. But unfortunately the Neerkatti is losing motivation to work as Neerkatti. Income is not important criteria for their de-motivation but caste discrimination.

Conclusion

The ability of the traditional institution to govern the tank water sustainably is tested with design principle offered by the Ostrom. The traditional irrigation system has well defined boundaries, is highly visible and user of the resources has clear spatial boundaries, is relatively small in size and the farmers are residing close to the tanks. Importantly, the most of the farmers share the same perception of risk of tank failure and wish to cooperate and use the resources appropriately (Wade, 1988). It seems that it has all the traits needed for the sustainability and possibly succeed in the management of tank water resources when compare to externally enforced institution. We are happened to believe that in many ways, the traditional tank institution would offer best possible solution for improving water resource management sustainably.

References

Ananth Pur, K. (2007). Rivalry or synergy? Formal and informal local governance in rural India. Development and Change, 38(3): 401-421.

Ananth Pur, K., and M.M000re. (2007). Ambiguous institution: Traditional governance and local democracy in rural India. Working paper No. 282. Institute of Development Studies, London.

Gomathinayagam P. and Ratnavel (2007). Traditional Irrigation System, Dhan Foundation. Madurai

Kahkonen, S. (1999). Does social capital matter in water and sanitation delivery? Social capital working paper No.9. Washington DC: World Bank.

Lam, W.F. (2001). Coping with change: A study of local irrigation institutions in Taiwan. World Development, 29(9): 1569-1592.

Mosse, D. (2003). The rule of water: Statecraft, ecology and collective action in south India. Oxford University Press, New Delhi.

Ostrom, E. (1990). Governing the commons. Cambridge, UK: Harvard university press.

Palanisami and Ruth S.Meinzen Dick (1999). The endangered tanks of south India: Who will pay for the survival? Forth coming, *Economic and Political Weekly.*

Reddy,V.R (1998). Institutional imperatives and cooperatives strategies for large irrigation system in India. Indian Journal of Agricultural Economics, 53(3): 440-445.

Shanmugam, C.R., and A. Gurunathan (2007). Irrigation tanks and their traditional local management: A remarkable ancient history of India. Proceeding of International history seminar on irrigation and drainage, May 2-5, Tehran, Iran. Pp: 173-182.

Sengupta, N. (1985). Irrigation: Traditional vs Modern. Economic and Political Weekly, 20(45,46 and 47): 1919-1938.

Swaminathan, M.S. (2006). Agriculture cannot wait, The Hindu, 26 May 2006.

Uphoff, N. (1992). Local institution and participation for sustainable development, Gate Keeper series No.31, SIDA, International Institute for Environment and Development, London.

Varughese, G and Elinor Ostrom (2001).The contested role of heterogeneity in collective action: Some evidence from community forestry in Nepal. *World Development,* 29(5) : 747-765.

Wade, R. (1988). Village republics: Economic conditions for collective action in south India. Cambridge: Cambridge university press.

30

Training Needs of Wetland Farmers in Organic Farming in Tamil Nadu

Noorjehan A., K.A.Hanif[1] and I. Mohamed Iqbal[2]

[1] Assistant Professor (Agrl.Extn), Dept of Social Sciences & Languages (SSL),

ADAC & RI, Trichy-620 009 and [2] Professor (Late), Department of Agricultural Extension, TNAU, Tamil Nadu

Introduction

Organic agriculture is defined by Food and Agricultural Organisation / World Health Organisation Codex Alimentarius Commission as "a holistic production management system which promotes and enhances agro-ecosystem health including biodiversity, biological cycles and soil biological activity. It emphasizes the use of management practices in preference to the use of off farm inputs. This is accomplished by using, where possible agronomic, biological and mechanical methods as opposed to using synthetic materials, to fulfill any specific function within the system".

Organic Farming is growing or cultivating crops without using chemicals or chemical fertilizers and producing crops by using only naturally available materials like organic manures, green and green leaf manures, biocontrol agents (so called biopesticides) and biofertilizers etc. (Swaminathan, 2002).

The State Department of Agriculture, Government of Tamil Nadu has taken up the challenge to achieve higher growth rate in agriculture by implementing several development schemes and also propagation of relevant technologies to step up the production; Organic Farming is one of the schemes. Soil health improvement through Bio-fertilizers including Green Manuring, adoption of Integrated Nutrient Management (INM) and Integrated Pest Management (IPM) technologies are given priority by the Government of Tamil Nadu to fetch better return and value addition to agricultural produce in order to improve the economic status of the farming community.

Organic Farming practices reduces the pressure on land, water and biodiversity without adverse effects on Agricultural production and nutritive value of food; it comprises, the judicious use of organic manures viz., farmyard manure, compost, crop residues, vermi-compost, etc., integrated in an efficient nutrient management practice, cropping systems conjunctive use of rain, tank and underwater, integrated pest management and conservation of genetic resources. Among them, soil fertility is given top attention due to its dynamic action with various physical, chemical and biological properties (Halingali, 2003).

The reduction in soil fertility and the residual toxicity in some of the food crops had led to increased adoption of Organic Farming by the farmers to a considerable extent. The farmers had been practicing Organic Farming in different crops over years with reported advantages as perceived by them. However, the large-scale adoption of Organic Farming practices in various crops is still being debated at macro level. At this juncture, the documentation of all the ongoing research efforts and findings on Organic Farming together with the scientists' views' on the feasibility Organic Farming for large-scale adoption in different crops will throw much light on Organic Farming from researchers' point of view. Through training, a farmer / individual improve his skill, knowledge and attitude towards a practice. Keeping this in view, a study was conducted with one of the specific objective to analyze the training needs of farmers on organic farming.

Methodology

The study was conducted in Lalgudi block of Tiruchirappalli district by focusing on two major wetland crops viz., Paddy and Sugarcane. By random sampling, a sample size of 60 was chosen (30 each for Paddy and Sugarcane) from six villages namely, Poovalur, Sirumayamgudi and Mettupatti for Sugarcane crop and Keelaperungavur, Melaperungavur and Valadi for Paddy crop. An interview schedule was developed for this purpose and pilot tested with non-sample farmers. The responses were collected by using this pre-tested and well-structured interview schedule. Within the sample size, two case studies were also conducted (one in Paddy and another in Sugarcane crop). The responses were tabulated and the data were analyzed by using statistical tools such as simple per cent analysis and cumulative frequency and precise, meaningful inferences were made. The data were collected on areas such as subject matter preferred for training, method of training and the suggestions were also collected to improve the training programmes.

In the present study, Training is conceptualized as the process of aiding an individual to gain effectiveness in his present or future work through the development of appropriate habits of thought and action, skill, knowledge and attitude.

Findings and Discussion

The findings of the study on training needs of the Paddy and Sugarcane farmers in Organic Farming, the methods of training preferred by the farmers and the suggestions for effective conduction of training programmes were discussed below.

1. Training needs of Wetland farmers in Organic Farming

The wetland farmers inclusive of both paddy and sugarcane growers preferred the following subject matter areas as the first five ranked ones for giving training as these were perceived as important ones by more than two-third of the respondents. They were biopesticides (100.00 per cent), bio control agents (100.00 per cent), biofertilizers (96.67 per cent), coirpith compost (95.00 per cent), vermicompost (93.33 per cent) and Sugarcane trash compost (86.67 per cent) (Refer Table 1 and Fig.1). Enriched Farm yard manuring (76.67%) and Biogas slurry (73.33 %) were next important areas of subject where the farmers expressed their interest for conduction of training so as to gain adequate knowledge and adopt the improved organic farming practices.

The subject matter areas like Panchakavya and Matka (50.00%each) are the farmer innovative technologies for pest and disease management and manuring, henceforth the farmers would have less interest on its training. The other areas like Farm yard manuring, Pressmud, Bone meal/fish meal, Oilcakes, and Indigenous Technical Knowledge in Organic Farming showed lesser percentage to be considered as a training need as these would more familiar to them by heredity occupation.

The finding of Maheshwari (2000) which revealed that the vegetable growers had more need about the use of biocontrol agents, biopesticides, vermicompost, coirpith compost, biogas slurry, sugarcane trash compost, enriched farmyard manure and biofertilizers was in coincidence with the present study.

The findings of Chawang and Jha (2010) revealed that majority of the farmers had medium level of training needs. Plant protection measures, subject matter relating to loan and intercultural operation were the top most training needs of the farmers and the least training need was identified in the subject related to nursery raising. The variables age and cultivation experience had negative and significant relationship with the training needs.

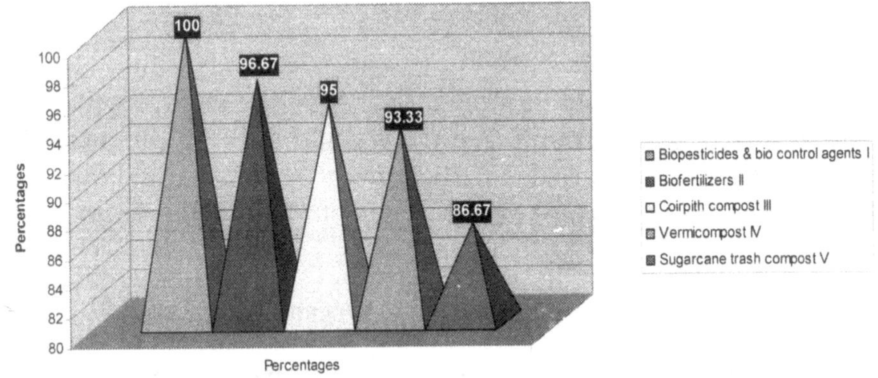

Fig.1.Training Needs of Farmers in Organic Farming

Table 1 : Distribution of respondents according to their Training Needs on Organic Farming

(n=60)

SI. No	Activities / Subject matter	Number	Percentage	Rank
1.	Farm yard manuring	25	41.67	
2.	Biogas slurry	44	73.33	VII
3.	Vermi compost	56	93.33	IV
4.	Sugarcane trash compost	52	86.67	V
5.	Coir pith compost	57	95.00	III
6.	Biofertilizers	58	96.67	II
7.	Pressmud	17	28.33	
8.	Bone meal / fishmeal	27	45.00	
9.	Oilcakes	23	38.33	
10.	Biopesticides	60	100.00	I
11.	Biocontrol agents	60	100.00	I
12.	Enriched Farm yard manuring	46	76.67	VI
13.	Panchakavya	30	50.00	
14.	Matka	30	50.00	
15.	Indigenous Technical Knowledge in Organic Farming	20	33.33	

2. Training methods preferred by the Paddy and Sugarcane farmers practicing Organic Farming

More than half of the respondents perceived first four ranked methods as the most preferred methods *viz.,* Demonstrations/Hands on experiences (100.00 per cent), Field visits (100.00 per cent), Study tours (100.00 per cent), Slides (100.00 per cent), Participatory trainings (78.33 per cent), Video lessons (75.00 per cent) and Exhibitions (60.00 per cent) for their training. (Refer Table 2). The reason would be that any training involving both lectures and practical demonstrations attract most of the farmers than mere lecture or discussion methods. Similar findings are reported by Raji (1991), Perumal (1994) and Maheshwari (2000).

Table 2 : Distribution of respondents according to their Identification of Training methods on Organic Farming Practices

(n=60)

SI. No.	Method of training	Number	Percentage	Rank
1.	Lectures	2	3.33	
2.	Discussions	27	45.00	
3.	Demonstrations/Hands on experiences	60	100.00	I
4.	Field visits	60	100.00	I
5.	Exhibitions	36	60.00	IV
6.	Study tours	60	100.00	I
7.	Video lessons	45	75.00	III
8.	Slides	60	100.00	I
9.	OHP	2	3.33	
10.	Participatory trainings	47	78.33	II

3. Suggestions offered by the Paddy and Sugarcane farmers to make the training programmes more effective

It could be enumerated from the Table 3 that the important suggestions offered by more than two-third of the respondents to make the training programmes more effective are viz., more hours be allotted to practical than mere theory, visit to successful farmers field practicing Organic Farming, sufficient printed materials be provided, need based trainings be arranged, venue of training be nearer, trainings be area / season bound, income oriented trainings be conducted, duration of training be short, interactive sessions be included and publicity about the conduct of the trainings be made.

The findings of the study undertaken by Chawang and Jha (2010) suggested that young farmers having less exposure in requisite training related to improved package of practices of paddy cultivation may be given preference for imparting training in the prioritized areas of training as identified.

Table 3 : Suggestions offered by the respondents to make the training programmes more effective

(n =60)

SI. No.	Method of training	Number	Percentage
1.	More hours be allotted to practical than mere theory	60	100.00
2.	Visit to successful farmers field practicing Organic Farming	58	96.67
3.	Venue of training be nearer	50	83.33
4.	Interactive sessions be included	45	75.00
5.	Sufficient printed materials be provided	57	95.00
6.	Need based trainings be arranged	56	93.33
7.	Trainings be area / season bound	50	83.33
8.	Duration of training be short	48	80.00
9.	Income oriented trainings be conducted	49	81.67
10.	Publicity about the conduct of the trainings be made	42	65.00
11.	Study tours be taken to other districts/state during the training period	35	58.33
12.	Off-season trainings be arranged	28	46.67
13.	Incentives be given during the training to encourage the participants	25	41.67

Conclusion

To ensure the latest methods being adopted by the farmers in Organic Farming, it is necessary that they must be trained in the improved methods of farming in the organic way. Training is essential to induce motivation, create confidence and inculcate efficiency in an individual. Training is also inevitable for imparting new knowledge and updating the skills of the farmers. Training of farmers had assumed further importance and urgency in the context of the high yielding varieties and improved practices in agriculture and allied fields. Thus training plays a very important role for human resource development. In order to make any training meaningful and effective, it is imperative on the part of the training organizers to identify the training needs of the farmers based on which a suitable training module can be developed so that the appropriate training is given to the right people, in the right form, at the right time so that higher degree of productivity and profitability can be achieved. The purpose of training programme is to keep the farmers abreast of the latest technologies in Farming and to hasten their adoption behaviour. Further to make the training more useful and purposeful they must be trained in such areas of farming wherein they really need training. The above study throws light on training needs on the subject matter areas, method of training and suggestions to improve the training programmes as preferred by the wetland farmers.

References

Chawang, J.K. and K.K. Jha. 2010. Training Needs of Paddy Cultivators in Nagaland, Indian Res. J. Ext. Edu . 10 (1), January, 2010.

Halingali, Jinnappa. 2003. Concept of Organic Farming. Kisan World, Aug.2003, p-38.

Maheshwari, M. 2000. Training Needs of Vegetable Growers in Organic Farming Unpub. M.Sc. (Ag.) Thesis, TNAU, AC & RI, Madurai.

Perumal, G .1994. Impact of TANWA Training on Participants. Unpub.M.Sc. (Ag.) Thesis, AC & RI, TNAU, Killikulam.

Raji, R.J. 1991. Impact of Training Programme on Adoption of Irrigation Management Practices in Paddy. Unpub. M.Sc. (Ag.) Thesis, AC & RI, TNAU, Madurai.

Swaminathan, Arul A. 2002. Need for Organic Farming. Kisan World, Oct.2002, p. 35-36.

●●●

31

Impact of Tsunami on the Farm Households of Coastal Tamilnadu State of India

M. Shantha Sheela[1]*, K. Palanisami*[2]*, Chieko Umetsu*[3] *and V. Ravichandran*[4]

Dept of Agricultural Extension & Rural Sociology, Tamil Nadu Agricultural University, Coimbatore

Introduction

On 26th December 2004, out of the 7516 km long coastline of India, more than 4500 km stretch was badly affected by the 9.0 magnitude earthquake-triggered tsunami, resulting in the total destruction of living environment along the coast. The worst affected areas along the Indian coast were in Tamil Nadu, Kerala, and Andhra Pradesh states. Tamil Nadu state suffered maximum loss with the damage concentrated in 4 districts.

It was reported that due to 26 December, 2004 Tsunami in Tamil Nadu state, 8.96 lakh people were affected, 376 villages had heavy damage, 7951 human lives lost, 1000 KM coastal length is affected, the sea water penetrated 1-1.5 KM distance into the main land, 128394 dwelling units affected, 9559 cattle lost, 10245 ha cropped area affected, 42655 boats damaged. (GOI, Ministry of Home Affairs, 09.01.2005). Many felt that impact will be very serious and it will take years to resume normal activities in the region. This paper presents an analytical study of the impact of tsunami on agricultural production, household income of the farm households on a continuous basis from 2005 to 2008.

Methodology

The study was conducted in Nagapattinam district of Tamil Nadu State, India. Two hundred and forty respondents from twenty four villages of coastal Nagapattinam district were selected. From 2004 onwards every year upto 2008 (consecutively 4 years) the same respondents were contacted to assess the impact of tsunami on agricultural production, household income of the farm households. Year 2004 represents the year of tsunami and the crop pattern during the period will represent before tsunami situation and the subsequent years will represent the after tsunami situation.

Findings and Discussion

Effect of Tsunami on Occupation

It could be observed from Table 1 that before Tsunami (2004), 77 per cent of the respondents involved in farming. Agriculture being the predominant occupation in villages, naturally agriculturists were in large number. After Tsunami it has been drastically reduced to 25 per cent, and it has increased to 37 percent during 2007.

Table 1 : Effect of Tsunami on Occupation of the Households

Occupation category	Before tsunami		After tsunami					
	2004		2005		2006		2007	
	No	%	No	%	No	%	No	%
Farming	184		58	24.17	61	25.42	89	37.08
Non farming	23		57	23.75	90	37.50	48	20.00
Farm labour	26		117	48.75	80	33.33	96	40.00
Fishing	5		5	2.08	5	2.08	5	2.08
Unemployed	0		1	0.42	1	0.42	0	0.00
Temple land	2		2	0.83	2	0.83	2	0.83
Total	240		240	100.00	240	100	240	100.00

The tsunami had left behind a thick (2 to 20 cm) layer of sea sediments as a slushy black layer over rice fields which affected rising of subsequent crops. Regarding crops, standing crops of rice (*Oryza indica*) and groundnut (*Arachis hypogea*), in different growth stages were dried up due to induced exosmosis (The passage of a fluid through a semipermeable membrane toward a solution of lower concentration, especially the passage of water through a cell membrane into the surrounding medium) inflicted severe damage. Even though farmers recovered from tsunami effects, flood occurred during October-November 2006 affected the standing crops. This may also one of the reasons for a slow recovery from the tsunami induced shock in the agriculture sector. Also, agriculture fields near to the coastal areas were affected by sea water intrusion made the field unfit for cultivation. Many have reported that to avoid taking risk in farming they were reluctant to take up farming as a sole occupation.

In the non-farm sector, 10 per cent of the sample households were involved in non farm activities such as small scale fish vending, shell collection and selling, fish net knitting and other similar activities before tsunami, and shift towards non-farm activities had increased after the tsunami, viz., 24, 38 and 20 per cent in 2005, 2006 and 2007 respectively. Later they slowly shifted to their original occupation of farming.

Before tsunami 11 per cent of the sample respondents were farm labourers. Subsequently, 50 per cent turned to be the farm labourers looking for relief from different agencies. Because of tsunami, many foreign agencies and Indian Government pumped money to the affected areas and to receive the relief packages, many discontinued the agriculture and defined as labourers. Hence, there is a close negative relationship between the number of households in farming and in labour categories.

During 2006, the percentage of farm labourers has reduced to 33 and in 2007 it has been increased to 40 per cent. In addition, few households got employment in the National Rural Employment Guarantee Scheme (NREGS) to sustain their livelihood requirements. This might be another reason for high percentage of respondents in this category. In fishing sector no change in the sample respondents' occupation. Similarly, in the case of unemployment and temple land cultivation, no difference was observed.

It could be revealed from Table 2 that only one member in each family involved in farming occupation. After tsunami single family member percentage was increased and reached to 91 per cent. In allied activities (livestock, poultry) and non-farm sector (shop), there is no change in percentage of respondents before and after tsunami.

In all the cases, one member from the family was involved in agriculture and allied activities. Only in the hired work category, more than two family members were involved indicating less importance given to agriculture and allied activities by the households.

Table 2 : Participation of Family Members in Different Occupation Categories

No. of family members	2004		2005		2006		2007	
	No	%	No	%	No	%	No	%
i) Farming								
1	105	78.36	5	55.56	63	80.77	132	91.03
2	24	17.91	3	33.33	14	17.95	12	8.28
3	4	2.99	1	11.11	1	1.28	1	0.69
4	0	0.00	0	0.00	0	0.00	0	0.00
5	1	0.75	0	0.00	0	0.00	0	0.00
Total	134	100.00	9	100.00	78	100	145	100
ii) Allied activities (livestock, Poultry)								
1	5	62.5	5	62.5	4	57.14	5	71.43
2	3	37.5	3	37.5	3	42.86	2	28.57
Total	8	100	8	100	7	100	7	100
iii) Non - farming (shop, etc)								
1	10	76.92	8	72.73	10	76.92	9	75
2	3	23.08	3	27.27	3	23.08	3	25
Total	13	100	11	100	13	100	12	100
iv) Hired work								
1	117	63.93	123	64.06	119	63.30	114	68.67
2	56	30.60	57	29.69	59	31.38	12	24.10
3	10	5.46	11	5.73	10	5.32	0	7.23
v) Old age pension scheme	1	0.00	1	0.52	0	0.00		0.00
Total	183	100	192	100	188	100	166	100

Effect of Tsunami on crop production

Total normal rainfall of the region is about 1341.7 mm. The North-east monsoon (October to December) contributes about 65% of the total annual rainfall. The South West monsoon (June to September) contributes about 20% of the total annual rainfall. The summer and winter rain accounts for the rest. Normally the cropping season coincides with the North-east monsoon season and if adequate water facility is available, farmers will raise the crop, otherwise the land will be kept fallow. The rainfall pattern shows that in 7 out of 13 Years, the Northeast monsoon was deficit and in 5 years, it was surplus thus indicating the climate vulnerability of the region.

Paddy is the main crop of the district and depending upon water availability and other factors, the farmers grow two crops viz., Kuruvai (April to July) and Thaladi (Aug to Nov) or Samba (Aug to Nov) crops. Other cereal crops like Cumbu, Ragi, Cholam, etc., account for a very small area only. Similarly, some pulses like Red gram, Green gram and Black gram are grown in small area (Statistical Handbook of Tamil Nadu, 2006).

Table 3 : Crop Production in summer season (Jan-May)

Category	2004		2005		2006		2007	
	No	%	No	%	No	%	No	%
Temple land	2	0.83	2	0.83	2	0.83	2	0.83
Not cultivating	229	95.42	229	95.42	236	98.33	235	97.92
Cultivating Cashew	1	0.42	1	0.42	0	0.00	1	0.42
Cultivating Coconut	3	1.25	3	1.25	1	0.42	2	0.83
Cultivating Mango	2	0.83	2	0.83	0	0.00	0	0.00
Cultivating Blackgram	0	0.00	2	0.83	0	0.00	0	0.00
Current Fallow	1	0.42	0	0.00	0	0.00	0	0.00
Leased out	2	0.83	1	0.42	1	0.42	0	0.00
Total	240	100.00	240	100.00	240	100.00	240	100.00

It could be inferred from Table 3, during summer season more than 95 per cent of the respondents had not cultivated annual crops such as paddy, cumbu, ragi, vegetables etc,. in their field in all the years. Only very meager percentage has grown perennial crops such as coconut, cashew and mango crops that exists in the summer season.

Table 4 : Crop Production in Season I (June- September)

Category	2004		2005		2006		2007	
	No	%	No	%	No	%	No	%
Temple land	2	0.83	2	0.83	2	0.83	2	0.83
Not cultivating	231	96.25	231	96.25	236	98.33	237	98.75
Cultivating Paddy	5	2.08	5	2.08	1	0.42	0	0.00
Cultivating Ragi	2	0.83	2	0.83	1	0.42	1	0.42
Total	240	100.00	240	100.00	240	100.00	240	100.00

Regarding Kharif (June- Sep) season crops, based on the availability of water only few farmers (2%) were able to cultivate paddy during 2004 and 2005 and this also reduced over years (Table 4).

Table 5 : Crop Production in Season II (October — January)

Category	**2004 (Oct 04— Jan 05)**		**2005 (Oct 05— Jan 06)**		**2006 (Oct 06 — Jan 07)**		**2007 (Oct 07- Jan 08)**	
	No	**%**	**No**	**%**	**No**	**%**	**No**	**%**
Temple land	2	0.83	2	0.83	2	0.83	2	0.83
Not cultivating	80	33.33	128	53.33	176	72.92	94	39.17
Cultivating Paddy	155	64.58	105	43.75	61	25.83	144	60
Cultivating Black gram	1	0.42	0	0	0	0	0	0
Cultivating Fodder Sorghum	1	0.42	1	0.42	0	0	0	0
Cultivating Cassuarina	0	0.00	1	0.42	0	0	0	0
Leased out	1	0.42	3	1.25	1	0.42	0	0
Total	240	100	240	100	240	100	240	100

However during Rabi (Oct-Jan) season, immediately after tsunami, 20 per cent reduction in paddy cultivation was observed. Drastic reduction in paddy cultivation during October 2006 to January 2007 was due to flood in November 2006 which washed away the standing crops (GoTN, 2006). Cyclonic storm brings havoc normally once in 3 or 4 years and heavy downpour during North-east monsoon leads to flooding of the district and damaged the field crops and wealth of soil. Hence, many farmers had reported that they could come back to normal cultivation during the Kharif season (October 2007 to January 2008) (Table 5).

Economics of cultivation

In order to see the economics of crop cultivation after tsunami, detailed cost of cultivation was worked out. The cost of cultivation has increased after tsunami due to the above agronomic practices even though the government also provided these inputs at subsidized prices. As indicated earlier, during tsunami year, the standing crop was totally devastated and the year after tsunami, about 70 per cent of the crop had failed due to poor soil quality.

Regarding the cost of cultivation, before tsunami 44 per cent of the paddy cultivating respondents had the expenditure upto Rs.7500/ha. The cost of cultivation of paddy has increased slowly from 2004 to 2007. Before tsunami, 27 per cent of the paddy cultivating respondents had a cost of cultivation of less than Rs.5000/ha and this percentage has reduced in the subsequent years. During 2004, about 28 percent of the farmers had a cost of cultivation of more than Rs.12000/ha, and it has gradually increased to 44 per cent during 2007 indicating the magnitude of cost increase in crop production Among the components of the cost of cultivation, fertilizer and manure accounted for more share, followed by seeds, machine power and human labour.

Table 6 : Cost of Cultivation of Paddy in Season II

(Rs/ha)

Category (Rs./ha)	2004 (Oct 04— Jan 05)		2005 (Oct 05— Jan 06)		2006 (Oct 06 — Jan 07)		2007 (Oct 07- Jan 08)	
	No	%	No	%	No	%	No	%
Up to 5000	42	27.10	10	9.52	0	0.00	17	11.81
5000-7500	27	17.42	27	25.71	5	8.20	24	16.67
7500-10000	25	16.13	23	21.90	18	29.51	22	15.28
10000-12500	18	11.61	13	12.38	7	11.48	18	12.50
> 12500	43	27.74	32	30.48	31	50.82	63	43.75
Total	155	100.00	105	100.00	61	100.00	144	100.00

Effect of Tsunami on Soil and Water

The average increase in soil EC was 4.58 dSm^{-1}, and the maximum increase was 13.3 dSm^{-1}. The average soil EC before the tsunami was 0.98 dSm^{-1}. Before tsunami, soil pH ranged from 6.9 to 8.6 with an average of 7.6. Afterwards, it ranged from 7.0 to 8.4 with an average of 8.0, and decreased to 7.7 in 2006. In 2007, average soil pH was 7.8, and ranged from 6.9 to 8.7. These results clearly showed the large scale increase in EC and the minor change in soil pH that resulted from tsunami.

The results of the soil and water analysis had shown that the salt deposited by seawater during the tsunami was rapidly leached out by rainfall, and the vegetation rapidly recovered. From these results, we conclude that the agricultural environment of the district recovered rapidly after the tsunami. (Kume et.al ., 2009)

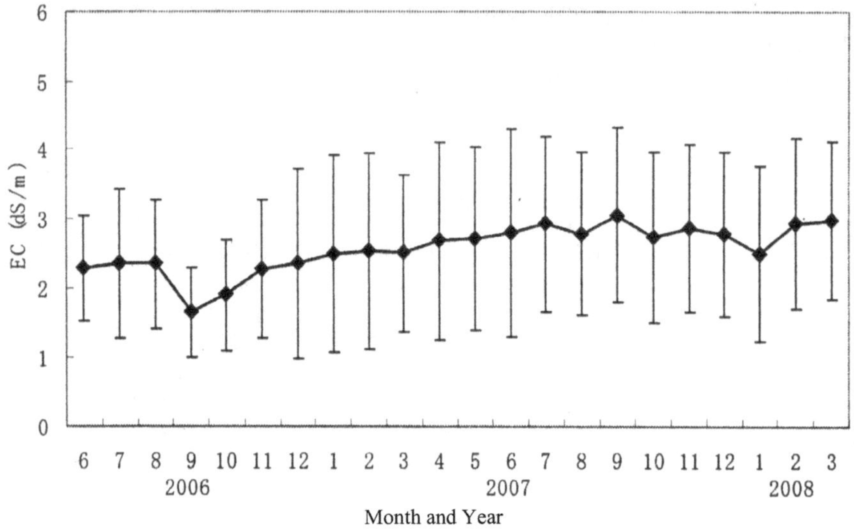

Fig. 1 : Monthly EC changes in Ground Water Level

Conclusion

Before tsunami 77 per cent of the respondents involved in farming. After Tsunami it has been drastically reduced to 25 per cent, and it has increased to 37 percent during 2007. Before tsunami almost 11 per cent of the sample respondents were farm labourers. Subsequently, almost 50 per cent turned to be the farm labourers looking for relief from different agencies. During 2006, the percentage of farm labourers has reduced to 33 and in 2007 it has been increased to 40 per cent. Regarding the cost of cultivation, before tsunami 44 per cent of the paddy cultivating respondents had the expenditure upto Rs.7500/ha. The cost of cultivation of paddy has increased slowly from 2004 to 2007. The effect of tsunami on soil and water results showed that the salt deposited by seawater during the tsunami was rapidly leached out by rainfall, and the vegetation quickly recovered.

References

Centre for Agriculture and Rural Development Studies (CARDS). 2007. Cost of Cultivation of Principal Crops. Unpublished reports. Tamil Nadu Agricultural University. Coimbatore.

Government of Tamil Nadu. 2006. Flood Damages in Coastal Districts of Tamil Nadu. Draft report. Agriculture Department. Chennai.

Kume, Takashi.. Chieko Umetsu, and K. Palanisami 2009. Impact of the December 2004 tsunami on soil, groundwater and vegetation in the Nagapattinam district, India . Unpublished paper. RIHN. Kyoto.

Statistical Hand book of Tamilnadu 2006 Directorate of Economics and Statistics. Chennai.

●●●

32

Contract Tree Farming in Industrial Wood Agro Forestry

H.Philip [1], K. T.Parthiban[2], Vennila[3] and P.Durairasu [4]

[1]Professor (Agri. Extn.), Directorate of Extension Education, TNAU, Coimbatore

[2]Associate Prof & Head,

[3]SRF, Dept. of Tree Breeding

[4]Dean (Forestry), FC&RI, Mettupalayam

Introduction

Indian forests are being denuded at an appalling rate of 1.5 million ha per year but only a tenth of this is being renewed. This has resulted in the accelerated dwindling of forest resources to a meager level of 23.55 per cent of the total geographical area against the mandated requirement of 33 per cent. This low forest cover coupled with poor productivity (0.5 - 0.7 m^3 ha-1 yr^{-1}) in comparative global statistics (2.1 m ^{3} ha-1 yr^{-1}) of Indian forest has ushered in a total mismatch between supply and demand of both domestic and industrial wood besides creating environmental de-stability and disequilibrium. The National Forest Policy of 1988 has resolved to phase out the supply of raw material to wood based industries and ultimately totally stopped the raw material supply from the forest (Anon., 1988). The policy also indicated that wood based industries have to become self reliance in meeting the raw material demand by establishing direct linkages with the farmers by providing lending facilitates and other input needs. This is going to pave the way for a strong industry-farmer nexus, with many farmers attracted towards tree husbandry due to assured marketability and income. As the activity gets intensified, location specific problems warranting solution are also found to be in the ascendancy.

Promotion of tree husbandry in the farmlands is a very big task faced by wood based industries in the state. The reason behind this is due to lack of superior genetic resources, poor adoption of precision and site specific silviculture techniques, imperfect linkage between industries and growers and lack of assured price information system. Hence it is essential to develop and promote site specific clones with higher productivity and amenable for short rotation under Agroforestry system Traditionally, trees are harvested manually which results in 20-30 per cent of logging waste during the process of extraction. After extraction and in the process of conversion also accumulation of huge industrial residues was evident. These residues neither used nor recycled in the field. The current study helped to address these issues by introducing mechanized harvesting to reduce the logging loss and facilitated value addition of plantation and industrial residues by converting them into briquettes. The current study aims to augment the supply chain through tripartite and quad-partite models and to develop entrepreneurship skill among the clusters of farmers with market driven approaches. Market and price information system is yet another bottleneck faced by the growers. Against this backdrop, the current study was planned to design and adapt a systematic value chain model to promote short rotation clones through contract farming and development of value addition technology for augmenting the Production to Consumption System (PCS) in pulp and paper industries.

Methodology

NAIP funded scheme on “A Value Chain on Industrial Agroforestry in Tamil Nadu” has taken a lead to conceptualize the contract tree farming model and facilitating linkages among various stake holders from growers to end users. The art of value addition to plantation and industrial residues and creating marketing linkages are the major agenda of the project objectives. Before promoting the concept, the existing pattern of forestry and agroforestry models coupled with their baseline status have been assessed through a pretested questionnaire and interview schedule. For this purpose, purposive sampling was carried among farmers, industries, traders etc. The quintessences of the base line survey are furnished.

Five districts viz Coimbatore, Erode, Karur, Namakkal and Vellore were purposively selected. The base line survey was conducted among 125 respondents distributed in 41 villages of the selected 5 districts of the Tamil Nadu state. For the selection of districts, taluks, villages and farmers, purposive sampling procedure was followed based on logistics.

The districts were selected based on logistics on development of industrial wood plantation coupled with its harvest and transport. In each district, two taluks representing as clusters were selected. Only the farmers who are willing to take up agro forestry were selected for the project as sample farmers and they represent different villages in the taluk. Since irrigation is essential for these tree species, a minimum of 1 acre to a maximum of 5 acre was identified for each farmers to cover a total of 20 ha per clusters. Using the pre-tested interview schedules, the baseline information at farmers, taluk and district level were collected. Interview was conducted among the randomly selected 125 farmers using well structured and pre tested interview schedule. The data collected were analysed and presented here.

Findings and Discussion

A. Distribution of Respondents based on Age

The distribution of respondents based on age ranged from young (less than 25 years) to old (more than 50 years). However, 50 per cent of the respondents were between the age group of 26 to 50 years who opt for industrial Agroforestry plantation. In total, around 92 per cent of the respondents are from the middle and old age category implying that farming experience is- the major driving force for adoption of farm forestry.

B. Literacy Status

In case of sampled villages, literacy level of 125 respondents indicated varying percentages with illiterate to collegiate. The respondents at village level recorded 25.6 per cent literacy at secondary level followed by primary (13.6%). The illiteracy at the sampled village was around 15 per cent. This shows that the level of literacy does not play a major role in farmers willingness to take a tree cultivation and reason for willingness may be the other factors viz., labourers scarcity for agricultural operations, income generation and marketability and prices for the produced commodities.

C. Working status

The working population status of sample villages in 5 districts was assessed through 125 respondents and the results are presented. The study indicated that 68.8 per cent of the respondents are involved in agriculture as the major occupation and 31.2 per cent in the other relied sector.

The percentage involvement of respondents in agricultural is more than twice the state average of employment as a agriculture labourers indicates that the reliance of the village population is more on agriculture. Hence the income generated through agricultural is the major source for the sustainability of the agricultural labour force for their livelihood.

D. Method of Irrigation

In case of irrigation system, 88 respondents out of 125 still follow conventional method of irrigation. However, the drip irrigation system has been slowly accepted by the respondents where in 8 respondents, in 72 — 90 per cent of the area used drip irrigation system followed by 14 respondents with 52 — 69 per cent of the area, 11 respondents with 32 — 49 per cent of the area and 4 respondents with 10 — 29 per cent on a multiple response mode (Table 1).

Table 1 : Categorization of Farmers based on per cent Area under Drip and Conventional Methods of Irrigation

SI.No.	Per cent area	Drip	Conventional
1.	90-100	0	88
2.	70-89	8	7
3.	50-69	14	13
4.	30-49	11	9
5.	10-29	4	8
6.	0	88	0

* Multiple Responses

The per cent area under irrigation showed that 68 out of 125 respondents had 100 per cent area under irrigation followed by 49 individuals with 50- 99 per cent area under irrigation and 61responents showed less than 50per cent area under irrigation on a multiple response basis (Table 2).

Table 2 : Classification of Farmers based on Per cent Irrigated Area

SI.No.	Irrigated Area (per cent)	No. of farmers
1.	100	68
2.	50-99	49
3.	<49	61

* Multiple responses

E. Farming Constraints

The farmers were asked to express their constraints in agriculture during survey. The findings are depicted in Table 20. The labour shortage (57.60%) is the major constraint experienced in the farming activities followed by irregular rainfall (33.60%), pest and disease outbreak (26.40%), increased fertilizer cost (20.80%), fertilizer shortage (16.80%) and non-availability of quality genetic stocks (4.00%) (Fig.4). The labour shortage and migration of rural labour are the major issues which increased the area under current fallows in the project operational districts.

Table 3 : Constraints Expressed by the Respondents

SI. No.	General Constraints	Respondents	
		No.	Percentage
1.	Labour shortage	72	57.60
2.	Irregular rainfall	42	33.60
3.	Pest and Diseases attack	33	26.40
4.	Price of fertilizer	26	20.80
5.	Fertilizer shortage	21	16.80
6.	No superior proper genetic resources available	5	4.00

* Multiple Responses

F. Family Size and Occupation

The type of family among the respondents is mostly of joined family nature with 63.2 per cent and 36.8 per cent of the respondents followed nuclear type.

G. Occupation

The status of identified respondents was recorded that more than 68 per cent were still practicing agriculture and only 3lper cent are involved in non-agricultural and business activities.

II. Farm power possession

This survey indicated that the respondent possessed various kinds of farm power viz., bullocks (57.60%), tractors (23.20%), pump sets — oil engines (54.40 %) and electric motors (84.80%) (Table 28). The maximum number of respondent possessed electric motors followed by bullocks and pump sets. However, 23.20 per cent of the farmers possessed tractors which indicate the richness of the respondents. The possession of pump sets-oil engines and electric motors by the respondents indicated that the selected respondents spread over 10 clusters of five districts have irrigation sources which are the basic requirement for the present project.

I. Farm size

The studies on farm size indicated the distribution of respondents in all types of Jand viz., dry, garden and wet lands. About 54 per cent of respondents had garden land followed by 48.8 per cent with dry land and 39.2 per cent with wet land. A maximum of 70 per cent of the respondents had dry land of less than 5 acres followed by 69 per cent with wet land. Out of 125 respondents, in multiple responses, 117 respondents have garden and wet land. This substantiates the table 18 on the irrigation facilities available to the respondents. (Table 4).

Table 4 : Distribution of Respondents based on Farm size (in acre)

SI. No.	General Constraints	Respondents*	
		No.	Percentage
Dry land (Acre)			
1.	Upto 2.5 to 5.0 acres	43	70.49
2.	Above 5.0 to 10.0 acres	18	29.51
Total		**61**	**48.80**
Garden land (Acre)			
3.	Upto 2.5 to 5.0 acres	31	45.59
4.	Above 5.0 to 10.0 acres	37	54.41
Total		**68**	**54.40**
Wet land (Acre)			
5.	Upto 2.5 to 5.0 acres	34	69.39
6.	Above 5.0 to 10.0 acres	15	30.61
Total		**49**	**39.20**

* Multiple Responses

J. Farm Income

The annual income of the respondents varied between Rs. 40,000/- and Rs. 80,000/-. In case of agriculture as a primary source of income around 64 per cent of the respondents had Rs. 80,000 and above as annual income and around 21 per cent have an annual income of below Rs.

40,000/-. In case of agriculture as a secondary employment, around 44 per cent of the respondents recorded low income of up to Rs. 40,000/- per annum. In case of non-agriculture as a primary source of income, nearly 49 per cent recorded higher income of above Rs. 80,000/- and around 18 per cent recorded low income. Around 40 per cent of respondents have an annual income of below Rs. 40,000/- followed by 37 per cent of the respondents with a high income of above of Rs. 80,000/- and 23 per cent with medium income (Table 5).

Table 5 : Distribution of Respondents based on Annual Income

SI. No.	General Constraints	Respondents	
		No.	Percentage
Agriculture-Primary			
1.	Low income (Upto Rs.40,000)	18	20.93
2.	Medium (Above Rs.40,000 to 80,000)	13	15.12
3.	High Income (Above Rs.80,000)	55	63.95
Agriculture–Secondary			
1	Low income (Upto Rs.40,000)	17	43.60
2.	Medium (Above Rs.40,000 to 80,000)	15	38.46
3.	High Income (Above Rs.80,000)	7	17.95
Non Agriculture–Primary			
1.	Low income (Upto Rs.40,000)	7	17.95
2.	Medium (Above Rs.40,000 to 80,000)	13	33.33
3.	High Income (Above Rs.80,000)	19	48.72
Non Agriculture –Secondary			
1.	Low income (Upto Rs.40,000)	34	39.53
2.	Medium (Above Rs.40,000 to 80,000)	20	23.26
3.	High Income (Above Rs.80,000)	32	37.21

However from the table it could be observed that the more number of respondents depend on agriculture as the primary source of income when compared to all other categories. Hence the income generated through the farm forestry project will contribute substantially for the development of livelihoods of these selected farmers.

K. Knowledge test

The knowledge test was conducted among 125 respondents using very well prepared knowledge instruments for 3 tree species viz, Casuarina,

Eucalypts (Pulp Wood) and Ailanthus (Match Wood). The result indicates that the knowledge on tree husbandry is ranged between 8 per cent and 15.2 per cent for Ailanthus, 8 per cent and 15.2 per cent for Eucalyptus and 8 per cent and 14.4 per cent for casuarina.

The result of knowledge test indicates that the knowledge level of farmers ranged between 8 and 15 per cent for both pulp and match wood species which needs to be improved through intervention by the current NAIP scheme to inculcate the knowledge on tree husbandry through capacity building, field visits and demonstration of precision pulp and match wood plantations in a decentralized manner across the various clusters (Table 6).

Table 6 : Knowledge Level of Farmers

SI. No.	General Constraints	Respondents	
		No.	Percentage
1.	0–39	125	100
2.	40–59	0	0
3.	60–100	0	0

Existing Supply Chain

Currently there exist two supply chain systems viz., Forest department supply and farmer's supply. The major raw material supply comes from the Tamil Nadu Forest Plantation Corporation (TAFCORN), which accounts for nearly 1.5 lakh tonnes of wood pulp supply (Parthiban *et al.*, 2010). The remaining raw material comes from the farm lands through the traders. The supply chain of farm forestry is depicted in figure 1.

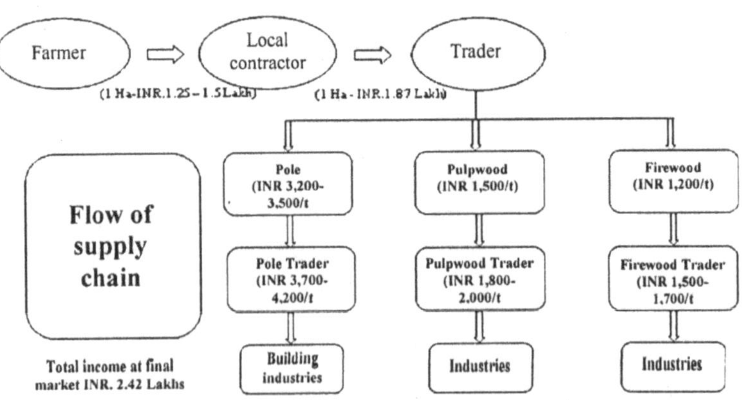

Fig 1 : Flow of Traditional Supply Chain

In the existing supply chain, growers sell the whole plantation on the acreage basis not taking into consideration the actual volume of the growing stock, which resulted in minimal returns to the growers. The local contractor's buy the plantations, fell them and classify them into poles, pulpwood and firewood. Only the wood size of 1 to 2 inches alone is supplied to the industries and other sizes are sold for pole purpose by the traders, which resulted in uncertain and erratic pulp recovery and quality (Parthiban *et al.,* 2009). This necessitated development of strong supply chain with a successful contract farming model.

Design and Dynamics of Value Chain Model

a) Design and Promotion of Supply (Value) Chain Model

To meet the above needs and increase the area under pulp wood plantation through farm forestry, Tamil Nadu Agricultural University has designed and developed a tripartite and quad-partite model for promotion of pulp wood based contract farming system in Tamil Nadu. Through the system, it is intended to produce quality and sustained raw material through a strong supply chain in a consortium mode.

i) Tri-partite Model

This model incorporates industry, growers and financial institutions. Under this system, the industry supplies quality planting material at subsidized rate and assures minimum support price of Rs. 2000 per tonnes or the prevailing market price which ever is higher. The financial institutions viz., Indian Bank, State Bank of India and Syndicate Bank provide credit facilities to the growers at the rate of Rs. 15000 to 20000 per acre in three installments. For credit facilities, a simple interest rate at 8.5 per cent is followed and the repayment starts after felling (Fig. 2). The contract farming system extends no collateral security for loan amount up to Rs.1, 00,000/- per farmer.

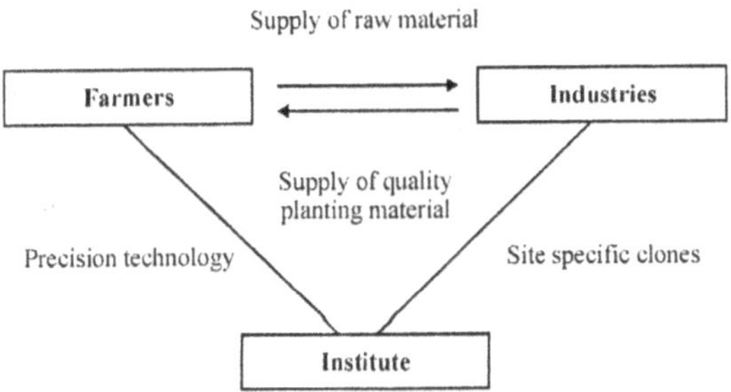

Fig. 2 : Tri-partite Model Contract Tree Farming

ii) Quad-partite Model

This system is similar to tri-partite model barring the involvement of financial institution (Fig. 3). In this system, the financial institutions provide credit facilities to the tree growing farmers which is unique in the country. A pre and post-plantation scientific advice helps to develop human resources through on and off institute mode to farmers and plantation staff of the industries.

Similarly, the industries mass multiply the potential genetic materials identified by the research institute in a decentralized manner and supply them at subsidized costs. The industry also facilitates felling and transport at their own costs, which resulted in strong linkage between industry and the farmers. The industry also help to repay the loan amount after felling of farm grown raw materials there by help the financial institutions for timely repayment, which resulted in strong institutional mechanism for sustainability of the contract tree farming system in the state.

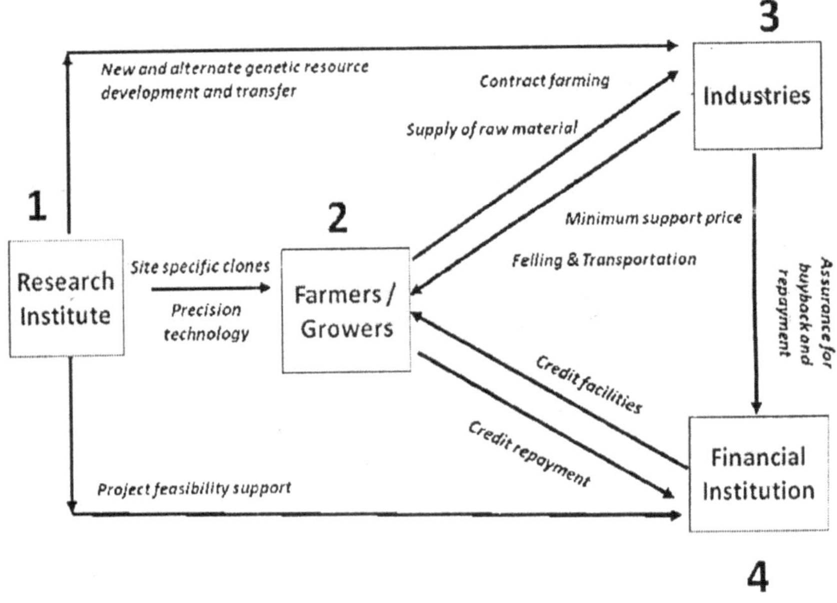

Fig. 3 : Quad-Partite Model

Impacts of Industrial Agroforestry Value Chain

a) Successful quad-partite model contract farming

A Quad-partite model contract tree farming has been designed and demonstrated in 10 clusters of 5 districts and covered an area of 200 ha which benefited 207 farmers. This successful demonstration has gained significant momentum both within and outside the project site which resulted in horizontal expansion of industrial agroforestry model to the tune of 11700 ha and benefited 4389 farmers distributed in 23 districts of the State of Tamil Nadu (Fig. 4). These promotional activities indicated successful adoption of Industrial agroforestry by the farmers of the State.

b) Income and employment generation

It is estimated that 300 man days are required for establishment of one ha of plantation from seedling up to harvesting and transportation. These flow of operations created 60,000 man days (0.13 million US D) directly in the project operational area and around 3.0 million man days (6.5 million USD) through horizontal expansion and created adequate employment and income generation to the landless labourers, small farmers and self-help groups, particularly women members.

The project also created enough scope and established farmer and industrial linked decentralized clonal propagation centres which created 0.1 million man days of work to land less women members. These activities have created an income generation of nearly ten million rupees (0.2 million USD) and the benefit was distributed to various stake holders viz., farmers landless labourers and women self-help groups.

c) Value addition of plantation residues

The plantation residues of Casuarina and the industrial residues of matchwood species promoted through this project have been successfully value added in to briquetting technology and new supply chain has been created by linking growers with processing and value addition industries. This value addition of plantation residues into briquettes resulted in a net profit of Rs.1750/tonne of briquette produced.

d) Carbon sequestration and climate mitigation

This current project has excellence scope of mitigating the climate change through carbon sequestration. It is estimated that 40 tonnes of carbon is sequestered/ha of plantation. The current plantation established through industrial agroforestry have the potential of sequestering 0.5 million tonnes of carbon. This carbon sequestration potential may help to augment the clean development mechanism.

Sustainability Plan

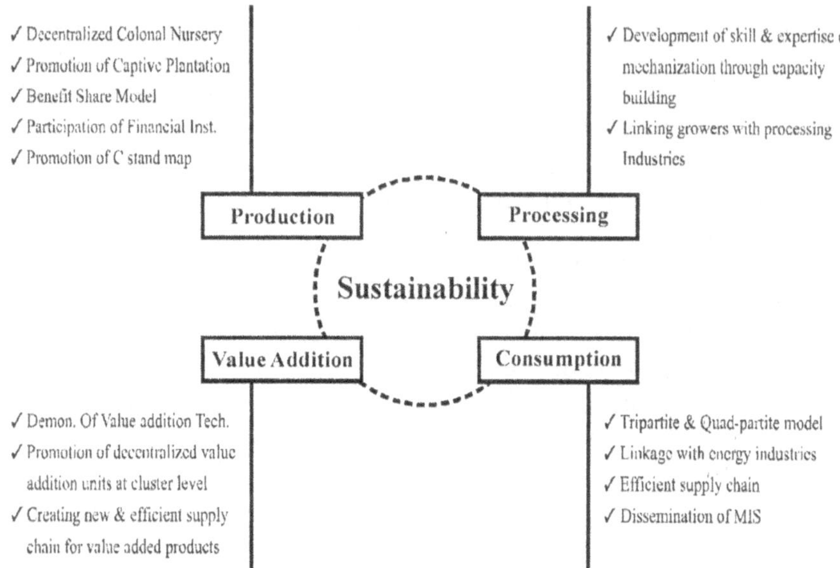

Fig. 4 : Sustainability Plan

Conclusion

The current value chain on Industrial Agroforestry model has been designed in a Quad partite consortium model and implemented in 23 districts of the State of Tamil Nadu by deploying short rotation and high yielding pulp wood clones. The demonstration of HYSC clones 200 ha of 5 districts of the State has made significant impact which resulted in horizontal expansion of more than 11700 ha involving 4389 farmer beneficiaries. This promotional activity was carried out through contract tree farming model which incorporated farm forestry, agroforestry, land lease and benefit share model. The farmers gained a potential gross income of INR 3.00 lakhs per ha in 3 years against less than INR 2.00 lakhs per ha. This besides value addition of plantation residues in the form of briquettes has given additional net revenue of Rs. 17.50 per tonne of briquette produced and extended enough scope for promotion of rural industries across the state of Tamil Nadu. In a holistic perspective, this value chain model has been recognized as a business model by the growers, industries and the other stake holders due to its impact on technological advancements, augmenting the supply chain, socioeconomic improvement and finally its impact on the environmental as a whole thereby this model has an excellent scope of adoption by other agro based industries.

References

Anon. 1988. National Forest Policy, MoEF, New Delhi.

Kulkarni, H.D.2008. Private farmer and private industry partnerships for industrial wood production: A case study. *International Forestry Review* 10: 147-155.

Parthiban, K.T, R. Seenivasan and M.Govinda Rao, 2010. Contract Tree Farming - A successful Industrial farm Forestry approach. Indian Forester. Vol. 136 (2) : 187-197.

Parthiban, K.T, S.Vennila, R. Seenivasan, G.P.Sararvanan and M.Govinda Rao, 2009. Industrial Agroforestry through Contract Farming. In: Agroforestry: Natural Resource sustainability, Livelihood and Climate moderation (eds. O.P.Chaturvedi et al., 2009). Satish Serial Publishing House, New Delhi pp:277-288.

Puri, S. and Nair, P.K.R. 2004. Agroforestry research for development in India. 25 years of experiences of a national program. *Agroforestry Systems* 61: 437-452.

Srivastava, M.B. 2005. Timber industries and non-timber forest products. CBS Publication, New Delhi. p. 518.

●●●

33

Rationality Analysis of Indigenous Tribal Agricultural Practices followed in Kolli Hills

P. Venkatesan[1] and M. Sundaramari[2]
1 Subject Matter Specialist (Agri. Extn), KVK, Gandhigram Rural Institute, Gandhigram, Dindigul
2 Professor (Agri. Extn), Faculty of Agriculture and Animal Husbandry, Gandhigram Rural Institute, Gandhigram, Dindigul

Introduction

The people in Kolli Hills were more traditional in nature believing the fith and practices of the local communities. They happened to manage their livelihood through agriculture and maintained an indigenous life with their own knowledge system. They used to maintain the longstanding traditions from their ancestors and spread the knowledge in different spheres of livelihood. Such socially generated knowledge is popularly called as local knowledge. The ongoing practice of using such knowledge for indigenous communities established the belief that such knowledge used in atraditional manner was fruitful for the people. The concept of indigenous knowledge gained its worldwide recognition through the United Nations Conference on Environment and Education in 1992, World Conservation Strategy of International Union and Conservations of Natural Resources in 1980, Brundtland Commission, and World Commission on Environment and Development, 1987. These events recognized the existence of indigenous knowledge in every country, society, culture. Since India has a long history and much enriched culture there is abundant reservoir of indigenous knowledge in every part of the country. Similarly Kolli Hills of Namakkal district in Tamil Nadu, is a historical land having enriched cultural heritage which has varied communities and immense resources. Its flora and fauna are vast and varied in nature. The tribal communities, of this Kolli Hills, "Malayali" appear to live on their own knowledge system. The paper is intended to unfurl the essence of Indigenous Tribal Agricultural Practices (ITAPs) in Kolli Hills and the rationality analysis ITAPs in the study area. According to Grenier (1998) indigenous knowledge is the traditional knowledge of the local community existing within and developed around the specific conditions of women and men indigenous to a particular geographical area.

Wang (1998) defined the ingidenous knowledge as the sum total of knowledge and practices that are based on people's accumulated experience in dealing with situations and problems in various aspects of life and such knowledge and practices are special to a particular culture. Indigenous Agricultural Practices were also defined as those practices developed and possessed by the farming community in a given geographical area and/or those practices which draw on the local inputs. Local ideas and internal solutions (Sundaramari, 2001). Further, incur low or no cost and do not require complex learning process on the part of farmers.

However, these indigenous practices are endanged ones and there is a possibility for them to become extinct particularly during this era of globalization, liberalization and commercialisation.

Hence, it is high time to collect, document these practices and to study their rationality so that the rational practices can be further experimented and recommended for better adoption by the farmers and the irrational ones may be discarded. Keeping these in view, this study was undertaken with the following objectives:

1. To collect and document Indigenous Tribal Agricultural Practices existing in the hilly tribal areas.
2. To study the rationality of selected Indigenous Tribal Agricultural Practices.

Methodlogy

Biodiversity and indigenous knowledge are interrelated phenomena (Warren 1993). Once the diversity of floral and faunal resources disappears, the knowledge associated with these resources also disappears. With this perspective in view, a study was conducted at the Kolli Hills in South India.

Collection and Selection of Indigenous Tribal Agricultural Practices

Seven cluster villages Ariyur Nadu, Bail Nadu, Gudini Nadu, Gundur Nadu, Selur Nadu, Thinnanur Nadu and Valappur Nadu were selected based on the geographical area covered under the respective farming systems. In each selected villages, 20 aged and experienced farmers were conducted through informal interview method for collecting indigenous practices associated with agricultural crops cultivated there. Thus, a total of 140 farmers were contacted. Indigenous Tribal Agricultural Practices were also collected from secondary source to a limited extent, apart from the above mentioned farmers. Thus, a total of 670 Indigenous Tribal Agricultural Practices were collected.

The collected Indigenous Tribal Agricultural Practices were then classified systematically into 27 crops containing 527 practices and into 13 general agriculture sub heading containing 143 practices.

Rationality Study of selected ITAPs

In this study, "Rationality" refers to the degree to which Indigenous Tribal Agricultural Practices can be explained or supported with scientific reasons or established based on long time experience. Likewise, "Irrationality" refers to the degree to which Indigenous. Tribal Agricultural Practices cannot be supported with scientific background or cannot be established based on experience.

For testing, the rationality, the selected 250 ITAPs were classified into three groups namely production aspect of field crops (125), production aspects of horticultural crops (57) and protection aspects of all the crops (68). Three questionnaires containing the three different list of ITAPs, as mentioned above, were prepared and presented to the scientists viz., Agronomists (43), Horticulturists (38) and Plant protection specialist (41) respectively for judging their rationality. Then the following weightages were assigned to the responses of scientists by adopting the procedure as enunciated by Rambabu (1997).

Findings and Discussion

1. Crop wise rationality of ITAPs

It could be constructed from the Table 1 that 84.80 per cent of the total of 250 ITAPs were judged as rational while 15.20 per cent were rated as irrational by the scientists. It is quite interesting to note that all the ITAPs of Red gram (100%) were rated as rational. Rational ITAPs were found to be more than 90 per cent in the case of Finger millets, Pineapple and Banana, and 80 per cent and more in rspect Little millet, Italian millet, Low land Paddy, Field beans and general agriculture and it was relatively low in Tapioca (76.47%).

Table 1 : Crop wise rationality of ITAPs

SI. No.	Crops	Rational		Irrational		Total	
		No	%	No	%	No	%
1.	Little millet (n = 84)	19	82.60	4	17.40	23	100.00
2.	Italian millet (n = 84)	13	81.25	3	18.75	16	100.00
3.	Finger millet (n = 84)	15	93.75	1	6.25	16	100.00
4.	Low land Paddy (n = 84)	53	80.30	13	19.70	66	100.00
5.	Red gram (n = 84)	14	100.00	0	0.00	14	100.00
6.	Field beans (n = 79)	8	80.00	2	20.00	10	100.00
7.	Tapioca (n = 79)	13	76.47	4	23.53	17	100.00
8.	Pineapple (n = 38)	11	91.67	1	8.33	12	100.00
9.	Banana (n = 79)	24	92.30	2	7.70	26	100.00
10	General Agriculture (n = 84)	42	84.00	8	16.00	50	100.00
		212	84.80	38	15.20	250	100.00

Most of the technologies recommended by the extension agencies in respect of Red gram, have been based on the indigenous technologies reported in this study and hence all the ITAPs in this crop were found rational. Crop such as Little millet, Italian millet, Finger millet, Low land Paddy, Field beans, pineapple and Banana are being cultivated over centuries and most of the ITAPs on these crops would have been tested and standardized over generations. Hence, more than 80 per cent of ITAPs have been found rational.

ITAPs on general agriculture comprises of Season, Traditional equipments, Preparatory cultivation, Seeds and sowing, Soil and water conservation, Weed management, Manuring, Live fencing, Pest and disease management, Rat control, Bird scaring, Storage pest control etc. which are common to all the areas and all the crops in hills and hence most of them would have been test verified. Thus, 84 per cent of the ITAPs in this aspect were rated as rational. The above findings are in conformity with those of Natarajan et.al. (2006), Ranjay K. Singh (2007) and Sakeer Husain et.al. (2011).

2. Unit wise rationality of ITAPs on general agriculture

It could be observed from the Table 2 that a large majority of total ITAPs related to general agriculture (84%) were found rational. Among the individual units all the ITAPs related to Traditional equipments, Preparatory cultivation, Soil & water conservation, Weed management, Manuring, Live fencing and Storage pest control (100%) were crated as rational. The rational ITAPs were 75 per cent in Seasonal aspect followed by 66.67 per cent in the case of Pest and disease management. Only half of ITAPs (50%) were rational related to Seeds & Sowing and Rat control. The rational of ITAPs were found to be low in the unit wise aspect of Bird scaring (33.33%).

Table 2 : Unit wise rationality of ITAPs on general agriculture

(n = 84)

SI. No.	Crops	Rational		Irrational		Total	
		No	%	No	%	No	%
1.	Season	3	75.00	1	25.00	4	100.00
2.	Traditional equipments	5	100.00	0	0.00	5	100.00
3.	Preparatory cultivation	5	100.00	0	0.00	5	100.00
4.	Seeds and sowing	1	50.00	1	50.00	2	100.00
5.	Soil and water conservation	9	100.00	0	0.00	9	100.00
6.	Weed management	2	100.00	0	0.00	2	100.00
7.	Manuring	3	100.00	0	0.00	3	100.00
8.	Live fencing	3	100.00	0	0.00	3	100.00
9.	Pest and disease management	4	66.67	2	33.33	6	100.00
10	Rat control	2	50.00	2	50.00	4	100.00
11.	Bird scaring	1	33.33	2	66.67	3	100.00
12.	Storage pest control	4	100.00	0	0.00	4	100.00
		42	100.00	8	16.00	50	100.00

Traditional equipments, Preparatory cultivation, Soil and water conservation, Weed management, Manuring, Live fencing and Storage pest control were the most important components of traditional agriculture and this would have been the reason for the cent per cent rational ITAPs in these units. Since the ITAPs on Season aspects have been followed by the farmers over the millennium, most of the ITAPs related to this unit would have been experimented several times and got standardized and hence more rational ITAPs. Though the ITAPs related to Pest and disease management are also followed for a long time, nearly 66.67 per cent of them alone have been rated as rational, because in many cases, different combinations of plant species have been used and the insecticidal properties of some of the species have not yet been verified scientifically.

Similarly for ITAPs applicable to Seeds and, sowing, Rat control Bird scaring, the scientific basis for many of them has not yet been proved through experiments and hence many were observed irrational. These results derive support from the study of Purusottam et. al. (2009), Balasubramanian (2009) and Sakeer Husain et. al. (2011).

3. Rationality of ITAPs on crop production

Out of the 250 ITAPs selected for further analysis, 183 were applicable to crop production aspects and their rationality has been presented in the Table 3 which reveals that nearly 90 per cent of the total ITAPs were found to be rational while only 10 per cent were rated as irrational. It is quite interesting to note that all the ITAPs of Red gram (100%) were rated as rational. Rational ITAPs were found to be more than 90 per cent in the case of Finger millets, Low land Paddy, Pineapple, Banana and general agriculture and nearly 80 per cent and more in respect of Little millet, Italian millet, Tapioca and it was relatively low in Field beans (75%).

Table 3 : Rationality of ITAPs on crop production

SI. No.	Crops	Rational		Irrational		Total	
		No	%	No	%	No	%
1.	Little millet (n = 43)	15	78.95	4	21.05	19	100.00
2.	Italian millet (n = 43)	12	80.00	3	20.00	15	100.00
3.	Finger millet (n = 43)	14	93.33	1	6.67	15	100.00
4.	Low land Paddy (n = 43)	34	94.44	2	5.56	36	100.00
5.	Red gram (n = 43)	7	100.00	0	0.00	7	100.00
6.	Field beans (n = 38)	5	71.43	2	28.57	7	100.00
7.	Tapioca (n = 38)	13	81.25	3	18.75	16	100.00
8.	Pineapple (n = 38)	11	91.67	1	8.33	12	100.00
9.	Banana (n = 38)	21	95.45	1	4.55	22	100.00
10	General Agriculture (n = 43)	31	93.45	2	6.06	33	100.00
		163	89.56	19	10.44	182	100.00

Crop production aspects mostly relate to Season, Traditional equipments, Preparatory cultivation, Seeds and sowing, Soil and water conservation, Weed management, Manuring and Live fencing, which are common to all the areas and all the crops in hills and hence most of them would have been test verified. Thus, 89.62 per cent of the ITAPs in crop production aspects were rated as rational. The above findings are in conformity with those of Natarajan et. al. (2006), Ranjay K. Singh (2007), Balasubramanian (2009) and Sakeer Husain et. al ., (2010).

4. Rationality of ITAPs on crop protection

Rationality of remaining 68 ITAPs applicable to crop protection was worked out and the same has been displayed in the Table 4.

About 72.06 per cent of the ITAPs were reported as rational. Of the individual crops Little millet, Italian millet, Finger millet, Red gram, Field beans have got all their ITAPs as rational. In case of Hill banana about three fourths of the ITAPs were rated as rational and was only 64.70 per cent in general agriculture and 63.33 per cent in Low land Paddy as rational.

Table 4 : Rationality of ITAPs on crop protection

(n = 41)

SI. No.	Crops	Rational		Irrational		Total	
		No	%	No	%	No	%
1.	Little millet	4	100.00	0	0.00	4	100.00
2.	Italian millet	1	100.00	0	0.00	1	100.00
3.	Finger millet	1	100.00	0	0.00	1	100.00
4.	Low land Paddy	19	63.33	11	36.67	30	100.00
5.	Red gram	7	100.00	0	0.00	7	100.00
6.	Field beans	3	100.00	0	0.00	3	100.00
7.	Tapioca	0	0.00	1	100.00	1	100.00
8.	Banana	3	75.00	1	25.00	4	100.00
9.	General Agriculture	11	64.70	6	35.30	17	100.00
		49	72.06	19	27.94	68	100.00

Since there was less number of ITAPs in the case of Little millet, Italian millet, Finger millet, Red gram and Field beans, they would have recorded cent per cent rationality. Through continuous trial and error the tribal farmers have preferred the ITAPs on rainfed cultivation of millets. These results derive support from the study of Narayasami (2006), Purusottam et al. (2009), Balasubramanian (2009) and Sakeer Husain et. al. (2011).

Conclusion

As more than four firths (84.80%) of the total ITAPs analysed in the study were reported to be rational, we cannot dispel the fact that the indigenous practices developed and tested by our ancestors over centuries are really an intellectual wealthy of our country. Hence, if these rational ITAPs are further validated, fine tuned and integrated with the present modern technologies and/or adopted in lieu of them; India will be marching towards the Farmers and Eco friendly, Real Second green Revolution.

References

Balasubramanian, P., 2009. Indigenous practices in dryland. Agropedia, posted in dryland indigenous practices portal.

Narayanasamy, P., 2006. Traditional knwoeldge of tribals in crop protection, Indian Journal of Traditional Knowledge, Vol. 5 (1): 64-70.

Natarajan, M., and Santha Govind, 2006. Indigenous agricultural practices among tribal women, Indian Journal of Traditional Knowledge, Voo. 5(1): 118-121.

Purusottam Mohapatra, N. Ponnurasu and P. Narayanasamy, 2009. Tribal pest control practices of Tamil Nadu for sustainable agriculture, Indian Journal of Traditional Knowledge, Vol. 8 (2): 218-224.

Ranjay K. Singh, 2007. Indigenous Agricultural Knowledge in Rainfed Rice based farming systems for sustainabvle agriculture: learning from Indian farmers, Indigenous Knowledge system and sustainable development, Tribes and tribals, special volume 1:101-110.

Sakeer Husain and M. Sundaramari, Scientific rationality and perceived effectiveness of indigenous technical knowledge on coconut cultivation in Kerala, Journal of Tropical Agriculture 49 (1-2): 78-87, 2011.

●●●

Section IV

Gender Issues

34

Rural Farm Women Knowledge Gain and Perception of Interactive Multimedia Compact Disc (IMCD)

N. Anandaraja[1], *T. Rathakrishnan* [2] *and M. Ramasubramanian* [3]

[1] *Asistant Profesor, e-Extension Centre, Directorate of Education,*[2] *Director, Students Welfare, Tamil Nadu Agricultural University, Coimbatore-641 003 and*

[3] *Assistant Professor, Home Science College & Research Institute, Madurai*

Introduction

Very few countries have experienced rapid economic growth without agricultural growth either preceding or accompanying it. Agricultural growth is a catalyst for broad based economic development in most of the low-income countries. Over the last few decades, the share of agriculture in GDP has steadily declined in India from over 19.7 per cent in the 1970-7 to 5.7 per cent in 2009-10. In contrast, the share of industry and services has increased to 26.4 and 55.1 per cent respectively due to greater focus on these sectors. But, agriculture still remains the backbone of Indian economy as nearly as 58 per cent of the rural population depends upon agriculture and allied activities for livelihood. The National Agricultural Development Policy (NADP) of India envisages a growth rate of 4 per cent per annum in the agriculture sector so as to achieve a target of over 300 million tonnes of food grain production by the year 2020. (Survey of Indian Agriculture, 2007).

Farm Women in India

Rural women form a very significant part of the productive work force in the agriculture sector in India and most of the developing countries.

Women are central to the selection, cultivation, preparation and harvest of food crops. About 80 per cent of all economically active women in the country are engaged in the agriculture sector. Also, about 48 per cent of India's self-employed farmers are women. There are 75 million women engaged in animal husbandry as compared to 1.5 million men (www. digitalopportunity.org). The findings of Chaudhary and Sangram Singh (2003) support the above statement. They reported that women played considerable role in various agricultural activities such as presowing, sowing and transplanting operations. Women's role was also reported in the preparation of compost and rising of vegetable nursery. Women's role was found to be maximum in hoeing and weeding practices on intercultural operations. Harvesting and post harvesting operations were reportedly more suited to the physical condition of women. Reaping of crops, storage of food grains, storage of seed and processing of grains were reported to be mostly done by women.

It is an acceptable fact that women possess talent which could be harnessed for the productive purpose. But it is most unfortunate that the role of women in agriculture has been neglected. A holistic development of women requires empowerment in all aspects involving them as managers, planners, scientists, and technical advisories. This is supported by Radha Singh (2005) as "A large part of women's contribution in agriculture is often not recognized and is also unpaid".

Knowledge Plays Crucial Role in Farm Sector

The world is in the midst of a knowledge revolution, complemented by opening up entirely new vistas in communication technologies. Recent developments in the fields of Information and Communication Technologies (ICTs) are indeed revolutionary in nature. Hundreds of millions of dollars are being spent on ICTs, reflecting a powerful global belief in the transformatory nature of these technologies. ICT, when used as a broad tool for educating the rural communities especially women, heralds the formation of a new class of society — the Knowledge Society. Knowledge thereby becomes the fundamental resource for all economic and developmental activities. The one resource that liberates people from poverty and empowers them is knowledge. *Possessing knowledge is empowering, while the lack of knowledge is debilitating.* "The World Bank organized a forum called "Voices of Poor" which got feedback from 60,000 people in 60 countries and concluded that people wanted access to knowledge and opportunities instead of charity to fight conditions leading to poverty", stated World Bank (2000).

A reorientation of existing strategies is essential to meet the new requirements of farming systems in the country and challenges in the areas of relevance, accountability and sustainability of agricultural extension. Ponnusamy and Jancy Gupta (2004) supports with their findings that "The paradigm shift is required to meet the challenges of the 21' century and to bring food, nutritional and environmental security and for serving all sections of the rural community including farmers, farm women and farm youth". "To give rural women visibility, they must be getting organized into Self Help Groups (SHGs). Group approach is a viable set up to empower women economically, socially and technologically for improved life", reported Suman Singh and Puja Mathur (2005).

Multimedia in Technology Dissemination

Interactive Multimedia Compact Disc (IMCD) is a modern electronic gadget which demonstrates complex farm technologies in simpler terms using text, graphics, audio and video. It is user friendly encompassing labeled instructions to learn the contents, structure and information using the mouse without depending on the help of a facilitator. It is highly self controlled which could be retained and viewed repeatedly as per the demand of the users. It is highly self controlled, comprehensive, userfriendly, highly reliable, pragmatic, innovative, cost-effective and simple tool which offers an unprecedented potential of providing knowledge to the users. It caters for individualized instruction and the contents could be easily updated and expanded. IMCD is a part of the vast ICT arena. This paper describes the experiences of using this IMCD as a transfer of technology tool to farm women.

Methodology

Most of the time, farmers or farm women or the farming system is not in a position to accept the latest research findings and technologies from organizations / Institutions. The reason may vary to context to context. But, by and large, misplaced and required priority between the farmers, field extension officials and research scientist are vary and not one and the same. Hence, this paper aims to overcome the mismatch between the research system and user system at Micro Level. For experimenting new idea, it needs a mutual consensus among the stakeholders who are part of the project. In this context, a strong baseline data is required to study and implement the project.

Identification of Village

Several discussions were held with Madurai district (Tamil Nadu) Krishi Vigyan Kendra (KVK) scientists and officials of the department of agriculture and horticulture, village administrative officers on the identification of a village under wetland ecosystem. The research team has identified and selected Mettuneerathan village which is located at three kilometers from Vadipatty block of Madurai district in Tamil Nadu with the following reasons.

- Compact village with 150 households
- Diversified cropping pattern
- Good irrigation source with crops grown all through the year
- Inquisitiveness of the villagers to acquire latest farming practices
- Highly approachable
- Existence of TANWA groups (The Tamil Nadu Women in Agriculture (TANWA)
- Good cooperation, support and response

Prioritizing Technological Needs

In order to elicit the felt technologies from the farm women, it was highly felt that only participatory techniques would help rather than applying conventional data collection methods. Hence, the PRA techniques were applied in the above village. Only select PRA tools namely Resource Mapping, Seasonal Calendar and Pair-wise matrix ranking techniques were selected.

Resource Mapping

The map shows the resources, infrastructure facilities available in the village and also depicts the social set up of the village. Further, it gives an idea about the resources like soil, water, forest etc. For the above exercise, charts which were pasted together to form a big sheet was given to the participants along with a variety of colour marker pens. The participants were given a briefing on the exercise and its purpose. The participants located few important structures of the village such as Temples, Schools, Water tank, Shops, Phone booths, Public latrines, Primary health centres, Milk society unit, etc. They differentiated each structure using different colours like green colour for paddy, yellow `colour for settlements, blue colour for water bodies so on and so forth. It took around one to one and a half hour to finish the exercise.

Seasonal Calendar

It shows the seasonal patterns in rural areas related to rainfall, farming practices, employment etc over the months in a year. The participants were explained about the seasonal calendar and its purpose. A pre-drawn seasonal calendar indicating months on one column was pasted in the wall for reference. The participants were given free space on the chart to enroll themselves on the crops, varieties and operations carried out in a particular season. They also involved themselves fully and brought out a complete seasonal calendar including the major crops being cultivated each season namely rice-rice- pulses. Various agricultural operations starting from sowing, transplanting, weeding, fertilizer application and harvesting of the crops grown and varieties were also collected with the application of the exercise. The outcome of the exercise is as follows.

Table 1 : Season wise Farm Activity of Mettuneerathan village

Month	Crop	Varieties	Operations
June	First crop - Paddy	ADT 43, ADT 45, Chinna Ponni	Sowing, Transplanting
July			Weeding
August			Second weeding
September			Harvesting
October	Second crop - Paddy	ADT 45, White Ponni, ADT 37, ADT 39, Sona, Karnataka Ponni	Nursery
November			Transplanting
December			First Weeding
January			Second Weeding
February			Paddy Harvesting (first week)
	Blackgram & Greengram	TN 4, TN 9 – BG; T2, Anjugam - GG	Sowing of Blackgram, greengram
March			
April	Banana		Harvesting of fruits
	Blackgram & Greengram	TN 4, TN 9 – BG; T2, Anjugam - GG	Harvesting of pods
May	Banana	Ottunadan	Planting of banana
			Summer Ploughing

Pair wise matrix ranking

According to Neela Mukherjee (1993), Pair wise ranking uses two items or attributes at a time for ranking to explore people's criteria for choosing one alternative over another. Group of villagers compare one pair at a time and give reasons for the choice made. At the end, the most favored choice is identified if it is a question of choice or, the major problem is identified in case of problems being ranked.

The number of times each technology was preferred was placed in total column. Ranks were assigned to the technologies according to the total number of times they were selected. The technology preferred more times was ranked first, similarly other ranks were assigned.

Table 2 : Pair wise matrix ranking for identifying felt agricultural technologies

Sl. No.	Agricultural Technologies	1	2	3	4	5	6	7	8	9	10	11
1.	SRI		1	1	4	5	1	7	8	9	10	11
2.	Land leveling			3	4	5	6	7	8	9	10	11
3.	Fodder cultivation				4	5	6	7	8	9	10	11
4.	Black gram — Package of Practices					5	4	7	4	9	10	11
5.	Vermicompost Production						5	7	5	9	10	11
6.	Mushroom production							7	8	9	10	11
7.	Rat control in paddy								7	7	7	7
8.	Mango — Package of Practices									8	10	11
9.	Coconut Eriophid mite control										9	11
10.	Amla -Package of Practices											11
11.	Identification & uses of medicinal plants											

Further, the research team intervened and shared the latest technological findings from Tamil Nadu Agricultural University. Finally the following felt agricultural technologies were finalized and the result is presented below:

Table 3 : Felt Technological Needs of Farm Women in Wet Land Eco System

Sl. No.	Prioritized and Identified Technological Needs of Farm Women
1.	Integrated Rat Management Practices
2.	Paddy Hybrid Seed Production Techniques
3.	Coconut Eriophid mite control
4.	Vermicompost Production
5.	Mushroom Production Technologies
6.	Drip Based Sugarcane Cultivation Technologies
7.	Paddy–Plant Protection (PP) measures
8.	Coconut Eriophyid mite Management

Development of Interactive Multimedia Compact Discs (IMCD)

Computer multimedia offer learners more complete and individual control over their learning. The main reason behind recommending the use of computer based multimedia system for farm women is to facilitate interactivity and better understanding between individual learners and the subject matter.

IMCD is a modern electronic gadget which demonstrates complex farm technologies in simpler terms using text, graphics, audio and video. It is highly self controlled which could be retained and viewed repeatedly as per the demand of the users. The steps involved in devising IMCD are given in Figure 1. The pre-production phase involved all preparatory works including review of literature, content preparation / script writing, collection of video / pictures / video capturing etc. in the second phase the software selection was given importance. Among the available different software, Microsoft Office 2003 Power Point was selected specifically for the study purpose. This software possesses number of unique features such as simple operation; create highly interactive multimedia presentation viz., possibility to add sound, photo, picture, animation etc., play movies directly with in the power point, create simple hyperlinks, convert the presentation into web pages, pack and wizard have the advanced facility to compress and save our presentation across the multiple disc etc.

Each IMCD differed in terms of designing and layout ie., font size, type, colour, background colour and animation. Audio recording was done in a digital audio recording studio through professional voice artists and research team with good vocal ability. The captured videos were edited into a very short duration movie of one to two minutes. In the third phase developed IMCDs were exposed to user groups (farm women) through orientation training. During the training, farm women groups were allowed individually to handle the computer including the monitor, mouse, CPU and CD so that they get rid of computer phobia. Before handing over the IMCDs to the user groups, pre testing of all IMCDs were done for content validation in non sampling areas and with experts. With necessary incorporations of valuable suggestions obtained from pre-testing, the final IMCDs were handed over to the user groups by installing computers at the respective villages.

Since the respondents of the study were farm women, the farm women of the entire village were considered as samples. The selected village had 150 households and the entire farming population families exposure and post exposure knowledge test. These 70 farm women constituted as sample size of the study.

The main aim of the study was to diffuse the farm technologies to farm women through IMCD and to assess the effectiveness of IMCD in terms of knowledge gain. The knowledge level of farm women was assessed as pre exposure and post exposure to the felt technologies through the IMCD.

Sample and Sampling Technique

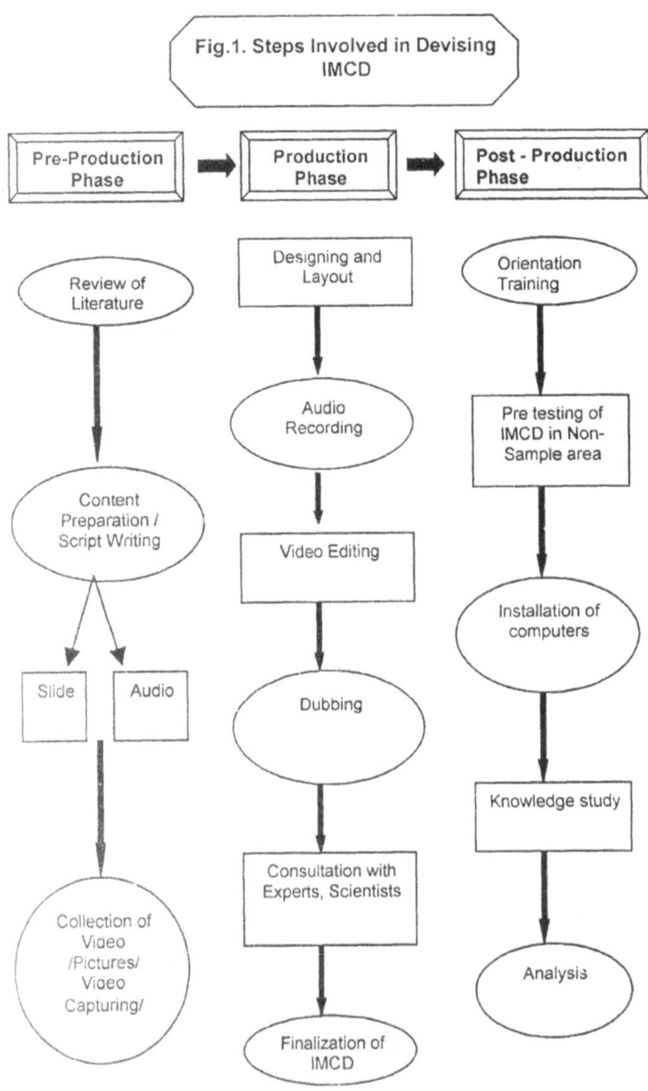

Hands on Exposure of IMCD

Farm women who are vulnerable to digital divide are felt to be trained on IMCD for effective utilization of the module in terms of knowledge gain. Hence, Orientation training was given to farm women on the usage of IMCD using laptop and LCD projector. The training was considered as a golden opportunity for the farm women of rural background and felt very happy and enthusiastic in learning more about them. Most of the participants handled the computer and independently started operating IMCDs immediately after the training by overcoming the computer phobia. Later a village kiosks was established in coordination with the panchayat representatives and local leaders to enable and empower the women farming community to access the latest agricultural information through developed IMCDs.

Measurement of Knowledge Gain

The effectiveness of the IMCD is measured by the knowledge gained by the farm women due to their exposure to the IMCD. Knowledge is defined as the extent of information or understanding a farm wom an has about the felt technologies expressed by them. Knowledge gain was operationalised as the quantum of information or messages newly learnt by an individual respondent due to the exposure of User-Friendly IMCDs on felt technologies.

Knowledge Test Questionnaires were prepared by reviewing pertinent literatures from various sources like journals, websites, thesis, books and so on. About twenty items were framed for each felt technologies in order to assess the existing knowledge of farm women on the identified technologies.Thesamesetsofitemswerefurtherusedtoassessthepostexposure knowledge after exposing them to the multimedia module on the select technologies. Such knowledge test questionnaires were prepared for all the eight technologies developed in the form of IMCDs. Questionnaires were prepared in Tamil (vernacular language) facilitating the farm women an easy understandability to fill them on their own.

Respondents Perception about the IMCD

Perception is the process of understanding sensation or attaching meaning based on past experiences to signs (Taneja, 1989). This variable helps us in assessing farm women's perception about Interactive Multimedia Compact Disc in transfer of technology i.e., Over all Perception of IMCD. The overall perception of IMCD is the degree to which the receiver perceives the message, text, audio, video and illustration communicated through the IMCD in a clear and understandable form.

Measurement of Perception

To measure the perception level of respondents on IMCD, a list of items seeking different contents were prepared. These listed items were administered to the respondents. The individual farm women was asked to state on a three-point continuum highly satisfactory, satisfactory and not satisfactory with the statements in his hand, with a score of 3, 2, and 1 for the responses respectively. Based on the scores the Perception Index was calculated using the following formula

$$\textbf{Perception Index (PI)} = \frac{\text{Individual respondents score}}{\text{Total score}} \times 100$$

The overall perception level was categorized based on the identified class interval. The data was collected using a well defined interview schedule.

Data Collection was done with carefully constructed interview schedule by reviewing various literatures, journals, theses, books and in consultation with experts and scientists. The knowledge gain was measured using percentage analysis. The Paired — SamplesT test were used to know the significant of knowledge gained by the respondents. The paired—samples 't' test procedure compares the means of two variables that represent the same group at different times.(i.e., before and after exposure to IMCD).

Findings and Discussion

Knowledge Gain of IMCD

In the wetland ecosystem the farm women were exposed to eight different IMCD's based on the priority set. The percentage of knowledge gain was calculated and the result is given below.

The knowledge test scores collected from the respondents for each module (IMCD) is pooled and the average given in Table 4.

Table 4 : Knowledge gain for the Eight IMCD modules

SI. No.	IMCD	Before Exposure	Immediately after Exposure	Mean Knowledge Gain	Percent of knowledge Gain	't' Value
1.	SRI Techniques	3.3	17.3	14	70	29.69**
2.	Rat management Techniques	6.1	16.6	10.5	52.5	16.05**
3.	Hybrid seed production technologies for Paddy	2.3	11.2	8.9	44.5	21.87**
4.	Eriophyid mite Management	2.5	14.1	11.6	58	29.00**
5.	Vermi compost Technology	3.6	14.1	10.5	52.5	21.00**
6.	Mushroom Production Technologies	2.8	14.3	11.5	57.5	17.58**
7.	Drip based Techniques	3	16.8	13.8	69	18.97**
8.	1PM in Paddy	8.1	17.1	9	45	11.84**

** Significant at 0.01% level of probability

From Table 4, it is evident that SRI IMCD had brought about a significant knowledge gain among the farm women respondents. The average per cent knowledge gain was 70 per cent which was found to be significant at one per cent level of probability.

The results indicate that there occurred a significant increase in knowledge level at the post exposure stage of System of Rice Intensification IMCD. Hence it could be concluded that the IMCD on System of Rice Intensification technology, acted as an intervening variable in changing and increasing the cognitive domain of respondents in the wet land eco system. The reason for the highest knowledge gain (70 per cent) on SRI IMCD as expressed by the respondents were newer technology, suitable to their cropping system, relevant and attractive video clippings, pictorial representation, number of relevant video clippings and clear voice.

The IMCD on Drip technology obtained 69 percent knowledge gain. This is attributed to newer technology, suitability to overcome water scarcity, higher yield and pictorial representation of field layout and field preparation. It was observed that the average knowledge gain was 58 per cent in respect on Eriophyid Mite Control. This is due to novelty, suitable as an alternate crop during water scarcity, clear package of practices, relevant video clippings and better combination of text and background colour. The IMCD's on Mushroom production technology, Vermi compost Technology, Rat management Techniques, Hybrid seed production in paddy and IPM in paddy followed in the level of knowledge gain.

Perception of IMCD by Farm Women

Respondent's extent of perception on different components of IMCD as perceived by the respondents is given in Table 5. It was observed that the perception of farm women was highly satisfactory in the case of the module on Mushroom production technology. it was observed that nearly cent (97.71 per cent) per cent of the respondents, were satisfied with the different components of Mushroom Production Technology IMCD. The detailed discussion for the satisfied level is here under on different components names i.e., message, text, video, audio, illustration.

Table 5 : Overall Perception about Different Technical Components of IMCD

Sl. No.	IMCDs	Message		Text		Audio		Video		Illustration		Average	
		S	NS	S	NS	S	NS	S	NS	S	NS	S	NS
1	Integrated Rat Management Practices	61 (87.14)	9 (12.86)	34 (49.57)	36 (51.43)	52 (74.29)	18 (25.71)	57 (81.43)	13 (18.57)	68 (97.14)	2 (2.86)	55 (77.91)	15 (22.09)
2	Paddy Hybrid Seed Production Technologies	45 (64.29)	25 (35.71)	35 (50.00)	35 (50.00)	51 (72.86)	19 (27.14)	12 (17.14)	58 (82.86)	12 (38.57)	43 (61.43)	31 (48.59)	36 (51.41)
3	Coconut Eriophyid mite management practices	66 (98.57)	4 (5.71)	57 (81.43)	13 (18.57)	50 (71.43)	20 (28.57)	62 (88.57)	8 (11.43)	49 (70.00)	21 (30.00)	57 (81.43)	13 (18.57)
4	Vermicompost Production Technologies	69 (98.57)	1 (1.43)	54 (77.14)	16 (22.86)	69 (98.57)	1 (1.43)	58 (82.86)	12 (17.14)	56 (80.00)	14 (20.00)	61 (87.30)	9 (12.70)
5	System of Rice Intensification	60 (85.75)	10 (14.29)	45 (6429)	25 (35.71)	56 (80.00)	14 (20.00)	63 (90.00)	7 (10.00)	38 (54.29)	32 (45.71)	53 (74.86)	17 (25.14)
6	Mushroom Production Technologies	68 (97.14)	2 (2.86)	65 (92.86)	5 (7.14)	63 (90.00)	7 (10.00)	69 (98.57)	1 (1.43)	70 (100)	0 (0.00)	67 (97.71)	3 (2.29)
7	Drip	67 (95.71)	3 (4.29)	56 (80.00)	14 (20.00)	58 (82.86)	12 (17.14)	49 (70.00)	21 (30.00)	39 (55.71)	31 (44.29)	54 (77.14)	16 (22.86)
8	Integrated Plant Protection Measures in Paddy	61 (87.30)	9 (12.70)	59 (84.29)	11 (15.71)	51 (72,86)	19 (27.14)	*		68 (97.14)	2 (2.86)	60 (85.39)	10 (14.61)

* Video clipping have not been incorporated. Since, relevant clippings are not available at the time of study. The number within parenthesis is percentage to the total

Nearly cent (97.14 per cent) per cent of the farm women were satisfied with the message component. Higher per cent of satisfaction among the respondents on message components of the Mushroom Production Technologies IMCD might be due to the reason of the following aspects production techniques of Oyster and milky mushroom, bed preparation Schedule for Oyster and milky mushroom , pest and disease management in mushroom production, preparation techniques of mother and bed spawn, maintenance of the mushroom sheds of Oyster and milky mushroom, climatic requirements of Oyster and milky mushroom and economic comparison for both the methods.

Majority (92.86 per cent) of the respondents were satisfied with the text component of Mushroom Production Technology. Nearly cent (98..57 per cent) of the farm women were satisfied with the video clippings of Mushroom Production Technologies. About 10 per cent of the respondent were not satisfied with the audio. This might be due to the reason of lengthy dialogue of text over voice. Respondents always give importance to illustration because learning from illustration was very easy than from text. It is interesting to note that cent (100 per cent) of the farm women were satisfied with the illustration.

The second component that has gained higher score is the IMCD on Vermicompost Production Technologies. The average satisfaction level was 87.30 per cent. Message and audio were the major contributors for this level of satisfaction. The next rank was obtained by the IMCD on Integrated Plant Protection Measures in Paddy (85.39 per cent). the major contributors were illustration and message in its presentation. The others in order are Coconut Eriophyid mite management practices (81.43 per cent), Integrated Rat Management Practices (77.91 per cent), Drip (77.14), System of Rice Intensification (74.86) and Paddy Hybrid Seed Production Technologies (48.59). Similar modules with perfect combination of audio, video, illustration, message and text will be perfect tools in transfer of technology.

Conclusion

Access to information can be seen as a central issue concerning empowerment of women. Women in developing countries, however have been traditional excluded from the external information sphere both deliberately and because of factors which inherently work to their disadvantages such as little freedom of movement, low education-levels etc. However, IMCD, a component of ICT opens up a direct window for women to the outside world. Information now flows to them without distortion or any form of censoring, and they have access to the same information as their male counterparts. This leads to broadening of perspectives, building up of greater understanding of their current situation and causes of poverty, and initiation of interactive processes for information exchange and in turn empowerment and transformation of rural women community:

The paper as a whole aimed at empowering the farm women in using the new communication advancements, helps in developing new education modules, an in exploring the feasibility of this new communication network established. Moreover this venture can be used for administrative convenience in a longer run. Educationalist, policy makers and administrators can take a lot from this project in developing new modules in educational, research and extension institutes in the near future. Development of locality specific IMCD's requires careful planning and preparedness. But its usage as an extension tool is extensive. Hence it is imperative for extension staff to plunge into the opportunity and make their works easier. Group approach in this effort will make better results as the dynamic interaction results in better and rapid adoption of technologies.

References

Chaudhary, H., and Sangram Singh. 2003. "Farm Women in Agriculture Operations". Agricultural Extension Review. 15(1): 21-23. http: www. digitalopportunity.org

Neela Mukherjee. 1997. Participatory Appraisal of Natural Resources. Concept Publishing Company, New Delhi.

Ponnusamy, K., and Jancy Gupta. 2004. "Agricultural Extension — New Paradigms". Agricultural Extension Review. 16(6): 3-6.

Radha Singh, 2005. Gender issues are central to the development dialogue. www.digitalopportunity.org .

Suman Singh and Puja Mathur. 2005. "Self Help Group — A Successful Approach for Women Empowerment". Agricultural Extension Review. 17(6): 27-30.

Aeron, R.K. 1998. What is multimedia, Multimedia Production Concepts? Summer School on Computer Multimedia Application in Agriculture and Allied Sciences. College of Technology and Agricultural Engineering, RAU, Udaipur, India.

Malik, B.S., Das, B.C., and V.K. Yaday. 2004. "Women Self Help Group — Disseminate Agri Information". Agricultural Extension Review. 16(1): 10-14.

Swanson. B.E., Farner, B.J., and Bahal, R. 1990. The current status of agricultural extension worldwide. In FAO, Report of the global consultation on agricultural extension, Rome: FAO, p.43-76.

Taneja, R.P. 1989. Dictionary of Education. Anmol publications, Ansari road, New Delhi. http://en.wikipedia.org/wiki/Multimedia.

●●●

35

Empowerment of Urban Women through SHGs under SJSRY in Coimbatore District

K.C.Leelavathy[1] and K.S.Selvanayaki[2]
[1]Director, Centre for Women's Studies, Avinashilingam Women's University, Coimbatore and
[2]Assistant Professor, Faculty of Women's Studies, Tamil Nadu Institute of Urban Studies, Coimbatore.

Introduction

Women empowerment is a global issue and it is the outcome of criticism and debates generated by the women's movements / feminists. Empowerment is the process by which the powerless gain greater control over the circumstances of their own levels relating to both resources and ideology.

Empowerment means the ability to assert oneself; striving for upward mobility or attaining the psychological status of being powerful. Empowering of women is necessary for sustainable development of a community. Sustainability implies a state of balance and equilibrium in factors related to human life which can be social, political, financial, environmental and even spiritual. (B.K.Singh,2006).

Women empowerment often involves the empowered developing confidence in their own capacities. Empowerment, probably, the totality of the following or similar capabilities (social welfare,2010).

- Having decision-making power of their own
- Having access to information and resources for taking proper decision

- Having a range of options from which you can make choices (not just yes/no, either/or.)
- Ability to exercise assertiveness in collective decision making
- Having positive thinking on the ability to make change
- Ability to learn skills for improving one's personal or group power
- Ability to change others' perceptions by democratic means
- Involving in the growth process and changes that is never ending and self-initiated
- Increasing one's positive self-image and overcoming stigma.

Women are an integral part of every economy. All round development and harmonious growth of a nation would be possible only when women are considered as equal partners in progress with men. Empowerment of women is essential to harness the women labor in the main stream of economic development. Empowerment of women is a holistic concept. It is multi-dimensional in its approach and covers social, political, economic and social aspects. Of all these facets of women's development, economic empowerment is of utmost significance in order to achieve a lasting and sustainable development of society. Self- Help Groups (SHGs) are the voluntary organizations which disburse micro credit to the members and facilitate them to enter into entrepreneurial activities (Rekha R. Gaonker, 2008).

Women in India

Women in India now participate in all activities such as education, politics, media, art and culture, service sectors, science and technology, etc. The Constitution of India guarantees to all Indian women equality (Article 14), no discrimination by the State (Article 15(1)), equality of opportunity (Article 16), equal pay for equal work (Article 39(d)). In addition, it allows special provisions to be made by the State in favour of women and children (Article 15(3)), renounces practices derogatory to the dignity of women (Article 51(A) (e)), and also allows for provisions to be made by the State for securing just and humane conditions of work and for maternity relief. (Article 42).

Women's Bureau was set up in 1976 by the government as a division within the ministry of culture and social services, to "uplift the status of women and to increase their involvement in the national development process" (Subhasini Mahapatra, 2006). In 1990s, grants from foreign donor agencies enabled the formation of new women-oriented NGOs. SHGs and NGOs such as Self Employed Women's Association (SEWA) have played a major role in women's rights in India. Many women have emerged as leaders of local movements. For example, Medha Patkar of the Narmada Bachao Andolan. The Government of India declared 2001 as the Year of Women's Empowerment (Swashakti). The National Policy For The Empowerment Of Women was passed in 2001. (Kabeer.N, 2001).

The role of SHGs in India

One has to believe that the progress of any nation is inevitably linked with social and economical plight of women in that particular country. For concrete results, we have to assert and act with our full might and what is needed most (Kay, 2002/03). Empowerment by way of participation in SHG can bring enviable changes and enhancement in the living conditions of women in poor and developing nations. SHG is a process by which a group of 12 - 20 women with common objectives are facilitated to come together voluntarily to participate in the development activities such as saving, credit and income generation thereby ensuring economic independence.

The principles underlying the SHGs are financing the poorest of the poor, and achieving holistic empowerment. SHG phenomenon certainly brings group consciousness among women, sense of belonging, adequate self confidence. What she cannot achieve as an individual, can accomplish as a member of group with sufficient understanding about her own rights, privileges, roles and responsibilities as a dignified member of society in par with man. When she becomes a member of SHG, her sense of public participation, enlarged horizon of social activities, high self-esteem, self-respect and fulfillment in life expands and enhances the quality of status of women as participants, decision makers and beneficiaries in the democratic, economic, social and cultural spheres of life. Thus undoubtedly SHG can be an effective instrument to empower women socially and economically by which the implication on the overall development of women is indisputably possible particularly for a country like India wherein still large segment of women population are underprivileged, illiterate, exploited and deprived of basic rights of social and economic spectrum. (Reddy, 2005).

The experiences of SHGs in many countries have been proving great success as an effective strategy and approach in recent years. Group-oriented efforts in the form of Micro-credit groups in different countries of Latin America, Africa and Asia are examples of current self-help efforts. The grameen groups in Bangladesh, Local self-help development efforts — harambee (Thomas, 1985) in Kenya, Tontines or Hui with 10 to15 members involved in financial activities through cash or kind in Vietnam, self help efforts through credit unions, fishermen groups, village-based banks, irrigation groups etc (Gaonkar, 2004) in Indonesia, the self-help groups (SHGs) in countries like Thailand, Nepal, and Sri Lanka and India are successfully proving forms of micro-credit groups or SHGs. (Ranjula, 2007).

Genesis of SHGs in SJSRY

After the 74th constitutional amendment Act, 1992 the Ministry of Urban affairs and employment, government of India have launched a rationalised poverty alleviation scheme namely the Swarna Jayanti Shahari Rozgar Yojana (SJSRY). It was functioning from 1st December197 in al urban towns India. It was functioning in al Urban Local Bodies (ULBs) (Corporation, Municipalities and Town Panchayats) of various states in India. For each group 12-20 members from Below Poverty Line (BPL) families eligible to join in group.

The main objectives of the study are

a) to enumerate the economic growth and functioning of SHGs under SJSRY.
b) to understand and analyse the empowerment of women participating in SHGs under SJSRY.

Methodology

The study is a descriptive in nature as it has attempted to describe social and economic benefits enjoyed by the members by participating in SHGs. The information collected directly by the researcher from 80 respondents as primary data and other details collected from various text books, reports, journals, newspapers, published literature, websites and records of SJSRY from randomly selected ULBs in Coimbatore district as secondary data. This study was conducted among women of 20 SHGs completed 5 years of existence. A total of 80 women (members of SHGs) were selected comprising four persons (one Resident Community Volunteers (RCVs), three ordinary members) randomly from every group purposively. As most of the respondents were semi - literate interview schedule was utilized as a tool of data collection.

Locale of study

This study was conducted in seven ULBs i.e., one Corporation, two Municipalities and four Town Panchayats of Coimbatore District, Tamil Nadu.

Findings and Discussion

Individual income of the Members

Women income in a family will determine her independence and freedom in all activities and become major determinant of the standard of living of the people. As the SHG members income has been increased after joining the SHGs, the women members of the groups are independent to meet their personal expenditure, and they contribute more to their household expenses. Many housewives (32 out of 80) did not earn anything before joining SHGs, but after joining the SHGs, they are also earning reasonably. Table 1 and Fig.1 present the data collected on monthly income level of the members.

Table 1 : Individual income of the Members

Monthly Individual Income	Before Joining SHG	After Joining SHG
Nil	32	0
< 500	18	19
500 - 1000	11	18
1000 - 1500	10	20
1500 - 2000	6	17
> 2000	3	6
Total	**80**	**80**

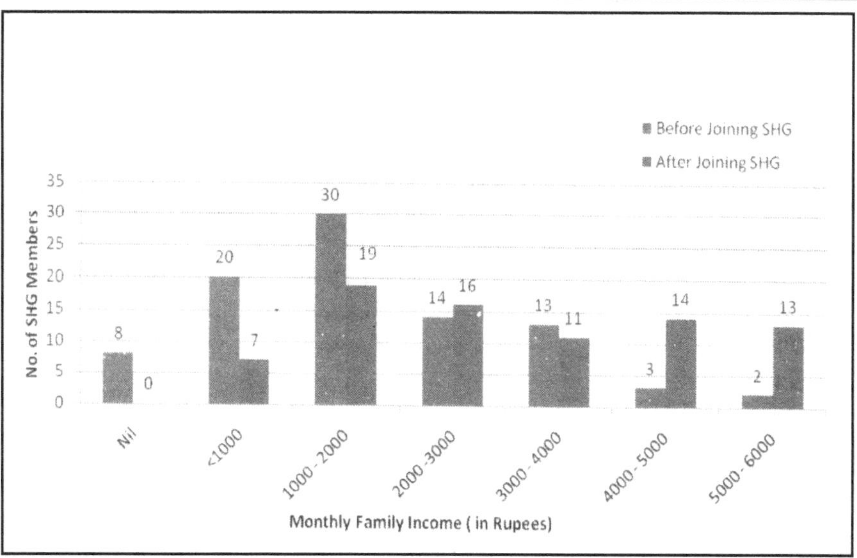

Fig. 1 : Distribution of Monthly Individual Income among 80 SHG Members

Considering,

H_0 : There is no significant increase in monthly Individual income among the SHG members.

H_1 : There is a significant increase in monthly Individual income among the SHG members.

Monthly Individual income	**Mean**	**SD**	**Observed value of Z**	**Table Value of Z**
Before Joining SHGs	493.75	356.26	Z_0 = -6.07774	Z_α=-1.645
After Joining SHGs	1006.25	664.77		

Since, the absolute value of Z_0 = 6.07774 is greater than the absolute value of Z_α = 1.645 for 5% level, the null hypothesis is rejected at 5% level and the alternative is accepted. We may conclude that there is a significant increase in monthly individual income among the SHG members after joining SHGs.

Monthly family income of the Members

The family income has been increased due to increase in the SHGs members' income. Usually working women are being respected by the household members and the society. Nowadays the women in the SHGs are also respected by the others, because they are independent in earning the income and they are contributing to household income, expenditure and savings. Therefore the above discussion clearly states that after joining in the SHGs, the members' well-being has been increased. (Table 2 and Fig.2).

Table 2 : Family incomes of the Members

Monthly Individual Income	**Before Joining SHG**	**After Joining SHG**
Nil	8	0
<1000	20	7
1000 - 2000	30	19
2000 —3000	14	16
3000 — 4000	13	11
4000 — 5000	3	14
5000 — 6000	2	13
Total	**80**	**80**

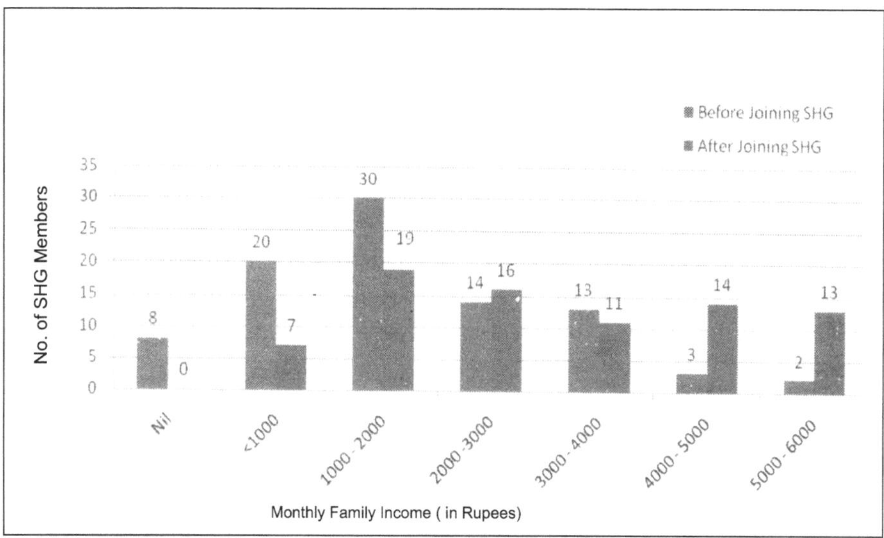

Fig. 2 : Distribution of Monthly Family Income among 80 SHG Member

Considering,

H_0: There is no significant increase in monthly family income among the SHG members.

H_1 : There is a significant increase in monthly family income among the SHG members.

Monthly family income	Mean	SD	Observed value of Z	Table Value of Z
Before Joining SHGs	493.75	356.26	Z_0 = - 11.5563	Z_α=-1.645
After Joining SHGs	1006.25	664.77		

Since, the absolute value of Z_0 = 11.5563 is greater than the absolute value of Z_α = 1.645 for 5% level, the null hypothesis is rejected at 5% level and the alternative is accepted. We may conclude that there is a significant increase in family income among the SHG members after joining SHGs.

Source of income for savings of members

The incremental income not only enhances the expenditure of the family but also promote the savings of the family after joining the SHGs and also the objective of the SHGs is fulfilled. This is an achievement of the women SHGs in the study area. Table 3 and Table 4 represent the source and the amount of monthly savings by the respondents. It clearly indicates that after joining SHGs, the saving habit has been motivated among the members.

Table 3 : Source of income for savings

SI.No.	Sources	No. of. Members
1.	Self-earning	60
2.	Husbands earning	20
3.	Children earning	0
4.	Borrow from other	0
	Total	**80**

Table 4 : Monthly savings of members

SI.No.	Savings (in Rs.)	No. of. Members
1.	Below 50	30
2.	51-100	24
3.	101-150	12
4.	151-200	8
5.	201 and above	6
	Total	**80**

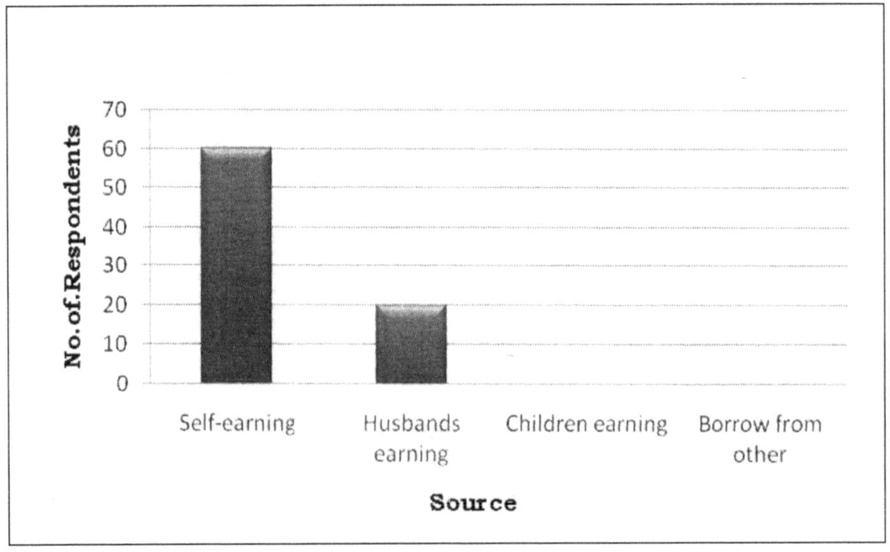

Fig. 3 : Source of income for savings

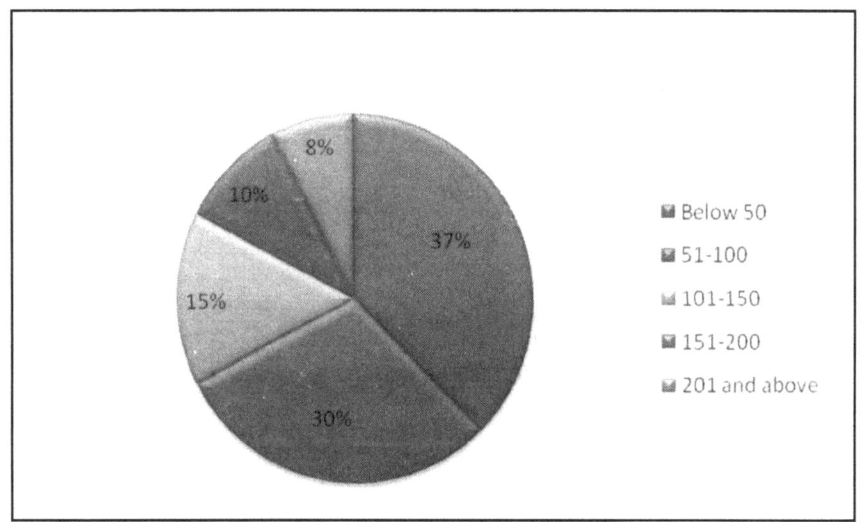

Fig. 4 : Monthly Savings of Members

Table 4 and Fig. 4 indicate that, the saving attitude of SHG members has increased enormously, but the amount of savings indicates that their savings still are very nominal. Only very few members were able to save more than 150 rupees per month due to their average income source.

Purpose of savings

The SHG members saved their hard earn money for different purposes. Even though they may have various reasons to spend their money, the priority is to their children's education. This change is appreciable that they realized the importance of educating their children. Table 5 and Fig. 5 indicate the various purposes represented by the members for savings.

Table 5 : Purpose of savings

Purpose	**No. of Members**	**Percentage**
Medical expenses	12	15
Children's education	35	44
Children's Marriage	13	16
Old age security	4	5
Celebrating festivals	2	3
To buy asset	7	9
For grandchildren	7	9
Total	**80**	**100**

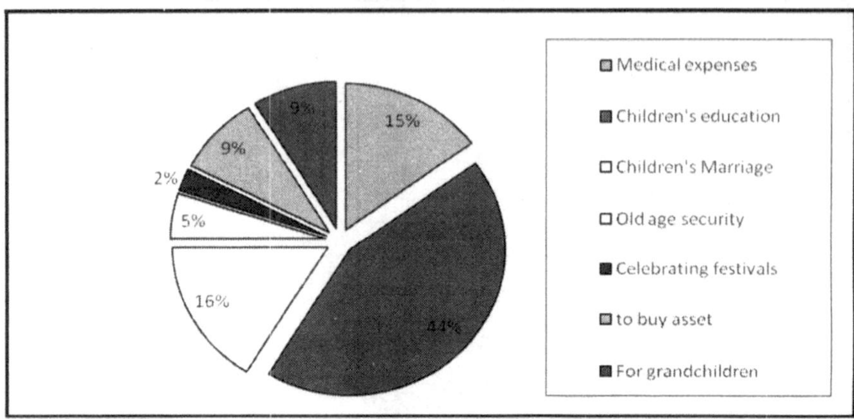

Fig. 5 : Purpose of Saving among 80 SHG Members

Internal Group Loan received from SHG

One of the reasons for joining SHGs is to avail loans. It is true in the present study area. Private money lenders charge very high rate of interest. In this situation SHGs are the boon to the urban poor women. The members of the group save their earning in the bank and it is accumulating in group account. These savings are utilized as an internal group loan for the needed members. The SHGs charge reasonable interest. In the study area, the prevailing interest rate is 2%. All the members are responsible to repay the loan at the given installment. Therefore, members repaid the loan in time. The SHGs in the study area grant the loan to their members for various purposes. The maximum loan amount per member and interest is decided by the general body meeting of SHGs. Almost all the members in the study area are availing the loan facilities in their SHGs.

Table 6 : Internal Group Loan received from SHG

Loan Amount	No. of Members	Percentage
Below 20000	6	8
20000 - 40000	13	16
40000 - 60000	17	21
60000 - 80000	21	26
80000 - 100000	15	19
Above 100000	8	10
Total	**80**	**100**

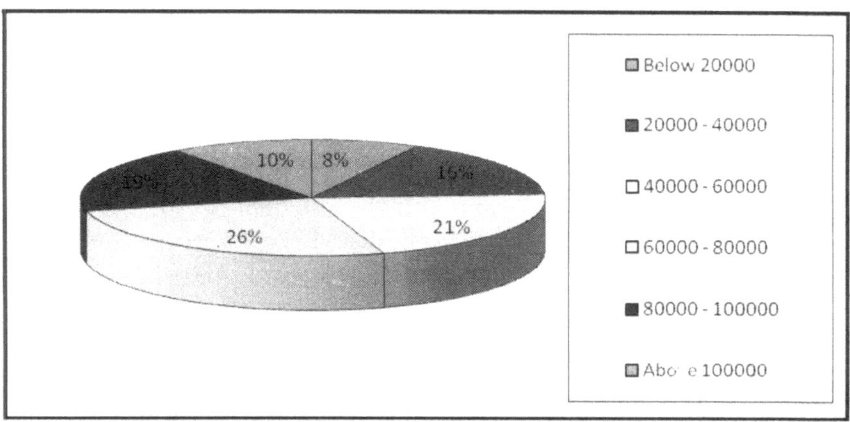

Fig. 6 : Internal Group Loan availed by 80 SHG Members

5.6. Purpose of Internal Group Loan Received by Members

All the members have got their loan from SHGs for different purposes. The various purposes for which loans obtained by the members are listed and ranked Table - 7.

Table 7 : Purpose of Internal Group Loan Received by Members

Sl. No.	Purpose of loan received	Rank 1	Rank 2	Rank 3	Rank 4	Rank 5	Rank 6	Rank 7	Rank 8	Rank 9	Rank 10	Total Number of Members	Average Ra nk	Rank
1	Purchase of House	20	14	17	6	10	3	2	4	1	3	80	3.45	Rank 2
2	Purchase of Household goods	16	10	8	14	9	17	2	1	2	1	80	3.91	Rank 5
3	Purchase of Jewels	5	3	6	9	12	11	16	10	3	5	80	5.75	Rank 9
4	Education expenses of children	30	10	8	12	4	3	5	2	4	2	80	3.34	Rank 1
5	Marriage expenses of children	22	17	9	10	6	4	2	3	3	4	80	3.53	Rank 4
6	Medical expenses	20	13	15	9	8	5	4	2	3	1	80	3.48	Rank 3
7	Delivery expenses	9	6	4	10	11	7	13	8	4	8	80	5.51	Rank 8
8	Starting self-employment	15	11	6	7	12	14	9	1	1	4	80	4.34	Rank 6
9.	Repay other loans	8	9	12	11	6	12	4	7	5	6	80	4.96	Rank 7
10.	Festival expenses	4	2	5	8	16	10	9	9	6	8	80	6.23	Rank 10

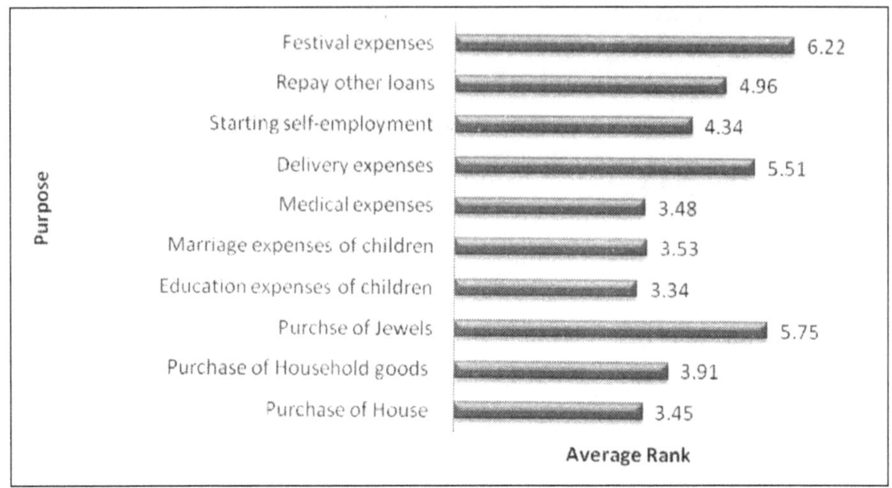

Fig. 7 : Purpose of availing Internal Group Loan among 80 Members

It is found from the study that the loan amount is being spent for various purposes in the hierarchy stated below:

Education, Purchase of House, Medical Expenses, Marriage expenses of children, starting self employment and repay other loans. The members rarely use the loan taken from SHG for festival expenses.

Benefits enjoyed by the member of SHG

SHGs have played valuable roles in reducing the vulnerability of poor, through economic freedom, consumption smoothing, provision of emergency assistance and empowering and emboldening women by giving them control over assets and increased self-esteems and knowledge. Table - 8 and Chart - 8 clearly picture the benefits enjoyed by the member being an SHG member.

Table 8 : Benefits enjoyed by the member of SHG

Benefits enjoyed by the member of SHG	**Rank 1**	**Rank 2**	**Rank 3**	**Rank 4**	**Rank 5**	**Total Number of Members**	**Average Ra nk**	**Rank**
Development of Self confidence	40	18	12	6	4	80	1.95	Rank 1
Social Recognition	24	32	9	7	8	80	2.29	Rank 2
Exposure to outside world	12	8	14	30	17	80	3.38	Rank 4

Economic development	10	11	15	25	19	80	3.40	Rank 5
Economic Freedom	20	16	28	7	9	80	2.61	Rank 3

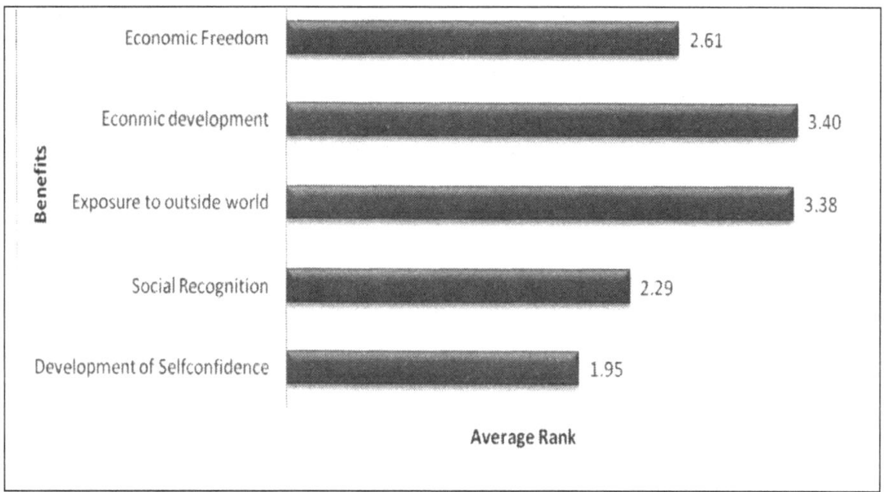

Fig. 8 : Benefits Enjoyed by the Member of SHG

The analysis realized our dream of increasing the self-confidence of women is achieved through SHGs, as the members ranked highly then other benefits.

Findings and Discussion

The members of SHGs are really privileged to enjoy many economic benefits including saving, liberation from debt trap, economic independence, economic skills for income generation, able to contribute for education of children, able to meet the healthcare needs, accumulate economic assets including house, land, jewels other valuable things, economically strong to meet unexpected financial challenges and so on. All the respondents were able to save reasonable amount for safety and security. It is found that the income of the women has been increased after joining SHGs. But the savings is increasing at slow rate, because the incremental expenditure is higher. Mostly they are spending for present consumption. The members should change it.

The SHGs are linked with the banks for external credit under the projects of urban development. The banks provide assistances for various entrepreneurial activities such as setting up small shops, vegetable shops, tailoring units etc. The groups undertake the responsibility of non-credit services such as literacy, health and environmental issues. The concept of

SHGs moulds women as responsible citizen of the country achieving social and economic status.

Thus the paper emphasizes that the SHGs are the effective instrument of women empowerment. The SHGs have also created better understanding between the member of the different religious group as the member of SHGs belong to different religious. This is the welcome change to have understanding and tolerance towards the members of other religions particularly in a country like India was there is a diversity of religions and castes.

The following suggestions and recommendations have been made based on the findings of the study and also in the interest of welfare of women in India in general. Government both state and Central should work hand- in -hand to promote the concept of SHG without political biasness and selfish motives taking into account the welfare of the womenfolk in general by incorporating as an important strategy for women empowerment in the national policy and allocating more grant in the budget and developing effective implementing machinery.

Conclusion

Women empowerment aims at enabling them to realize their identities, potentiality and power in all spheres of there lives. The real empowerment of women is possible only when a women as increased access to economic resources, more strength and course for entering into the power structure, more involvement through social relationships and participation, more self-motivation and confidence, and more say in the family matters. Women should be willing to take additional efforts for building a mind set which is suitable for their overall empowerment. It is a gradual and consistence process which requires the support of many who are related to them in one way or other.

References

Kabeer, N. (2001), "Resources Agency Achievements : Reflections on the Measurements of Women Empowerment - Theory & Practice", SIDA Studies, No.3.

Ranjula Bali Swain & Fan Yang Wallentin, (2007), "Does Microtinance Empower Women. Evidence from SI IGs in India, Working paper (August), Department of Economics, Uppsala University, Sweden.

Rekha R.Ganoker, (2004), Role of SHG in Empowerment of Women Social Welfare (2010), India Development Gateway

Rajammal P. Devadas, (1999), "Empowerment of Women through SHGs"

CS Reddy, (2005), "Self-help groups : A keystone of Microfinance in India - Women Empowerment and Social Security, Mahila Abhivruddhi Society, Andhra Pradesh

Thelma Kay, (2002/03), Bulletin on Asia-pacific perspective.

B.Singh.(2006), Women empowerment through SHGs"; Adhyayan publisher & Distributors, Delhi, 94.

R.K.Sahoo, and S.N.Tripathy (2006), Self help groups and women empowerments, Anmol Publication Private Limited, New Delhi, 02.

N.Jerinabi, (2006), Microcredit Management by women self help groups, Discovery Publishing House, New Delhi, 02.

SJSRY Guidelines, (1998), Ministry of Urban Affairs & Employment, Government of India, New Delhi

V.P.Sriraman, (2007) Microfinance, SHGs & Women Empowerment Current issues & Concerns

Zaman, H,(2001), "Assessing the poverty and vulnerability impact of micro-credit in Bangladesh : a case study of BRAC", unpublished background paper for world bank, world development report 2000/2001 (Washington, world bank).

●●●

36

Impact of Self Help Groups on Development of Entrepreneurship among Farm Women

***S.K. Meti*[1] *and S. Aparana*[2]**
[1]*Professor & Head and Dean (PGS)*
[2]*Ph.D., Scholar, UAS, Raichttr*

Introduction

In India, the marginal, small and semi-medium land holding constitutes more than 90% of the land holdings in the country. They are caught in a vicious cycle of low investment, low productivity and low income with little chance to get out of it. In general, majority of farm families possess limited and scattered land holding with a low productivity and they have very limited accessibility to the improved agricultural information because of their illiteracy and lack of participation in extension programmes. The development of entrepreneurship in subsidiary activity like vermicompost production and Dairying will definitely help the small and marginal farmwomen to increase the farm income. The entrepreneurship among farmwomen in vermi composting and dairying will definitely encourage them to start self employment in addition to the agricultural income. Farmwomen perform -multifaceted role in various fields such as production, processing and marketing. In spite of their significant contribution to production they have virtually remained invisible as farmers, home makers and the ultimate by protecting the whole family. Their role has become critical in the process of moving their family out of the vicious 'circle of poverty' Nataraju and Perumal, (1990). Farmwomen generally have lower income and less leisure time than men and seldom have equal opportunities to develop their capabilities. Farmwomen have been considered unorganized and inferior to men, Lelamma (2001).

Methodology

The research study was conducted in Manvi and Dedurga taluks of Raichur District, to know the knowledge status and adoption level of dairy and vermi compost production technologies. In each taluka 75 respondents were selected by following the random sampling procedure and training was given to those selected respondents based on their needs on vermi compost and dairy production technology. The information was gathered in pre-tested interview schedule by asking the open end questions. The gathered information was analyzed by using appropriate statistical tools and presented in tabular form.

Findings and Discussion

I. Impact of Vermicompost Training

Table 1 : Overall Impact of Vermicompost Training on knowledge status of Farmwomen

(n = 150)

SI. No.	Category	Knowledge		Adoption	
		Figures	Percentage	Figures	Percentage
1.	High Category	46	30.66	42	28.00
2.	Medium Category	90	60.00	88	58.66
3.	Low Category	14	9.33	20	13.33

Table 1 indicates that, majority (60.00%) of farmwomen belonged to medium knowledge category about vermi compost production technology through education, training and demonstrations. Further (30.66%) and (9.33%) of farmwomen belonged to high and low knowledge category about vermi compost production. Further, the results revealed that, the (58.66%) of farmwomen were having positive attitude towards the vermi compost production and adopted the technology. This might be due to impact of training and demonstration and also low investment with less effort and higher returns motivated the farmwomen to adopt vermi compost production technolo .gy in toto. Further, (28.00%) and (13.33%) of the farmwomen were belonged to high and low adoption category of vermi compost production technologies. This might be due to non-availability of time and lack of interest towards vermi compost production and less scope from the family leader. Further, it was clear that less than 80% farmwomen were having knowledge through education, training and demonstration regarding vermi compost production and nearly 65% of farmwomen given up their labour work and solved dairy and vermi compost. entrepreneurship because of earning additional income.

II. Impact Dairy Training

Table 2 : Overall Impact of Dairy Training on knowledge status of farm women

(n = 150)

SI. No.	Category	Knowledge		Ado Aion	
		Figures	Percentage	Figures	Percentage
1.	High Category	44	29.33	47	31.33
2.	Medium Category	85	56.66	78	52.00
3.	Low Category	21	14.00	25	16.66

It is clear from the Table 2 that, the majority (60.66%) of farmwomen belonged to the medium knowledge category whereas (29.5%) and (14.00%) of farmwomen belonged to high and low knowledge category, with respect to dairy management technologies. The possible reason for this might be due to the earlier experiences with a local cattle management and impact of dairy training. Further, it was observed from table - 3 that, majority (52.00%) of the farmwomen belonged to medium adoption categories of dairy management practices followed by high and low level of adoption category were (31.33%) and (16.66%) respectively. The reason for majority of farmwomen adopted dairy management practices due to their interest to earn additional income.

III. Knowledge level of Farm Women

The results from the Table 3 shows that the practice wise knowledge level of the farmwomen regarding vermi compost production practices were, the Advantages of use of FYM in Agriculture (70%), Knowledge about earth worms rearing (62%), Use of FYM in crops (56.66%). Further, less than 50% of the practices were known by the farmwomen were Use of chemical fertilizers in crops (54.66 %), Size of normal vermi compost pit (53.33 %), the benefits of vermi compost in agriculture (48.66 %), and Different types of waste materials used in vermi compost production (43.33 %). The quantity of vermi compost produced in one pit size (10 ft x 3 ft x 1 1/2 ft) (36.00%), quantity of earthworms required / pit to produced vermicompost (33.33%) and the price of one quintal of vermi compost (30.00 %), The reason for having practice wise knowledge regarding vermicompost production is due to impact of training and demonstration and effect of SHGs to motivate and encourage the farmwomen.

Table 3 : Practice wise knowledge level of farmwomen about vermi-compost

(n=150)

SI. No.	Vermi compost production practices	Figures	%
1.	The advantages of use of FYM in agriculture	105	70.00
2.	Knowledge about earth worms rearing	93	62.00
3.	Use of FYM in crops	85	56.66
4.	Use of chemical fertilizers in crops	82	54.66
5.	Size of normal vermi compost per pit (3ft*10ft*11/2ft)	80	53.33
6.	The benefits of vermi compost in agriculture	73	48.66
7.	Different types of waste material used in vermi compost production	65	43.33
8.	The quantity of vermicompost produced in one pit size (3ft *10ft*11/2ft)	54	36.00
9.	Quantity of earthworms required / pit for vermi compost production	50	33.33
10.	The price of one quintal of vermi compost	45	30.00

Table 4 : Distribution of farmwomen according to their source consultancy pattern

(n= 1 5 0)

SI. No.	Sources	Frequency	Percentage
1.	Friends / progressive farmwomen (PFW)	93	77.50
2.	Andra farmwomen	85	70.83
3.	Radio	78	65.00
4.	TV	60	50.00
5.	Live stock inspector	57	47.50
6.	Extension personnel	45	37.50
7.	Animal health centre	43	35.83
8.	Opinion leaders	40	33.33
9.	Magazine	20	16.67
10.	Films	18	15.00

The data in Table 4 depict that, the majority (77.50%) of respondents depends on Friends / progressive farmwomen (PFW), (70.83%) further on Andra farmwomen. A significant number of farmwomen depends on Radio and television. A medium level of farm women depends on veterinary live stock inspector (47.50%), extension personnel (37.50%)

and animal health centre to seek information (35.83%). It was interesting to note that, though about (95%) of farmwomen using radio and television, but because of lack suitable educational programmes only less percentage of respondents showed using radio and television to seek information regarding dairy management practices. This finding is confirmed by the findings of (Kempson, 1986, Tucker and Napier, 2002) that, when rural people are in information seeking using different level of sources like friends, neighbours and colleagues.

IV. Involvement of Farm Women

The study reveals that, 50% of the farmwomen were found to be involved continuously in vermicompost production whereas (23.33%) of the farmwomen were discontinued during crop season and again continued during off season. Further, only 26.66 % of the farmwomen discontinued the vermicompost production due to non-availability of water and higher temperature. The reason for more than 50% of farmwomen adopted the vermi compost production continuously due to because of effectiveness of group approach and impact of shelf help group on farmwomen which were motivated them for continuously adopting the technology to earn additional income.

V. Constraints faced by Farm Women

Table 5 : Constraints faced by farmwomen in adoption of Dairy technology and vermi compost production technologies

(n= 150)

SI. No.	Particulars	Frequency	Percentage
1.	Lack of knowledge about dairy and vermi compost production	98	65.33
2.	Lack of motivation among farmwomen	83	55.33
3.	Lack of participation in extension programme and mass media participation	80	53.33
4.	Lack of financial support from Bank in time	73	48.66
5.	Lack of facilities available in the villages	75	50.00
6.	Lack of support from the village people	64	42.66

It is clear from the Table 5 that, the farmwomen faced major constraints during adoption of vermi compost production and dairy management technologies were "lack of knowledge about dairy and vermi compost production (65.33%), lack of motivation among farmwomen (55.33%), lack of participation in extension programmes and mass media participation (53.33%), lack of financial support from Bank in time (48.66%), lack of facilities available in the village (53.33%) and lack of support from the village people (42.66%), as expressed by the farmwomen.

Conclusion

It is concluded that majority of farmwomen were middle age group and illiterate with low income group. Further, the social participation and extension participation were very weak among farmwomen in the rural area. Therefore, the selected farmwomen were motivated and changed their attitude towards vermicompost and dairy management entrepreneurship through series of education, training and demonstration. Further, majority of farmwomen had medium knowledge and adopted the vermicompost production and dairy management production technologies with the help of group approach and SHGs efforts. Hence group approach and exposure visit has to be increased to the farmwomen to motivate and encourage the farmwomen for development of entrepreneurship in rural area.

References

World bank (1992). Designing and implementing agricultural extension for women farmers. Technical note Washington DC: World bank, women arid development division.

Sherin, Muller, J. (1998). An analysis of characteristics of women groups and their role in rural developments M.Sc (Agri) Thesis, KAV. Vellayam.

M.H. Renjita and Premlata Singhs. 2006, Effectiveness of self help groups, IJEE 42 (1-2) : 66-70

●●●

37

Self-Help Groups Approach for Agri-Rural Environment : Need of the Hour

B. Narayanaswamy[1], K Narayana Gowda[2] and R.K. Naika[3]
[1]Senior Scientist/ Prog.Coordinator, KVK (IIHR-ICAR), Gonikoppal, Kodagu,
[2]Vice Chancellor, UAS , GKVK, Bangalore and [3] Associate Professor of Sericulture-UAS (B), Chinthamani

Introduction

Since six decades after independence, the existence of people in a state of social, economic, political and knowledge disempowerment is considered as major threat to overall development. There is an essential demand for the integrated development, which should cover up development in all aspects, specifically in status, equality, security, complete physical, mental and social well-being of the target group. One sided emphasis on economic achievements by policy makers has set in motion so for But the social cleavage in rural areas has deepened. The power to the people signifies a new movement, which has been born out of the realization that society's traditional arrangements for voicing their problems are very meager. The economic growth does not mean that only creation of wealth, but also creation of people's capacity to create wealth and this resides in their health, education, knowledge and skills. It has been proven fact that, the economic development in any country should be preceded by social development which is interdependent. To combat the poverty at present juncture, efforts should be through the group approach only. So the new paradigm like Self - Help Groups (SHGs) can attain sustainability because of the clear cut understanding among them. Almost failure of co-operatives to mitigate unemployment problems and combating rural poverty, it is essential to think over self help group approach to empowering the rural poor.

Accordingly many Governmental Organizations and NGOs have followed the emphasis on self help groups since one and half decades by considering the efforts of other agencies in our country and abroad. The savings and credit activities promoted by some NGOs and departments help members to take up income generating enterprises and there by, contribute to their economic development as well as social development. All the governmental efforts and NGOs initiatives lead to spend millions of rupees in this effort.

The World Bank Aid, NABARD's credit system emphasis, concerned State Government's new initiatives have been followed through SHGs. The Karnataka Government also allocated some portion of its budget to the Watershed development programmes. Sujala Watershed Project is one of the major developmental projects designed by the Government. It has distinct features in bringing people together on area groups, watershed associations and self help groups through the involvement of NGOs at all levels. Accordingly since few years, we could observe such developments in Karnataka's selected districts, which have enhanced the rural community to take up self employment activities in many areas especially through farming enterprises.

So far no studies have been reported in developing a separate scale to measure the performance of farming related SHGs , In this background an effort has been made to know the impact in terms of performance, usefulness and to analyze the overall constraints of selected SHGs under watershed project in Karnataka. In this, study the performance of SHGs was analyzed in three sub dimensions viz. Individual level, Group level and Community level.

Methodology

The study was taken up in Kolar and Tumkur Districts of Karnataka, coming under National Watershed Project area. The unit analysis of the study was individual member of exclusive Women Self Help Groups (SHGs) concentrated in farming activities. The SHGs who have undertaken one farming related activity and those completed minimum of three years of functioning was chosen. The total number of SHGs covered for the study were 50, giving equal representation for both the districts and four taluks. From each SHG, a random sample of four members was considered for the study. The criteria followed were: [i] They should be existing office bearers and members of the SHG; [ii] Out of the four members, two (50%) office bearers and other two (50%) were ordinary members. The total number of respondents for the study was 200, which constituted the total sample size.

Findings and Discussion

The findings revealed that (Table 1), a majority of the SHG members were small and medium (52.5%) farmers, followed by one third of them marginal land holding (32.5%), where as one sixth (15%) belonged to big farmers category. The results showed that, a great majority (62%) of the SHG members had medium family dependency ratio, where as only 24.0 percent of them with low family dependency ratio and only a small percent (14%) of them had high family dependency ratio. In case of faming experience, 37.5 percent were belonged to medium farming experience category. Whereas almost equal percentage of them (32 and 30.5%) were in low and high faming experience category. Majority of the SHG members (50%) enjoyed medium level of family encouragement, followed by 32 percent of them had high family encouragement; where as only 18 percent of them had low family encouragement. The results revealed that 42.5 percent of them had medium level of cosmopoliteness, where as 29 percent had high, followed by 27.5 percent of low cosmopolite category. In case of mass media participation, it was revealed that 36 percent of them had low mass media participation followed by high (34%) and medium (30%). It was found that the deferred gratification of the SHG members was medium (38%) followed by equal percentage (31%) of both low and high deferred gratification category.

Table 1 : Characteristics of the SHG members and category

(n = 200)

Characteristics	**Category**	**Respondents**	
		Number	**%**
Age Group	18-35 Years	119	59.5
	≥ 36 years	81	40.5
Education	Illiterate	81	40.5
	Primary and Middle	63	31.5
	High School and above	56	28.0
Land holding	Marginal	65	32.5
	Small and Medium	105	52.5
	Big	30	15.0
Family Dependency Ratio	Low	48	24.0
	Medium	124	62.0
	High	28	14.0
Farming Experience	Low	74	32.0
	Medium	75	37.5

	High	61	30.5
Family Encouragement	Low	36	18.0
	Medium	160	80.0
	High	04	02.0
Cosmo politeness	Low	57	28.5
	Medium	85	42.5
	High	58	29.0
Mass Media Participation	Low	72	36.0
	Medium	60	30.0
	High	68	34.0
Deferred Gratification	Low	62	31.0
	Medium	76	38.0
	High	62	31.0
Risk Willingness	Low	61	30.5
	Medium	86	43.0
	High	53	26.5
Achievement Motivation	Low	68	34.0
	Medium	84	42.0
	High	48	24.0
Management Orientation	Low	72	36.0
	Medium	60	30.0
	High	68	34.0
Extension Participation	Low	75	37.5
	Medium	65	32.5
	High	60	30.0
Extension Contact	Low	78	39.0
	Medium	72	36.0
	High	50	25.0
Participation in Training Programme	Low	45	22.5
	Medium	67	33.5
	High	88	44.0

The relationship of independent variables of SHG members with different performance levels is analysed and presented in the Table 2. A close look at the results indicated that all the five independent variables such as capacity building, economic activity, social status, communicability and self monitoring of members have established highly significant and positive relationship with the performance of SHGs at individual level. The findings were in accordance with the results of Ahmed(1999). With respect to the performance at group based activities the independent variables such as economic activity, social status, functional linkage, conflict management and transparency in functioning have shown highly significant relationships. Also, they have positively related. The results also reveled that the independent variables viz.

economic activity, social status and functional linkage of SHG members were positively and significantly related to the performance of SHGs at community level. The plausible reasons might be that the homogeneity, optimum size of the group to discuss effectively and the punctuality might be attributed more. It could be generalized that each of the subcomponents has profound relationship with the performance at different levels. The good performance at community level might be due to their understanding of gravity of their common problems of their villages, their needs and cordial relationship in the community. Also, the collective decision of the SHG members and leadership in the group and contact with other agencies and cooperation from outsiders might have contributed more to these performances.

Table 2: Correlation co-efficient between performance levels with independent variables

(n= 200)

Performance Level	**Independent Variables**	**Correlation co-efficient (r)**
Individual Level	Capacity Building	0.686 **
	Economic Activity	0.762 **
	Social Status	0.695 **
	Communicability	0.612 **
	Self-Monitoring	0.652 **
Group Level	Economic Activity	0.623 **
	Social Status	0.695 **
	Functional Linkage	0.716 **
	Conflict Management	0.621 **
	Transparency in Functioning	0.714 **
Community Level	Social Development	0.746 **
	Economic Development	0.846 **
	Functional Linkage	0.675 **

* * Significant at 1 % Level

A cursory look at the results indicated in the Table 3 reveled that all the interdependent variables together contributed to the extent of 99.4 per cent of total variation in the performance of SHGs at individual level. Also the ‘t’ value was highly significant. All the variables showed significant contribution. The possible reason might be their exposure to interactions, collective decisions and trainings led to better performance. Also, in the performance of SHGs at group level all the five variables together contributed to the extent of 99.6 per cent of total variation. It shows clearly that the significant characteristics have influenced on the performance at group level, each of the sub components profoundly contributed for performance of the activities.

Table 3 : Multiple regression analysis of independent variables with performance at different levels

SI. No.	Independent variables	Regression Co-efficient (b)	SE of Regression Co-efficient (SEb)	't' Value
Individual level				
1	Capacity Building	0.9725	0.0248	39.24 **
2	Economic Activity	0.9899	0.0169	58.58 **
3	Social Status	1.0096	0.0284	35.54 **
4	Communicability	0.9807	0.0321	30.58 **
5	Self-Monitoring	1.0010	0.0215	48.97 **
Group level				
1	Economic Activity	0.9890	0.0132	74.87 **
2	Social Status	0.9465	0.0206	45.96 **
3	Functional Linkage	1.0729	0.0179	60.03 **
4	Conflict Management	1.0118	0.0222	45.55 **
5	Transparency in Functioning	0.9834	0.0146	67.29 **
Community level				
1	Social Development	0.9790	0.0293	33.37 **
2	Economic Development	1.0022	0.0257	39.07 **
3	Functional Linkage	1.0039	0.0428	23.46 **

Individual level $R^2 = 0.994$, Group level $R^2 = 0.996$, Community level $R^2 = 0.974$
** Significant at 1 % Level

The probable reasons might be due to their experience in groups and good decisions towards existing problems, awareness to achieve economic benefits and aspiration to improve their living conditions. The results seek support from the studies of Snehalatha(1994), Prasad(2000), Puhazhendi (2000). An insight, into the Table 3 revels that all the three variables together contributed to the extent of 97.4 per cent of total variation in the performance of SHGs at community level. It clearly shows that significant characteristics are involved to know the performance of SHGs at community level. The possible reasons might be due to aspiration and zeal to come up in the society and motivational level to acquire social status and improve their economic status. Apart from this their recognition of their activities in the village and support from outside might have created an impact in the mind set of respondents to go for more activities. Similar results have been reported by Prasad (2000), Puhazhendi(2000) and Raghavendra (2002).

Conclusion

The results of the study indicate the potential for further enhancing the performance of SHGs. Also, Self Help Group approach is an effective extension perspective in changing agricultural — rural environment in the present juncture. While analyzing performance of any SHGs, one has to consider the overall dimensions rather than economic dimension alone. Also, farming related enterprises should be encouraged through group approach for the sustainable and long lasting services in the society. Also developmental activities through SHG can be effectively implemented for the overall up liftment of the rural families. Further, to ensure sustainable development of SHGs in future, Federation of SHGs can be thought of by the implementing, as well as promoting agencies. This would facilitate a forum for experience sharing and recognition of successful SHGs.

References

Ahmed, M.A., 1999, Women's empowerment: Self help groups. Kurukshetra, 47 (2) : 69 72.

NABARD, 1995, Role of self help groups and their linkage with formal credit institutions. NABARD, Mumbai.

Prasad, V., 2000, Self employment women set to change face of Kolar. The Indian Exp. May 21, 2000. p. 9.

Puhazhendi, V., 2000, Evaluation study of SHGs in Tamil Nadu. NABARD, Mumbai.

Raghavendra, H.G., 2002, Comparative study on performance of Women Self Help Groups organised by two NGOs in Kolar District. M.Sc. (Agri.) Thesis, University of Agricultural Sciences, Bangalore.

Snehalatha, 1994, A study on the impact of thrift and credit groups in improving the status of rural women. M.Sc. (Agri.) Thesis. Andhra Pradesh Agricultural University. Hyderabad.

●●●

38

Fortifying Farm Women through Entrepreneurial Training: Problems and Prescriptions

***S.R. Padma*[1] *and T. Rathakrishnan*[2]**
[1]Ph.D. Scholar, Dept. of Agrl. Extension and Rural Sociology and
[2]Director, Students Welfare, TNAU, Coimhatore - 641 003

Introduction

In the existing socio-economic scenario, economic independence is considered as one of the main status indicator in the society. Women especially in rural areas are the deprived section of the society on account of their dependence on others for economic reasons. This necessitates generation opportunities so as to make them economically independent. Moreover, women's contribution to national development is crucial. The process of development would be incomplete and lopsided, unless women are fully involved in it.Many innovations have failed due to non-satisfaction of women on the perceived attributes of innovations. So, it is necessary that they should be informed, trained, convinced of a technology primarily.

Farm women can't emerge as a good entrepreneur, if they are not equipped with any skill, knowledge, education and information. So, research studies are essential to assess the training needs of them to become a successful entrepreneur. In order to know the training needs of farm women for entrepreneurial development, and the constraints faced by them to become a good entrepreneur.

Methodology

This study was focused on training needs of farm women for entrepreneurial development. Hence, the TANWA farm women who had been given sufficient exposure on entrepreneurial activities by the programme operators served as the sample for the study. Among the six agricultural divisions in Madurai District TANWA beneficiaries groups are more in Thirumangalam Agricultural division and hence the same has been selected. A list of farm women was obtained from the office records maintained by the concerned TANWA personnel. They were arranged in alphabetical order. From this list, one hundred and twenty respondents required for this study has been selected by using random sampling method. The respondents were interviewed personally. A well structured, pre-tested interview schedule and suitable statistical techniques were used to analyse the collected data.

Training need was assessed in specific items under each enterprise by the use of a four point rating scale, the points being 'Very much needed', 'Needed', 'Less needed' and 'Not needed' scores of 4, 3, 2 and 1 respectively. This measures were exclusively developed for this study.

Findings and Discussion

Training need

Training need can be defined in terms of gap between requirements and performance. The gap could be in terms of knowledge, skill and attitude. Training is considered to be an essential factor that has its own influence in gaining knowledge and skill in respect of the subject matter that is taught during the training. Hence, attempts were made to identify the specific areas in which the farm women required training under each enterprise. The training needs under various subject matter areas of layer production were assessed and presented in Table 1.

Table 1 revealed that the top preference of training needs was given to medical care by cent per cent of the farm women due to the fact that in poultry faming, even if one or more birds are infected by certain diseases, the flock gets affected within no time and the disease spreads like wild fire. If they undergo training how to deal with such situations they will be able to identify the symptoms early and take appropriate preventive measures to avoid high mortality of birds.

The next order of priority noticed was with regards to feeding as expressed by more than three fourths (86 per cent) of the respondents, since the cost of feed in poultry production accounts to the extent of 70 per cent of the total production cost. But by own feed mixing, the cost of production of feed would be minimized. Therefore proper knowledge and skill on all dimensions of feeds and feeding are of paramount importance to the poultry growers.

Table 1 : Training need areas in layer production

(n=106)

SI. No.	Items	Very much needed		Needed		Less needed		Not needed	
		No	%	No	%	No	%	No	%
	Housing								
1.	Cleaning and disinfecting the house	-	-	29	27.36	23	21.70	54	50.94
2.	Cleaning and disinfecting the equipment	21	19.81	23	21.70	19	17.92	43	40.57
3.	Temperature maintenance	8	7.54	23	21.70	32	30.19	43	40.57
4.	Fumigating the house	-	-	23	21.70	39	36.79	44	41.51
	Feeding								
5.	Feeding and watering	56	52.83	20	18.87	10	9.43	20	18.87
6.	Selection of breed	7	6.60	23	21.70	35	33.02	41	38.68
7.	Grading of eggs	-	-	23	21.70	44	41.50	39	36.80
8.	Storage of eggs	4	3.78	23	21.70	42	39.62	37	34.90
9.	Removal of culled birds	45	42.26	10	9.43	26	24.52	25	23.59
Litter management									
10.	Removal of old litters, spreading of old litter, maintaining the moisture level of litter, rating of litter	-	-	8	7.54	48	45.28	50	47.18
Medical care									
11.	Vaccination	67	63.21	35	33.01	4	3.78	-	-
12.	Providing	73	68.87	35	31.13	-	-	-	-

	medicated feed								
13.	Disease control measures	67	63.20	39	36.80	-	-	-	-
14.	Deworning	66	62.26	20	18.87	20	18.87	-	-
15.	Delousing	61	57.54	20	18.87	25	23.59	-	-
Marketing aspects									
16.	Selling of eggs and culled birds	45	42.46	10	9.43	26	24.52	25	23.59

The next preference was in favour of culling, since the unproductive, unhealthy and weaklings in a flock consume the feed without any benefit to the farmers and bring forth economic loss. The farmers also preferred training in the area of marketing. Since the farmers were dependent on traders for marketing, they wanted to know about the marketing strategies to realize remunerative prices. This was expressed by 81 per cent of the farm women.

The housing contributed the indirect effect on the production and productivity of the birds. This might be the reason for the less preference of training needs,in housing (27.60 per cent). These findings are in line with the findings of Peru Mathialagan and Subramanian (1998).This is the typical unfelt need of the farmers which have to be brought to the surface, so that the farmers can feel and realize the real need for further adoption.

Training needs in specific items under mushroom production

From the Table 2, it could be concluded that 'contamination of beds', 'sterilization of spawn bottles', 'inoculation of spawn bottles', 'avoiding contamination in bottles', 'preparation of substrate for mushroom bed preparation', 'maintaining mushroom beds in cropping room' were the areas in which cent per cent of the farm women expressed the need for training. It is worth noting the cent per cent of the respondents felt that they needed training. This calls for organising a massive training programme on the felt need areas of farmers in mushroom production.

Table 2 : Training need areas in mushroom production

(n=71)

SI. No.	Items	Very much needed		Needed		Less needed		Not needed	
		No	%	No	%	No	%	No	%
1.	Sterilization of spawn bottles	39	54.93	15	21.13	17	23.94	-	-
2.	Inoculation of	32	45.07	23	32.40	16	22.53	-	-

	spawn bottles								
3.	Avoiding contamination in bottles	27	38.03	28	39.44	16	22.53	-	-
4.	Preparation of substrate for mushroom bed preparation	27	38.03	31	43.66	13	18.31	-	-
5.	Preparation of bed	28	39.44	30	42.25	13	18.31	-	-
6.	Avoiding contamination in beds	42	59.16	16	22.53	13	18.31	-	-
7.	Maintaining mushroom beds in cropping room	24	33.80	32	45.07	14	19.72	1	1.41
8.	Harvesting	6	8.45	26	36.62	20	28.17	19	26.76
9.	Packing	-	-	18	25.35	34	47.89	19	26.76

Mushroom cultivation involves relatively more simple techniques and small investments. Mushroom production is of such activities which can easily be taken especially by farm women as a leisure time activity which provides substantial income. This might be the reason why the farm women wanted training in the above mentioned areas. If trainings are given to them, that will enable them to carry out mushroom production in a successful way.

Harvesting and packing were the areas in which nearby three - fourths (73.24 per cent) of the farm women required training. Since harvesting and packing plays a very important role in preserving the produce. The respondents wanted to get training in these aspects, however 26.76 per cent were not opted for that probably those group might had possessed skill on those areas.

Training needs in specific items under value addition of fruits and vegetables

From the Table 3, it could be inferred that all the respondents (100 per cent) wanted to get training in all critical aspects of value addition of fruits and vegetables. Since every step in preservation and processing plays a vital role in getting the quality produce, they preferred to get training in the areas like selection of fruits and vegetables, preparation, bottling, marketing and storage.

Table 3 : Training need areas in value addition of fruits and vegetables

(n=78)

SI. No.	Items	Very much needed		Needed		Less needed		Not needed	
		No	%	No	%	No	%	No	%
1.	Selection of fruits and vegetables	12	15.39	48	61.54	18	23.07	-	-
2.	Washing	12	15.39	21	26.92	45	57.69	-	-
3.	Peeling	12	15.39	21	26.92	44	56.41		
Preparation									
4.	Pulping	14	17.95	31	39.75	33	42.30	-	-
5.	Testing sugar content (refractometer - brix)	60	76.92	16	20.51	2	2.57	-	-
6.	Preparation of sugar : pulp : water	61	78.20	14	17.95	3	3.85	-	-
7.	Preservatives	62	79.49	16	20.51	-	-	-	-
8.	Artificial colour	59	75.65	19	24.35	-	-	-	-
9.	Essence	49	62.83	29	37.17	-	-	-	-
Bottling									
10.	Sterilization of bottles and vessels	48	61.54	30	38.46	-	-	-	-
11.	Avoiding contamination while filling	50	64.10	24	30.76			-	-
Marketing and storage									
12.	Marketing outlet	43	55.12	32	41.03			-	-
13.	Storage	30	38.46	42	53.85			-	-

Constraints of farm women in starting layer production

It could be observed from the Table 4 that 'high cost of feed' (83.33 per cent), 'problem of disease' (75 per cent), 'high capital investment' (70.83 per cent), 'high price for construction of cages' (66.66 per cent) and marketing (66.66 per cent) were perceived as major problems as expressed by more than two - thirds of the sample.

Sl.No.	Items	No	%
1.	High cost of feed	100	83.33
2.	Problem of disease	90	75.00
3.	High capital investment	85	70.83
4.	High price for construction of cages	80	66.66
5.	Marketing	80	66.66

* Multiple responses

The problem such as high cost of commercial feed was expressed by 83.33 per cent of the farm women. The respondents had to depend upon the traders for commercial feed. The mounting cost of the feed was the major problem. Feed is a daily input in poultry unit which takes the lions share in the production cost. A bag of 75kg feed costed more than Rs.600. Farmers who had a very large unit, can afford to maintain a feed mixing unit. But for small and medium farmers, they could not afford a feed mixing unit because it required heavy capital investment. The next important problem was the disease incidence. Several diseases are very common in poultry birds. The most important is the infectious Bursal disease. The major epidemic IBD had ravaged the poultry farms a few years ago. There were cases where people had even lost almost all the birds. Even though vaccines had been identified, due to huge growth of population of birds and poor hygiene, the virulence of the disease had also increased. So, the farm women (75.00 per cent) were so reluctant to carry out layer production. The same was quoted by Taneja (1990).

Construction of cages is very important in poultry. It is more convenient for providing water, feed and collection of eggs. More birds can be accommodated in the cage system than the deep litter system also possible to maintain a better hygiene in cage system. But, installation of the cage system required a high capital investment. Without the shed, for cage alone, it would cost around Rs.50 for a bird. It was felt as too costly for average farmers to afford it. Even without a cage system, starting a poultry unit itself was a capital intensive one as expressed by the farm women (70.83 per cent). It required huge sheds and also materials for keeping feed, water, collection of eggs etc., High price for construction of cages stated by the sample (60.66 per cent). This is in conformity with the research finding of Thenamudha (1996).

Marketing of poultry products was one of the problems of the sample (66.66 per cent). The farmers, because they got feed from the traders, to compensate the feed cost they have to sell the eggs to those private traders. So the private traders enjoyed more profit by the egg production.

Problems of farm women in starting mushroom production as an enterprise

It could be seen from Table 5 that, marketing of mushroom was the foremost problem as expressed by cent per cent of the respondents. Mushrooms being highly perishable with less keeping quality can cause great losses to growers if they are not lifted for sale immediately. Some of the suitable methods were evolved commercially for increasing shelf - life of mushroom include canning, dehydration and freeze drying. These techniques are sophisticated and costly and are not practicable at grass root level. (Mamoni Das, 1997). Therefore it needs quick marketing. The next important problem of 'lack of financial assistance' was expressed by 83.33 per cent of the farm women. For rural women to start an enterprise financial assistance is a must. Even though financial assistance is given by various organizations, they are not getting it at proper time because of the long procedures in sanctioning loans. This result derived the support from the finding of Jeyalakshmi *et al* (1997).

Table 5 : Problems of farm women in starting mushroom production as an enterprise

(n=120)

SI.No.	Items	No	%
1.	Marketing problem	120	100
2.	Lack of financial assistance	100	83.33
3.	Non-availability of spawn	95	79.16
4.	Preservation of produce	20	16.68
5.	Problem of pests and diseases	20	16.68

* Multiple responses

The basic input in mushroom production is the spawn. It is not possible for the rural masses to produce their own spawn as it requires sophisticated and automatic machinery and other facilities. Shortage of spawn growing laboratories is one of the main constraints to this industry. So, more than three - fourths (79.16 per cent) of the farm women expressed 'non availability of spawn' as a constraint. `Preservation of produce' and 'pests and disease attack' were the minor problems expressed by only 16.68 per cent of the respondents.

Problems encountered by the farm women in value addition of fruits and vegetables

The results presented in Table 6, revealed that marketing was the major constraint as expressed by cent per cent of the respondents. Usually the produce of the rural women have no brand name. So, they find it difficult to sell their produce in cities and towns. This finding is in line with finding of Jeyalakshmi *et al* (1997).

Table 6 : Problems encountered by the farm women in value addition of fruits and vegetables

(n=120)

SI.No.	Items	No	%
1.	Marketing problem	120	100
2.	Lack of training	100	83.33
3.	High cost of fruits	98	81.66
4.	Lack of financial assistance	90	75.00
5.	Seasonal availability of fruits	80	66.66

* Multiple responses

More than three - fourths (83.33 per cent) of the farm women expressed 'lack of training' was the next major constraint in value addition of fruits and vegetables. Even though the farm women got trained they preferred further more training in some critical areas.

'High cost of fruits' and 'lack of financial assistance' were the major threatening expressed by the farm women (81.66 per cent and 75.00 per cent respectively) in this enterprise. Since majority of the farm women were coming under the low to medium annual income group, they could not able to invest more money. This result derived the support from the finding of Himachalam (1990),In one season, fruits and vegetables are available in abundance. The rural women are not able to convert and preserve these fruits and vegetables into marketable products within a limited time. And they didn't have any improved facility to store the excess produce and make it available during the period of glut. So, more than half (66.66) of the farm women expressed that 'seasonal availability of fruits and vegetables' as the important constraint.

Conclusion

This study revealed that even though the rural women have trained in various technologies, they required further more trainings in major areas. So, this identification of the area in which they are really lack in training, is essential so that training can be imparted in the required areas to the farm women to become a good entrepreneur.

The need for training and guidance at all stages could be one of the most effective ways of dealing with problems of farm women that hinder taking up self employment avenues. The major problems of credit, non-availability of inputs, lack of marketing facilities need to be rectified to build up the farm women as a successful entrepreneur.

References

Jeyalakshmi, G. S. Shilaja, and G. Sobhana, 1997. Constraints experienced by women Entrepreneurs in Kerala. Journal of Extension Education. 8(3): 1752-1754.

Himachalam, D. 1990. Entrepreneurship Development in Small Scale sector. Yojana. 34(3): 16-18.

Mamoni Das, 1997. Mushroom cultivation for rural women. Social Welfare. 44(3): 34-35.

Peru Mathiyalagan and R. Subramanian, 1998. Training needs of poultry farmers in scientific poultry management. Journal of Extension Education. 9(3): 2099-2101.

Taneja, B.S. 1990. Prospects and problems of poultry farming in Punjab. Poultry Guide. 27(12): 81-84.

Thenamudha, V. 1996. Entrepreneurial behaviour of farmers-A critical analysis. Unpub. M. Sc.(Ag.) Thesis, TNAU, Coimbatore.

●●●

39

Perception of Women SHG Members on Training Programmes

***Santha Govind*[1] *and T. Mukesh*[2]**
[1]*Profesor in Agrl. Extension and*
[2]*PG Student (Agri. Extension), Annamalai University, Tamilnadu*

Introduction

Women are vital part of the Indian economy, constituting one-third of the national labour force and are considered to be the major contributor to the survival of the family. Despite progress in several indicators, gender analysis of most social and economic data demonstrate that women in India continue to be relatively disadvantaged in matters of survival, health, nutrition, literacy and social status. More than 90 percent of rural women in India are unskilled thus restricting them to low-paid occupations. Women generally have no control over land and other productive assets, which largely excludes them from access to institutional credit and renders them dependent on high-cost informal sources of credit to secure capital for consumption and for productive purpose.

Among 621.1 lakh population in Tamilnadu (according to 2001 census) 308.4 lakh are females. The people below poverty line in Tamilnadu are about 21.12 percent, out of this, women share equal percentage of poverty. In a patriarchal society, women have inadequate knowledge on various aspects and depend largely on the male siblings or husbands. They also do not have much access to credit or income as most of the expenditure is controlled by men. Even where they have some measure of control they spend most of their earnings on family needs. Estimates shows that more women and girls experience the vigour of poverty than men in poor holds because of inequalities in access to food, healthcare and education.(www.undp.org.in).

Kuhn (1985) defined SHGs as organisations whose members have united on the basis of common interest to improve their economic and social conditions in order to pursue their paramount long term goals through joint action and self help.

The members of the SHGs are provided systematic training to bring about qualitative changes in their attitude and to promote cohesion and effective functioning of the group. All the SHG members are imparted training in 4 modules for 4 days. The primary objective of this training is to orient all members to the SHG concept and bring out the hidden talents and capacity of all the members. In addition, SHG members who are interested in starting economic activities or to develop skills to get self employment are provided skill training.

The main objectives of SHG trainings are to develop leadership quality, self confidence, increase social awareness, improve states of women in family and society, improve health and family welfare, functional literacy, increase assets, and inculcate the habit of saving and to develop economy. SHGs can become an extraordinary tool for women empowerment provided the nurturing agencies take care to design their interventions to improve the confidence level of members. Keeping this in mind, a study was taken up to assess the usefulness of the training programmes as perceived by women SHGs members.

Methodology

The study was conducted in Thanjavur district of Tamil Nadu state. Thanjavur district has fourteen blocks, among them Kumbakonam block was selected as more number of SHGs were involved in farm oriented activites in this block. There were eight NGOs functioning under Tamil Nadi' Women Corporation for Development Women (TNCDW) in the block. Among the eight NGOs, KMSSS had the maximum number SHGs were engaged in farm and allied activities. The NGO named Kumbakonam Multipurpose Social Service Society (KMSSS) promoting . agriculture and alied activities was working sucesfuly towards the development of SHG women members. Hence, the SHGs functioning under the NGO KMSSS was considered for the study.

It was decided to select the top three enterprises wherein more number of SHGs were involved in farming and allied activities. Accordingly two SHGs involved in paddy cultivation, two SHGs involved in dairy management and two SHGs involved in mushroom cultivation were selected based on the maximum loan utilized in the respective income generating activites. Thus, finally six SHGs comprising of 120 SHG members were selected as sample for the study. Perception was defined as a set of realization of an individual toward the technologies in the form of their usefulness towards higher productivity. Similar operationalisation was done by Vengatesan (2001).

The subject matter areas of the women Self Help Group trainings were listed from the Mahalir thittam working manual. The response on perceived usefulness of the training on these technologies was obtained from the Self Help Group women respondents as either useful or not useful. The percentage analysis was used to analyse the data.

Findings and Discussion

Perceived usefulness of training programmes

Results on perception on the usefulness of the training programmes of the beneficiaries towards 'Self Help Group members training', 'communication' and 'activity based skill training' were analyzed and discussed in the tables from 1-3.

Usefulness of Self Help Group members training programme

The various training areas offered under Self Help Group members were listed and the perception of women respondents towards the training were studied and the results are given in Table 1.

Table 1 : Usefulness of Self Help Group members training programme *

Sl.No.	Subject matter	Number	Per cent
1.	SHG orientation	76	63.33
2.	Leadership	62	51.66
3.	Accounts maintenance	35	29.16
4.	Literacy	73	60.83
5.	Health and nutrition	78	65.00
6.	Activities for village development	48	40.00
7.	Importance of group reserve fund	54	45.00
8.	Group income generation programme	62	51.66
9.	Women development	74	61.66
10.	Legal rights	26	21.66
11.	Better/Sustained development of family	57	47.50

12.	Definition of environment and ecology	36	30.00
13.	Afforestation	39	32.50
14.	Soil and water conservation	36	30.00
15.	Watershed management (Basics)	35	29.16
16.	Ecological balance	10	8.33
17.	Towards a green village programme	80	66.66
18.	Financial independence	36	30.00
19.	Independence from Govt. and NGOs	59	49.16
20.	Need for sustainability of SHGs	86	71.66
21.	Strategies for sustainability of SHGs	39	32.50
	Mean percentage	-	**45.55**

* Multiple response

It may be seen from Table 1, that out of the 21 subject matter areas offered, the subject matter areas viz., need for sustainability of SHGs, health and nutrition, SHG orientation, literacy, women development, towards a green village, independence from government and NGOs, leadership and group income generation programme were perceived as useful ranging from 51.66 to 71.66 per cent. In general, majority of the SHG women were motivated to join SHG groups and were aware about its impact on their socio-economic status, its education generating confidence and countering difference. These have been viewed as useful due to the motivation and enthusiasm created among the target groups and also due to imaginative approaches to course content making them relevant to the day-to-day needs of women target. Leadership training improved the self-confidence and self-reliance among the group members and changed the SHG womens' attitude positively towards a green village. These findings are in line with the findings of Chaturvedi *et al.* (1988).

The trainings wherein only less than fifty per cent of the respondents perceived them as useful were viz., on sustained development of family (47.50 per cent), importance of group reserve fund (45.00 per cent), activities for village development (40.00 per cent), strategies for sustainability of SHGs (32.50 per cent), afforestation (32.50 per cent), soil and water conservation (30.00 per cent), watershed management (29.16 per cent), financial independence (30 00 per cent), definition of environment and ecology (30.00 per cent), accounts (29.16 per cent), legal rights (21.66 per cent) and ecological balance (8.33 per cent). Hence, it may be inferred that most of the subject matter areas was perceived as useful by less than fifty per cent because Self Help Group members already knew about sustained development of family, group reserve fund and activities for village development.

Lack of awareness about legal rights was perceived to be less useful to women in the rural areas found it difficult to practice legal provisions. The respondents were aware of the other training areas viz., ecological balance and legal provisions and were practicing them. The items namely soil and water conservation, watershed management and accounts maintenance were perceived as not useful as they needed a greater degree of skill and literacy among the members.

Radharani and Laxmidevi (1992) had studied the problems of biogas beneficiaries. They inferred that the beneficiaries expressed the training programme as less useful to them. The results of the study are in conformity with the outcome of the relevant study.

Usefulness of training programme on communication

The various subject matter areas of training under communication were listed and the perception of women respondents towards the training were studied to understand the usefulness of training on communication and the results are presented in Table 2.

It could be seen from Table 2, that only four subject matter areas were perceived as useful by three-fourth of the Self Help Group members. They were trainings on thrift and savings, kitchen garden, group unity and collective bargaining.

Of the four, the training on thrift and savings (95.00 per cent) has been practiced by all the women and hence they might have mentioned it as highly useful. This might be possible as Self Help Groups inculcated the habit of thrift and savings operations among themselves and another reason being that more than just being a credit programme this intervention is targeted at bringing about intrinsic attitudinal changes within the communities for the effective use of savings and thrift towards their livelihood strengthening. However, the training on group unity was perceived as useful by 82.50 per cent of the SHGs members.

Group unity training improved group activities and group communication with each other to a greater extent and generally behave in ways designed to promote integration.

Table 2 : Usefulness of Training on Communication*

Sl.No.	Subject matter	Number	Per cent
1.	Literacy	96	63.33
2.	Thrift and savings	114	95.00

3.	Group unity and achievements	99	82.50
4.	Gender sensitization, status of women gender and development	75	62.50
5.	Education of girl child	75	62.50
6.	Child labour	79	65.83
7.	Personal hygiene	84	70.00
8.	Environmental sanitation	62	51.66
9.	Water borne diseases	97	22.50
10.	Low cost sanitation	79	65.83
11.	Soak pits	15	12.50
12.	Kitchen garden	64	53.33
13.	Safe motherhood	72	60.00
14.	Small family	76	63.33
15.	Common property resources and their protection	62	51.66
16.	Early marriage	86	71.66
17.	Evils of dowry	72	60.00
18.	Evils of female infanticide	27	22.50
19.	Evils of alcoholism	78	65.00
20.	Abolition of bonded labour	40	33.33
21.	Communal harmony/Caste harmony/ Human dignity/ Human rights	52	43.33
22.	Legal rights of women	27	22.50
	Mean percentage	-	**54.09**

* Multiple response

Chouhan (1997) had studied the eradication of poverty under million wells scheme. He had found that most of the respondents viewed the million wells training programme as more useful and was eradicating the poverty in rural areas which is in line with the observed findings.

The next group of subject matter areas to have emerged as useful in the perception of SHGs members were personal hygiene (70.00 per cent), early marriage (71.66 per cent), child labour (65.83 per cent), low cost sanitation (65.83 per cent), evils of alcoholism (65.00 per cent), small family norm (63.33 per cent), gender sensitization (62.50 per cent), education of girl child (62.50 per cent), safe motherhood (60.00 per cent) and evils of dowry (60.00 per cent). This might be due to the following reasons. Despite the prevalence of child marriage, registration marriage of girls under eighteen is still common in rural areas, partly because of lack of awareness and partly because of socially entrenched customs, but the women after undergoing Self Help Group training became aware about the evils of early marriage, evils of alcoholism, benefits of small family norms and evils of dowry system. Hence, this training might have been perceived as useful by the respondents of the study.

The status of women in society as individuals in their own right was consistently given priority in all development programmes. Hence, the above trainings would have been perceived as useful as they were oriented towards the development of women.

The training on kitchen garden was perceived as useful among 53.33 per cent of the respondents. Only fifty per cent of the women perceived the trainings viz., protection of common property resources and environmental sanitation to be useful. The above trainings were oriented towards social development and economic upliftment. The SHG members were more interested because of their social and economic backwardness.

Laharia and Bhatti (1991) reported that the most of the beneficiaries viewed the biogas technology to be highly useful. The report of the study agrees to this finding. The next group of training to have emerged as useful in the perception of SHG members were communal harmony (43.33 per cent), abolition of bonded labour (33.33 per cent), water borne diseases (22.50 per cent), evils of infanticide (22.50 per cent), legal rights of women (22.50 per cent) and soak pits (12.50 per cent). Respondents lacked awareness and there was poor participation on the above said training programmes.

Erappa (1988) had studied the Integrated Rural Development Programme as catalyst for upliftment of schedule castes and schedule tribes. He had inferred that the members perceived less usefulness of IRDP training programmes. The report of the study is in conformity with the present findings.

Usefulness of activity based skill training programme

The various trainings under activity based skill training were listed and the perception of women respondents towards the training was studied and is given in Table 3.

Table 3 : Usefulness of activity based skill training programme*

Sl.No.	Subject matter	Number	Per cent
1.	Theoretical and practical input on activity	40	33.33
2.	Marketing and other linkages	30	25.00
3.	Economics of the activity	45	37.50
4.	Loan and repayment	104	86.66
5.	Field visit	15	12.50
6.	Insurance	82	68.33

7.	Clarifications	76	63.33
8.	Care and management of assets	99	82.50
9.	Repayment and insurance	99	82.50
10.	Linkages	95	79.16
	Mean percentage	-	**73.41**

* Multiple response

It could be seen from Table 3, that more than 75.00 per cent of the respondents perceived seven subject matter areas to be useful viz., loan and repayment (86.66 per cent), economics of the activity (82.50 per cent), care and management of assets (82.50 per cent), repayment and insurance (82.50 per cent) and linkage (79.16 per cent).

Most of the trainings were oriented towards economic development and to improve their socio-economic status. Hence, the women would have been interested to participate and to adopt them very easily, resulting with high perception. SHG are fast emerging as a promising tool for promoting income generating enterprises. Further, SHGs can contribute towards improving the quality of lending by offering loan in prompt and simple manner, ensuring need based loan and keeping the loan size within the repaying capacity of the borrowers.

For the remaining two trainings, more than fifty per cent of the respondents perceived them as useful namely, insurance (68.33 per cent) and clarifications (63.33 per cent). Regular and sustained income obtained from the above trainings might be the possible reason.

The training on theoretical and practical inputs on activity (33.33 per cent), marketing and other linkage (25.00 per cent), field visit were perceived as useful only by a very few respondents (12.50 per cent) as most of the activity based skill trainings were conducted' at the institution with less field visits. This finding is in agreement with the findings of Erappa (1988), Chouhan (1997) and Vengatesan (2001).

Conclusion

The subject matter areas of training programme should be tailor made for SHG women focused on income generating programmes, as most of the respondents perceived the training on income generation programme as not useful. The other subject matter areas need to be revised and updated to make it highly useful for the SHG women members. Further, the subject matter areas viz., linkage, clarification, insurance and field visit need to be modified to make it highly useful for the SHG members.

References

Chaturvedi, Y.S., Naidu, K.K. and M.S. Sridhar. 1998 "Beneficiaries of IRDP in Gujarat, Karnataka and Rajasthan: A study, "Kurukshetra, 36 (12): 25-28.

Chouhan, S. 1997. "Eradication of Poverty: Million Wells Scheme", Yojana, 41(7): 15-16.

Kuhn, J. 1985. The Role of Non-Government Organizations in Promoting Self Help Organizations. Seminar papers, Druckavi Franz, Paffenholz Bomheins (FRG): 265.

Erappa, S. 1988. "IRDP as catalyst of Upliftment of SCs and STs", Kurukshetra, 40 (12): 25-29.

Laharia, S.N. and S.K. Bhatti. 1991. "Constraints in the Choice of Traditional Rainfed Low Land Rice Varieties", IRRI Research Paper Series, 154: 5-17.

Radharani, N and A. Laxmidevi. 1992. "Problems of Biogas Beneficiaries", Indian Journal of Extension Education, 28 (3&4): 44-48.

Vengatesan, D. 2001. Impact of Women Self Help Groups Organised by NGOs Under TNCDW of Cuddalore District, Unpublished M.Sc. (Ag.) Thesis, Annamalai University, Annamalai Nagar.

●●●

40

Positive Therapy for Enhancement of Self-Esteem, Self-Efficacy and Management of Stress among Women in Local Governance in Coimbatore District

M. Shanmugavani[1] *and K.C. Leelavathy*[2]
[1]*Ph. D, Scholar and*
[2]*Professor, Department of Home Science Extension Education, Avinashilingam University for Women, Coimbatore-43, Tamil Nadu.*

Introduction

Women in India had a high and glorious tradition. All nations have attained their pinnacle of glory only when women have been free, cultured and pure. Women have been the transmitter of culture in all societies. Modern living has brought with it not only innumerable means of comfort but also a plethora of demands that tax human body and mind. Stress is an inescapable part of human life is manageable to a large extent. With proper understanding of the processes that cause stress, the situation can be well managed. The stress management is very easy if we sincerely analyze the reasons of stress. Physical stress management techniques have been used since a long time.

The investigator had undertaken the study entitled "Positive Therapy for enhancement of Self-Esteem, Self-Efficacy and Management of Stress among Women in Local Governance in Coimbatore District".

Methodology

The area selected for the present study is Coimbatore District in Tamil Nadu State. The Coimbatore District comprises of two corporation and ten municipalities with 130 women councillors. Among 130 women councillors, 63 had attended the programme. The tools used for conducting the research were questionnaire, Rosenberg self-esteem scale, General self-efficacy scale, Stress inventory and Anxiety inventory.

The Coimbatore district comprises of two corporation and ten municipalities with 130 women councillors. Among 130 women councillors, 63 alone attended the programme.

Finding and Discussion

The findings of the study are discussed under the following:

Mean and t-values of Self-esteem and Self-efficacy

The mean and t-values of self-esteem and self-efficacy are presented in Table 1.

Table 1 : Mean and t-Values of Self-Esteem and Self-Efficacy

n=63

Variables	Before training	After Training	Difference	t — value
Self-esteem	75.79 ± 9.16	85.44 ± 9.96	9.65	5.66
Self-efficacy	86.67 ± 10.77	97.14 ± 3.69	10.47	7.30

**Significant at 0.01 level

The above table shows that mean values of self-esteem and self-efficacy were assessed. The mean values of self-esteem and self-efficacy were increased after training programme. The mean difference before and after training as revealed by t-value is significant at 0.01 level for both selfefficacy and self-esteem. It is evident that the programme has a good impact on the personality development of women councilors.

Mean and t-values of Stress and Anxiety

The mean and t-values of stress and anxiety are given in Table 2.

Table 2 : Mean and t-values of Stress and Anxiety

n=63

Variables	Before training	After Training	Difference	t — value
Stress	12.24 ± 4.79	5.73 ± 4.24	6.51	14.58
Anxiety	9.94 ± 7.08	4.06 ± 6.05	5.88	10.93

**Significant at 0.01 level

The mean values of stress and anxiety level of women councilors were assessed. The mean values of stress and anxiety were found to be increased after training programme. The mean difference before and after training as revealed by t-value is significant at 0.01 level for both stress and anxiety level. Hence it is clear that the programme has a good impact on the personality development of women councilors and has reduced the stress and anxiety among women councilors.

Symptoms of Stress Experienced by the Sample Before and After Training

Table 3 explains the symptoms of stress experienced by the women councilors before and after training.

Table 3 : Symptoms of Stress Experienced by the Women Councilors

Aspects	Percentage of Women Councillors (N:63)	
	Before Training	After Training Lack of
Lack of sleep	92	44
Carelessness	92	30
Incapable of decision making	89	25
Lack of appetite	87	25
Disturbed sleep	75	16
Disturbed breath	65	22
Confusion	62	22
Headache	56	25
Stomach ache	56	24
Day dreams	40	-
Ill health	40	6
Restlessness	40	5
Giddiness	35	5
Indigestion	32	3
Loss of weight	22	10
Sweating	17	8

*Multiple responses

The Table 3 depicts the symptoms of stress experienced by the women councillors before and after training. Majority (92 per cent) of them experienced lack of sleep and carelessness followed by struggle in decision making (89 per cent) lack of appetite (87 per cent), indigestion (32 per cent), loss of weight (22 per cent) and sweating (17 per cent) before training as a symptoms of stress experienced by women councillors. After training programme 44 per cent experienced lack of sleep, 30 per cent carelessness five per cent restlessness, giddiness and three per cent indigestion.

Advantages Expressed by Women Councillors Through Music and Dance Therapy

Music has a harmonizing effect on pain regulating neuro endocrine functions, important opiods of the pain inhibiting systems. Music gives enormous stimulation to the secretion of endorphins and encourage faster healing. It can be perceived by the patient. Music strengthens the immune activity and paves to good health. Music increases the activity of immune system too.

Dance therapy is the therapeutic use of movement to improve the mental and physical well being of a person. It focuses on the connection between the mind and body to promote health and healing. Benefits expressed by women councillors through music and dance therapy are given in Table 4.

Table 4 : Advantages Expressed by Women Councillors through Music and Dance Therapy

Benefits	% of Momen Councillors (N:63)
Improvement in concentration and memory power	81
Creating peace of mind	78
Relaxation	70
Stimulation of creativity	62
Reduced stress	60
Leads to good health	54
Creation of happy atmosphere	49
Sleeping well	46
Avoid fear	40
Developed self conf	33
Reduced Anxiety	29

*Multiple responses

The above Table expressed that most of the women councillors benefited through music and dance therapy. Eighty one per cent reported that music and dance therapy helped to improve their concentration and memory power, while 78 per cent of them perceived therapy as a capable of creating peaceful mind in one's life, 70 per cent felt that it helped the individual as a relaxation, stimulation of creativity(62 per cent), reduced stress (60 per cent), leads to good health (54 per cent), avoid fear (40 per cent) developed self confidence (33 per cent) and reduced anxiety (29 per cent).

Conclusion

Women today are storming all the male bastions and proving themselves to be equally good in all most all walks of life. Women in India today are also poised to take off. They are standing on the threshold of a new era. It should be realized that every issue is a woman's issue from water to militarization, violence to economic planning, ecology to economic development and from kitchen to parliament. Women are making efforts round the world to see that their rights are respected, their voices heeded, their opportunities widened. The "Voice from the kitchen" is being heard in international forum.

References

Harshipinder and Anjla Paramjit, 2001. "Physical Stress Management among Women", Psychological Studies, National Academy of Psychology, India, 46 (1-2) : 69.

Janaki, D., 2001. "Women issues- Perspectives from Social History", Dhana Publication, Chennai,p.11.

41

Role of Rural Women in Farming, Allied and Off-Farm Activities

P. Sumathi[1] and M. Senthilkumar[2]
[1& 2]Krishi Vigyan Kendra, ARS, Virinjipuram — 632 104
Vellore dt, Tamil Nadu

Introduction

Women constitute half of the Indian population. Around 80 % of the total female population lives in rural and out of 31 million women work force of the country, 20 million are living in rural areas. The role played by women in agriculture is more enormous and fully recognized. Women accounted for more than 76 per cent of the marginal workers and about 16 per cent of them are main workers. Among main workers as well as marginal workers more women belong to the category of agricultural labourers. The role of women and their involvement is more in production, processing and storage of grains. In general, 60-70 per cent of labour input is provided by womenfolk which increases upto 80 per cent in rice. Women formed part of a highly valuable human resources which with appropriate training and education, can bring about phenomenal changes in the desirable direction. By keeping this view, the main focus of this paper was, to find out the role played by the farm women in farming , allied and off-farm activities.

Methodology

The study was conducted in Tirupattur taluk of Vellore district. Ten villages were randomly selected. A list of farm women families were obtained from Village Administrative Officers. A sample of 360 respondents was selected using proportionate random sampling method from 10 villages of Tirupattur block of Vellore district. Out of which 210 were engaged in farming, 90 were in farm and allied activities and 60 were in farm and off-farm activities. The data were collected by using a pre-tested structured schedule through personal interview method. The data were tabulated and interpreted with suitable statistical tools.

Findings and Discussion

A. Role of women in farm activities

Table 1 : Role of farm women in farm activities

(n=210)

SI. No.	Role performance	Farm women	
		No.	%
	I. Land preparation		
i)	Ploughing, puddling and levelling	-	-
ii)	Stubble collection	55	26.19
iii)	Application of manures	25	11.90
iv)	Cleaning of field boundaries	30	14.28
v)	Farming ridges and furrows	-	-
	II. Nursery preparation		
i)	Selection of seeds	167	79.52
ii)	Seed treatment	152	72.38
iii)	Sowing the seeds	210	100.00
iv)	Irrigating the nursery	-	-
v)	Plant protection in nursery	-	-
	III. Transplanting		
i)	Pulling out the seedlings from the nursery	180	85.71
ii)	Bio-fertilizer application	62	29.52
iii)	Transporting the seedlings	91	43.33
iv)	Transplanting the seedlings	210	100.00
	IV. Inter-cultivation		
i)	Irrigation	-	-
ii)	Cleaning the irrigation channels	60	28.57
iii)	Thinning and gap filling	210	100.00
iv)	Earthing up	192	91.42
v)	Detrashing	102	48.57
vi)	Weedicide application	-	-
vii)	Top dressing	51	24.28
viii)	Plant protection measures	-	-

	V. Harvesting		
i)	Draining the water	-	-
ii)	Harvesting	210	100.00
iii)	Collection and heaping	152	72.38
iv)	Bundling	150	71.42
v)	Carrying to the yard	162	77.14

It could be ascertained from the Table 1 that the tasks performed by the farm women under land preparation is very negligible. It must be due to the fact that these activities would have necessitated arduous effort with more physical strain on the part of the doers. Thus the farm women had performed this task in a lesser magnitude.

Majority of the respondents have performed the job of sowing the seeds. About three-fourth majority of them have performed the tasks like selection of seeds and the treatment of those seeds in an effective way. This finding is in line with the findings of

Looking into the result, it may be concluded that most of them have played the role in only two activities viz., thinning and gap filling and earthing up. The role performed by the respondents with respect to other activities was less. The probable reason might be due to more risk involved in these roles and should have made the farm women physically weak to take up the roles. This finding is in conformity with the findings of Balaji (1990).

All the respondents were engaged in harvesting. More than seventy per cent of the respondents have performed the tasks like collection and heaping the produce, bundle out and carry off to the yard. None of them were engaged in draining the water. This finding is in agreement with the findings of Santhy (1991) and Kumari (1998).

Most of them have performed the tasks like drying and storage. The other practices viz., threshing (96.19%), winnowing (77.14%), bagging (76.19%) and transporting (24.76%) were carried out by majority of the respondents. The findings are contrast with the findings of Kumari (1998). Very meagre per cent of the women have been involved in marketing the produce and transporting the produce to market centers. It might be due to the fact that they were men-oriented practices / activities. The finding is in line with the findings of Sujatha (1996), who reported that most of the respondents have not performed the role in marketing.

Apart from agriculture there are innumerable specialized operations attended by the farm women in allied activities. They are dairy management, poultry keeping, goat rearing, sericulture and horticulture respectively.

Dairy activities

Table 2 : Role of farm women in dairy activities

(n=90)

Sl. No.	Role performance	Farm women	
		No.	%
i)	Purchase of animals	62	68.88
ii)	Cleaning the cattle shed	90	100.00
iii)	Cleaning the animals	72	80.00
iv)	Fetching water for animals	65	72.00
v)	Collecting fodder for animals	90	100.00
vi)	Preparation of cattle feed and feeding the animals	62	68.88
vii)	Milking	41	45.55
viii)	Taking care of sick animals	30	33.33
ix)	Collecting cowdung	190	100.00
x)	Making curd, butter and ghee	80	88.88
xi)	Marketing the milk products	55	61.11

It could be inferred from the Table 2 that cent per cent of the respondents have performed certain tasks like cleaning the cattle shed, collecting fodder for animals and collecting cowdung. More than seventy per cent of the respondents had been engaged in making curd, butter and ghee, clearing the animals and fetching water for animals.

It may be inferred that most of them have performed almost all the activities related to dairy farming. This might be due to the fact that they could earn continuously throughout the year for their daily livelihood by the way of getting additional income generated from the dairy enterprise. This finding is in agreement with the findings of Rexlin (1984) and Sujatha (1996).

Poultry keeping

Table 3 : Role of farm women in poultry keeping

(n=90)

Sl. No.	Role performance	Farm women	
		No.	%
i)	Purchasing birds	52	57.77
ii)	Preparation of feed	80	88.88
iii)	Nurturing the birds	70	77.77
iv)	Taking care of sick birds	35	38.88
v)	Cleaning the cages	80	88.88
vi)	Rearing layers	35	38.88
vii)	Collecting of eggs	65	72.22
viii)	Selling of birds / eggs	42	46.66

The above Table 3 reveals that in poultry keeping, all the activities were being carried out by the farm women by themselves. About 88.88 per cent of them were engaged in the preparation of feed and clearing the cages followed by 77.77 per cent in nurturing the birds and 72.22 per cent in collection of eggs. This might be due to the fact that they knew the practices from their childhood through experience.

Sericulture

About one fourth of them have been engaged in picking the leaves, feeding the larvae, disease management and cleaning the chandrikai and rearing room.

Horticulture

Table 4 : Role of farm women in horticulture

(n=90)

SI. No.	Role performance	Farm women	
		No.	%
i)	Establishing kitchen garden	62	68.88
ii)	Nursery management	10	11.11
iii)	Grafting	10	11.11
iv)	Vegetable seed production	15	16.66
v)	Production of vegetable seedlings	20	22.22
vi)	Flower cultivation	50	55.55
vii)	Preparation of pot mixtures	5	5.55
viii)	Harvesting	32	35.55
ix)	Post-harvesting operations	60	66.66

Perusal of the Table reveals that more than fifty per cent of the respondents have been engaged in establishing kitchen garden, post harvest operations and flower cultivation. It may be concluded that more than half of the respondents has been engaged in kitchen gardening. This might be due to the fact that they could have acquired skill through proper trainings and should have got adequate technical know-how on the kitchen gardening. To meet the day to day vegetable requirement for them and also to cater for a minimum local sales at demand they would have gone for this activity.

Role of farm women in off-farm activities

The farm family performing other than farm activities for earning is called as off-farm activity. In the study area there are small scale industries and income generating activities such as (a) small scale industries and (b) income generating activities.

Table 5 : Role of farm women in small scale industries

SI. No.	Activities	Farm women	
		No.	%
i)	Coir pith industry	44	73.33
ii)	Broom making	35	58.33
iii)	Juice factory	25	41.66
iv)	Pot making	5	8.33
v)	Mat making	15	25.00
vi)	Basket making	9	15.00
vii)	Agarbathi making	35	58.33
viii)	Beeding making	47	78.33
ix)	Match box making	37	61.66
x)	Building construction	15	25.00
xi)	Shoe factory	15	25.00

It could be ascertained from the Table 5 that 78.33 per cent of the respondents were found to have been engaged in beedi making jobs followed by 73.33 per cent in coir pith industry, 61.66 per cent in match box making, 58.33 per cent of them in broom and agarbathi making. Less than fifty per cent of them have been engaged in juice factory and building construction work. A meagre per cent was engaged in mat, basket and pot making and also jobs in shoe factory.

It may be stated that most of them had been engaged in small scale industries for getting additional income for the family maintenance.

It was found out from study that cent per cent of them have been engaged in vadagam preparation as well as curd, butter and ghee making. The higher percentage might be due to the fact that these were the activities traditionally being done by the farm women in their home without going outside in search of any jobs. Seventy per cent of them were found to be in rearing cocks for meat. They were rearing the birds not only for their household purpose but also for selling to neighbours in local areas to get additional income. About 41.66 per cent were engaged in pickle and vathal making.

It could be inferred that most of the farm women were found to be engaged in all these activities to get additional income.The finding is in conformity with the findings of Savarimuthu (1981) and Shanthy (1991).

Conclusion

On the basis of above findings, it could be concluded that most of the farm activities viz., nursery preparation, transplanting, inter-cultivation, harvesting and post harvest operations. Majority of the farm women were engaged in tasks of allied activities like dairy, poultry keeping and goat rearing and less respondents in sericulture and horticultural operations. Some of the respondents were engaged the role in small scale industries and other income generating activities to get additional income for family maintenance. Hence, women extension agents should take effort to create awareness and to impart knowledge and skill on production, processing and marketing technologies through periodical training, field trips, group discussion, demonstration and campaign. It would definitely help to increase the role performance of farm women in farm activities.

References

Balaji, S.R. 1990. Role of Farm Women in Groundnut Farming and Allied Agro Enterprises, Unpub. M.Sc.(Ag.) Thesis. TNAU, Coimbatore.

Premavathi, R. 1997. Rural Women in Farm and Home Decision-making, Unpub. M.Sc.(Ag.) Thesis, TNAU, Coimbatore.

Rexlin, R. 1984. Women's Participation in Decision-making on Farm Practices, Unpub. M.Sc.(Ag.) Thesis, TNAU, Coimbatore.

Suguna, K. 1994. A Study on Gender Issue — Pattern of Gender Responsibility in Agriculture, Unpub. M.Sc.(Ag.) Thesis, TNAU, Coimbatore.

Sujatha Jane, J. 1996. Gender Analysis in Different Farming Systems, Unpub. M.Sc.(Ag.) Thesis, TNAU, Coimbatore.

●●●

42

Empowerment Level of Farm Women Beneficiaries of Karnataka Community Based Tank Management Project

B.S. Swetha[1] *and N. Narsimha*[2]
[1]*Ph.D. Scholar, Department of Agricultural Extension*
[2]*Professor Department of Agricultural Extension, GKVK*
University of Agricultural Sciences, Bangalore

Introduction

Agriculture is the base of Indian economy on which the fortunes of over 70 million families directly or indirectly rest. The women in these families are also partners in crop and food production as managers, decision makers and skilled farm workers. The socio economic advancement of a country can be best judged by the status and position, which is bestowed on its women. Women as a class, play a vital role in the process of economic development. According to census 2001, women constitute 48.26 per cent of the total population of India and 25.67 per cent of the female population are designated. Amongst the female workers, 32.50 per cent are identified as cultivators and 39.43 per cent as agricultural labourers.

There is increasing realization of the critical role of women in agriculture and of the fact that empowerment of women is necessary for bringing about sustainable development at a faster pace. Government of Karnataka (GoK) with the financial support from the World Bank, launched tank systems improvement programme entitled Karnataka Community Based Tank Management Project (KCBTMP) during July 2002. The project development objective is to improve rural livelihoods and reduce the poverty by developing and strengthening community-based approaches through proper management of selected tank systems. To achieve this, the project conducted various On-Farm demonstrations (OFD) in water management, agricultural and horticultural development, fisheries, forestry and fodder production to help and ensure that improved water storage and efficiency is translated into increased household incomes.

Hence, Present study was conducted to know the impact of on farm demonstration on level of empowerment of farm women and to know the relationship between Socio-Psychological Characteristics and Empowerment level of the Farm Wotnen Beneficiaries in Kolar district of Karnataka state.

Methodology

Present investigation was carried out in Kolar district of Karnataka state. Two taluks Kolar and Mulbagal were purposively selected for the study where OFD has been organized in the year 2005-06. List of villages where OFD has been organized was collected from Karnataka Community Based Tank Management Project head quarters and the villages were selected randomly and 90 respondents were selected from the villages. Ex-post facto research design was employed for conducting the study. Data was collected by using a detailed interview schedule employing personal interview method. The responses were scored, quantified, categorized and tabulated using statistical methods like percentage, mean and standard deviation, frequencies, chi-square, correlation and regression test.

Findings and Discussion

Empowerment Level of Farm Women Beneficiaries of OFD

Empowering level of farm women measured in terms of process empowerment, product empowerment is presented in the Table 1.

Table 1 : Empowerment Level of Farm Women Beneficiaries

(n=90)

SI. No.	Variable	Categories	Respondents	
			Number	Per cent
1.	Empowerment Mean = 84.03 Standard deviation = 7.11	Low < 79.87 Medium 79.87-86.98 High >86.98	15 42 33	16.67 46.67 36.66

The results revealed that 46.67 per cent of the farm women had medium level of empowerment followed by 36.66 per cent of them having high empowerment level and 16.67 per cent of farm women belonged to low level of empowerment because of their changed attitudes, increased critical consciousness, better self perception and desire for the control in the changing scenario over a period of time.

The findings of the study are in line with observation of Tayde and Chole (2006) and Yavana Priya (2010) and contradictory to that of Jyothi (1998).

Process empowerment

Results showed that Cent percent of the farm women agreed that education is essential to women as much as men which indicates that they have realized the importance of education for their socio-economic development. Large majority of women agreed that Managerial skills of rural women will be enhanced through collective approach which shows that they have gained courage to manage people and skills related to it (Table 2).

Table 2 : Response analysis of statements of Empowering Farm Women beneficiaries of Karnataka community based tank management project.

(n-90)

SI. No.	Dimensions and Statements of Empowering	Level of Agreement /Disagreement among Respondents	
		Agree	Disagree
I.	**Process empowerment**		
1.	Do you feel that rural women have equal access to different sources of information as men and also socio-political participation is very crucial for their development	61 (67.78)	29 (32.22)
2.	Do you agree that education is as essential to women as men	90 (100)	0 (0.00)
3.	Women are not equal to men in performance of farm and entrepreneurial activities	21 (23.33)	69 (76.67)
4.	Managerial skills of rural women will be enhanced through collective approach	78 (86.67)	12 (13.33)
5	I confidently believe in my abilities for achieving success and can influence and convince others easily about my ideas	35 (38.89)	55 (61.11)

II.	Product empowerment		
1.	Do you feel that you have a good managerial competency?	60 (66.67)	30 (33.33)
2.	Do you feel that your family members help to reduce your drildgery in carrying out the farm activities?	75 (83.33)	15 (16.67)
3.	Do you feel that your urge for access to information and resources is good?	59 (65.55)	31 (51.67)
4.	Are you critically aware of development programes like MSY, PMRY?	29 (32.22)	61 (67.78)
5	Do you feel efficiency achievement will be better in group activity than in individual activity	73 (81.11)	17 (18.89)

Most of the women felt that rural women have equal access to different sources of information as men and also socio-political participation is very crucial for their development, whereas 61.11 per cent of women disagreed for the statement "I confidently believe in my abilities for achieving success and can influence and convince others easily about my ideas". However 76.67 per cent of women disagreed for the statement "Women are not equal to men in performance of farm and entrepreneurial activities" because of their confidence in performing any job with equal efficiency as that of men.

Product empowerment

About two-third of women agreed that their urge for access to information and resources is good Majority of the farm women lack access to information and resources which may be because of non-availabiliiy of information to them, lack of interest in accessing information and resources, lack of exposure and lack of social mobility. Farm women should be made to understand and recognize the concept that intellectual property (information) can defend her if challenged, and can be profitably incorporated for her benefit. Farm women should be an ethical consumer and should understand how free access to information, and free expression, contribute to a democratic society of empowered farm women. Farm women should be aware of the breadth and depth of information, synthesize and integrate information from a variety of sources, draw appropriate conclusions, and should be able to clearly communicate ideas to others. So, to have farm women the access for information and resources, developmental programmes and educational programmes should be conducted by the concerned departments to strengthen the women group by accepting the information and resources.

Majority of women felt that they have a good managerial competency. About 81.11 per cent of farm women agreed that efficiency will be achieved better in group activity than by individual activity and 83.33 per cent feel that their family members help to reduce their drudgery in carrying out the farm activities. Reduction in drudgery will facilitate to reduce fatigue, increase participation in productive work that ultimately leads to increase in social empowerment. Reduction in drudgery not only ensures equal participation of farm women at every level but also encourages them in taking up useful activities. Hence, a close interaction with farm women can help to identify suitable solutions for their problems like designing improved agricultural implements/techniques which can reduce the drudgery involved for farm women in farming. Only 32.22 per cent of farm women are critically aware of development programmes like Mahila Samakya Yojana (MSY) and Pradhana Mantri Rozghar Yojana (PMRY). It may be due to their low level of awareness and lack of general exposure.

Relationship between Independent variables and Empowerment level of the Farm Women Beneficiaries

The variables like education, social participation were found to have significant and positive relationship at one per cent level with the level of empowering and the variables like Management orientation, Extension contact, Innovative proneness, Mass media exposure, Extension participation were found to have significant and positive relationship at five per cent level with the empowering level and other variables were found non-significant (Table 3).

Table 3: Relationship between Independent Variables and Empowerment level of the Farm Women Beneficiaries

(n=90)

SI.No.	Independent Variables	Correlation coefficient (r)
1.	Age	0.174^{NS}
2.	Education	0.281**
3.	Family size	-0.021^{NS}
4.	Land holding	0.183^{NS}
5.	Family income	0.142^{NS}
6.	Cosmopoliteness	0.088^{NS}
7.	Social participation	0.268**
8.	Mass media exposure	0.225*
9.	Extension contact	0.231*
10.	Extension participation	0.221 *
11.	Innovative proneness	0.251 *

Contd. ...

12.	Management orientation	0.219*
13.	Risk orientation	0.064^{NS}
14.	Deferred gratification	0.072^{NS}

NS- Non significant, **- Significant at 1%, *- significant at 5%

Empowerment level and Education

Education was having highly significant association with process, product and overall empowerment. In most realistic, cases, education makes it possible to have an exhaustive understanding of an information domain. Empower means "putting power into". So the first step in empowering people is to refrain from doing anything that disempowers them or reduces their energy and enthusiasm for what they are doing. Thus improving critical consciousness, attitude formation, role perception, attitude towards collectivism, self perception and desire for control of farm women would empower them which can be brought through education and thus the above mentioned result.

Empowerment level and Social Participation

Empowerment is a multi-dimensional social process that helps people gain control over their own lives. It is a process that fosters power in people for use in their own lives, their communities and in their society, by acting on issues they define as important. Better social participation enhances an individual to deconstruct problems and find solutions. It helps the individual to deconstruct problems in agriculture, in society, and in communities and helps to focus on finding solutions to problems faced in agriculture and society. Also social participation helps to create a new era of critical awareness and understanding thus promoting constructive changes that lead to empowerment thereby explaining a significant association of social participation with process and product empowerment.

Empowerment level and Management orientation

Management orientation had positive and significant relationship with the extent of empowerment. A common explanation that would fit in for observed relationship was that management orientation is a basic character upon which other motives, drives and other attributes are built; it psychologically conditions an individual to orient herself to achieve higher income. Agriculture enterprise, being a highly remunerative enterprise, there are chances for excellent management orientation. Thus, the farm women would strive hard and seek to herself about different aspects of managing enterprise besides aiming at profit maximization thus empowering herself as a potential manager. Hence, it is quite natural to expect this type of relationship.

Empowerment level and Extension contact

Empowerment is a construct shared by many disciplines and arenas: community development, psychology, education, economics (agriculture) and organizations. It is a process that fosters power (that is, the capacity to implement) in people, for use in their own lives, their communities, and in their society, by acting on issues that they define as important. It goes without mentioning that farm women with higher level of contact with extension agency, are predisposed to acquire more information, skills and other factors, thus empowering themselves. The extension offices are staffed by one or more experts who provide useful, practical and research-based information to be used by rural farm women contributing to their empowerment and hence the result mirrors the significant association of extension contact with empowerment.

Empowerment level and Innovative proneness

On Farm Demonstration endeavors technology dissemination through indigenous technologies and innovative efforts to bring about sustainable development. It also emphasizes on the involvement and active participation of the target groups from its designing, planning to actual implementation stage and maintenance of activities on their own even after the project duration is over. A highly innovative individual is highly-motivated and is ready to embrace change and does the work necessary to create the wanted result. OFD empowers the groups and thereby the individual through skill development and capacity building which is well absorbed and put to use only if the farm women is innovative. When viewed from this angle the present finding of this study confirms the proposition that increased innovative proneness leads to increased empowerment.

Empowerment level and Mass media exposure

Mass media empowers farm women by keeping them "in the know," by informing on everything that promotes and affects their livelihood. Empowered ones' has a want and need to feel that they are aware of everything that is going on. Mass media participation was found to have a number of advantages. It evinces keen interest and makes individuals more receptive to ideas which interfere with survival or well-being. And in the borderless world or global village that we are currently living in, mass media has played an integral role in our daily lives and routine. Thus higher participation in mass media would facilitate the individual to develop habits of gathering more information, updating them leading to better empowerment.

Empowerment level and Extension participation

Farm women empowerment could be possible only by accomplishment of various activities accurately on farm and home which is possible only if necessary intelligence is obtained by the farm women. Hence, it is emphasized that farm women by participating in extension programmes can acquire knowledge and overcome barriers they face due to bias and discriminatory behavior, common in society. Hence, arrangements Should be made to encourage farm women to come out of house and to participate in different level of extension activities. Empowerment refers to increasing the political, social or economic strength of individuals and extension participation provides scope for it.

Conclusion

Women are very important segment in developing at local to global levels. Economic independence and education of women will go a long way in attaining self —reliance. Empowering women enhances their ability to influence the direction of social change, to create a better social and economic order. It gives autonomy and enables them to be strong enough to challenge and change their subordinate position in the society.

References

Jyothi, K.S., 1998, Employment Pattern and Empowerment of Rural Women-A Study in Kolar District. M.Sc.(Agri.) Thesis, (Unpublished), University of Agricultural Sciences, Bangalore.

Manga Sri, K., 1999, Empowerment of DWCRA Groups in Ranga Reddy District of Andhra Pradesh. Ph.D. Thesis, (Unpublished), ANGRAU, Hyderabad.

Sarada, 0., 2001, Empowerment of Rural Women in Self-Help Groups In Prakasham District of Andhra Pradesh — An Analysis. M.Sc. (Agri.) Thesis (Unpub.), Univ. Agri. Sci., Bangalore.

Selvaraj, A., 2007, Empowerment of Women, Kisan World, 34 (7):59-62.

Shrivastava, N.K., Bareth, L.S. and Sarkar, J.D., 1996, Impact of Farmers Training Centre on Rural Women. Maharastra J. Extn. Edu ., 15: 141-148.

Tayde, V.V. and Chole, R.R., 2006, A Scale to Measure Women Empowerment. Asian J. of Extn. Edu .,25:77-83.

Yavana Priya, D., 2009 Impact of Farmer Field School on Women Participants In Karnataka Community Based Tank Management Project. M.Sc.(Agri.) Thesis, (Unpub.), Univ. Agri. Sci., Bangalore.

●●●

Section V

Policy Issues

43

Credit Management Pattern of Farm Families

***K. Mahandra Kumar*[1] *and T. Rathakrishnan*[2]**

[1]*Assoc. Professor, Dept. of Agrl. Extension. Agrl. College and Research Institute, Madurai*

[2]*Director, Student Welfare, TNAU, Coimbatore*

Introduction

For uplifting the socio economic conditions of the poor mass, government enacted many programmes to make available of institutional credit flow to the farmers almost at their doorstep at a reasonable rate of interest. Hence, the mind set of borrowers has changed. credit is no more regarded as "hangman's rope" but are considered as an economic "ladder or elevator"

However, Credit can serve useful purpose only when it is used for productive purpose to generate additional income. Otherwise, its deviation for non productive purposes would affect the repaying capacity of borrowers and create over due and defaulters. The increasing rate of non repayment has been one of the major problems for all the financial institutions. Increasing default in the repayment of loan led to very serious implication, for instances, it discourages the financial institutions to refinancing to the defaulting members. Many eminent economists like Desphande and Amartya sen (2008), suggested that the present complex credit support system and situation has to be explored and research studies on institutional aspects and credit management pattern of farm families should be carried out. Hence, an attempt has been made in the study to explore the credit management pattern of farm families related to Turmeric cultivation.

Methodology

Turmeric crop was chosen because the turmeric growers are in need of credit support as it consumes more skilled labour for planting, irrigation, weeding, harvesting and processing compared to any other food crops cultivated.

For this study, Ex-post – facto research design was followed. The study was conducted at Bhavani block of Erode District. The district and block was selected purposively for its maximum area under turmeric crops cultivation. A sample size of 120 respondents consisting of farm families of 60 nationalised bank borrowers, and 60 cooperative bank borrowers had drawn from nine villages, were randomly selected for the study..

Findings and Discussion

Relationship between characteristics of farm families and their credit management pattern

Institutional credit management pattern reveals that the three important behavioural patterns of the borrowers namely borrowing, utilization and repayment. As there was no significant relationship between characteristic of farm families and borrowing behavioral pattern in the analysis, only utilization and repayment behavioural pattern are explained below.

Relationship between the characteristics of farm families with utilization behaviour

In order to find out the relationship between socio-economic, socio-personal characteristics of respondents and their utilization behaviour pattern with respect to institutional sources viz., Co-operatives and Nationalized banks were discussed here under. In order to find out the relationship, the simple correlation co-efficient was worked out and the results were presented in the following Table 1.

Out of eight variables analyzed five variables such as Caste (X3), Family annual income (X4), farm holding size (X5), Family norms(X7) and religious belief (X8), had significant and positive relationship and

Family educational status(X1) Family Occupational status(X2), had negative and significant relationship with the utilization of institutional loan. However, when the same set of variables analyzed independently with respective of the utilization behaviour of borrowers of nationalized and cooperatives, different sort of relationship pattern were observed in the behaviour pattern of cooperative borrowers. In the cooperative category, out of eight independent variables studied, only five variables exhibited significant and positive relationship and Occupational status had negative and significant relationship with utilization behaviour.

Table 1 : Correlation of characteristics of farm families with utilization behaviour

SI. No.	**Independent variables**	**Nationalized bank (n= 60)**	**Cooperatives (n=60)**	**Total (n=120)**
1	Family educational status	- 0.319*	- 0.224	- 0.260*
2	Family occupational status	-0.449**	-0.255*	-0.359**
3	Caste	0.322*	0.300*	0.304**
4	Family annual income	0.746**	0.763**	0.556**
5	Farm holding size	0.402**	0.703**	0.687**
6	Self reliance	0.136	0.147	0.127
7	Family norms	0.495**	0.739**	0.590**
8	Religious belief	0.583**	0.699**	0.625**

* - significant at 5 percent level of probability
** - significance at 1 percent level of probability
NS - Non- Significant

From the analysis, it could be inferred that Family with higher educational status divert the loan more, as they have more social commitments such as giving higher education to their children. Farmers who are earning only through farming utilize the loan properly, since there is no chance for diversification in other economic activities. The finding of negative and significant relationship between educational status as well as occupational status with utilization behaviour was supported by the study of Antonya (2004).

Those turmeric growers, who earned more income utilized the borrowed loan to intended purpose. As they already have sufficient money, the diversification was restricted. Farm holding size found to have positive and significant relationship with utilization behaviour. Such positive significant relation between farm size and utilization behaviour was also reported by Antony (2004), Anand Kumar Singh, *et al* .(2005), and Mishra (2006).

Adherence to the Family norms and Religious belief exert influence on moral behaviour of human being. Hence here too those turmeric growers who had adherence to family norms and religious belief utilized the loan for crop cultivation purpose.

Relationship between the characteristics of farm families with repayment behaviour

The relationship of profile of respondents and their repayment behaviour was studied and the results are presented below.

Table 2 : Correlation of characteristics of farm families with repayment behaviour

SI. No.	Independent variables	Nationalized bank (n= 60)	Co-operatives (n=60)	Total (N=120)
1	Family educational status	0.391**	0.387**	0.364**
2	Family occupational status	-0.421**	-0.280**	-0.324**
3	Caste	0.377**	0.467**	0.410**
4	Family annual income	0.692**	0.859**	0.551**
5	Farm holding size	0.326**	0.729*	0.693**
6	Self reliance	0.014^{NS}	0.085 ^{NS}	0.023 ^{NS}
7	Family norms	0.476**	0.817**	0.636**
8	Religious belief	0.535**	0.768**	0.275**
9	Utilization behaviour	0.857**	0.841**	0.790**

* - significant at 5 percent level of probability
** - significance at 1 percent level of probability
NS - Non- Significant

All the variables except self reliance had significant relationship with the repayment behaviour. Among them Family educational status (X2), caste (X4), Family annual income (X5), farm size (X6), adherence to family norms (X7) and religious belief (X8) had positive relationship. and occupational status (X2) had negative relationship.

Educational status would enhance the knowledge of latest improved practices and the advantages of adopting such practices. Educated loanees would have utilized the loan for agricultural purposes which in turn give more profit that leads to higher repayment. Moreover there was formal organized linkage between education and values, education and behaviour. Repayment of loans, promptly and regularly also a part of value system and moral behaviour. Further education has thus come to have some consequences for rational thinking, decision and action. In repayment of loan too, higher the educational status of a borrowers, greater the possibility of prompt repayment of loan. Hence educational status found to have positive and significant relationship. The findings of positive and significant relationship between educational status and repayment behaviour was supported by the study of Antony (2004), Sanjay Kumar (2005), Oni, Oladele and Oyewole (2005) and Dubey, M.K *et.al* (2006).

Farmers, who are earning their income only through farming, utilize the loan properly, since there is no chance for diversification of borrowed money hence we are findings occupational status exhibited negative and significant relationship. Such negative significant relation between occupational status and repayment behaviour was also reported by Antony (2004).

Those who have bigger farm could earned more farm income and hence they had ability to repay the loan. For the reason annual income as well as farm size had positive and significant relationship with repayment. Such positive significant relation between annual income as well as farm size with repayment behaviour was also reported by Sanjay Kumar (2005), Oni, Oladele and Oyewole (2005), Mishra (2005), Reddy,D.R, and Reddy, K.A (2006).

Adherence to the family norms and religious belief exert influence on moral behaviour of human being,. In turmeric cultivation too, those farm families who had strong adherence to the family norms and religious belief had prompt repayment behaviour, Similarly Prompt utilization leads to more profit that in turn influenced prompt repayment behaviour.

References

Anand kumar Singh, Anil kumar singh and Singh, V.K. 2005, Repayment performance of Borrowers with Respect to Agricultural Loan of Ranchi Kshetriya Gramin Bank: A Micro Analysis, Indian Journal of Agricultural Economics.

Antony.M.P. 2004, Institutional Finance of Agriculture in Kerala, Unpub, M.Sc., Thesis, Mahatma Gandhi University, Kottayam.

Desphande and Amritya Sen, Loan waiver scheme as Moral Hazards, Business line, March 2008.

Dubey, M. K., Abdul Rasheed Khan and Saxena.K.K. 2006. Factors Associated with utilization and Repayment behaviour of borrowers, Indian Research Journal of Extension Education, Vol .6 (3).

Mishra, R.K. and Dhal Samant, V, 2006, Utilization and efficiency of credit in Agriculture in Bank Block of Cuttack District, Indian Cooperative Review, January 2006, Vol.43 (3).

Mishra, R.K. and Pattanaik.S, 2005, Repayment Performance of borrowers with respect to Agricultural loans in Khurda block of khurda District, Orissa.

Oni, O.A, Oladele, O.I, and Oyewole, I.K.2005. Analysis of Factors Influencing loan default among poultry farmers in Ogun state Nigeria, Journal of Central European Agriculture, Vol.6 (4).

Reddy, D.R. and Reddy, K.A. 2006. Overdue and Recovery performance of Institutional credit, Indian Journal of Agricultural Research, Department of Agriculture, Andhra Pradesh.

Sanjay Kumar, 2005, Problems of Overdue in Tribal Area of Jharkand, Indian Journal of Agricultural Economics.

●●●

44

Strategies to increase Castor Production in India through Effective Resource-Use Management (RUM) Behaviour

R. Venkattakumar[1] *, M. Padmaiah*[2] *and C. Sarada*[3]
[1-3]*Directorate of Oilseeds Research, Hyderabad*
Andhra Pradesh

Introduction

Castor (*Ricinnus communis* L.) is an important non-edible oilseed crop and occupies an important place in the country's vegetable oil economy. The crop is mostly confined to Gujarat, Andhra Pradesh and Rajasthan. Although other states like Tamil Nadu, Karnataka, Orissa, Maharashtra, parts of Madhya Pradesh, Bihar and Chhattisgarh cultivate castor, their contribution to either area or production is limited. Presently, castor is grown over an area of 8.7 lakh hectares with a production of 11.7 lakh tonnes and productivity of 1352 kg/ha in the country (2008-09). Castor oil finds application in manufacture of a wide range of ever expanding industrial products such as nylon fibres, jet-engine lubricants, hydraulic fluids, soaps, varnishes, paints, medicines and a host of similar other products. Castor cake is used as manure and also as nematicide. Gujarat and Andhra Pradesh put together contribute to 68.5% of the total castor area and 80% of the total castor production in the country. The crop has been cultivated under resource-rich irrigated conditions in Gujarat with very high productivity (1963 kg/ha), while in Andhra Pradesh, the crop has been cultivated under rainfed and poor resource management conditions, with very low productivity (509kg/ha). There exists a wide gap between actual and potential yield levels of castor in both Gujarat and Andhra Pradesh, which is due to improper and ineffective RUM behaviour of the castor growers (Prasad 2002, Ramanjaneyulu and Padmaiah 2003, Raghavaiah *et al,* 2006 and Padmaiah 2007). Such gap resulted in near stagnation of castor production in the country over a past decade (Fig. 1). The reasons for yield gap in castor and castor production constraints can be overcome through efficient RUM behaviour (Ramanjaneyulu and Padmaiah, 2003). Thus, there is a need to improve the RUM behaviour of castor growers under both irrigated and rainfed situations in order to overcome production constraints and increase castor productivity, for which, "assessing the actual RUM behaviour of the castor growers", is obviously, a pre-requisite. Hence, a project was formulated to assess the resource-use management behaviour of castor growers in the country, so that pragmatic strategies can be suggested to improve the castor productivity.

Castor Area CON ha) Production (`000 t) and Productivity (kg/ha) in India

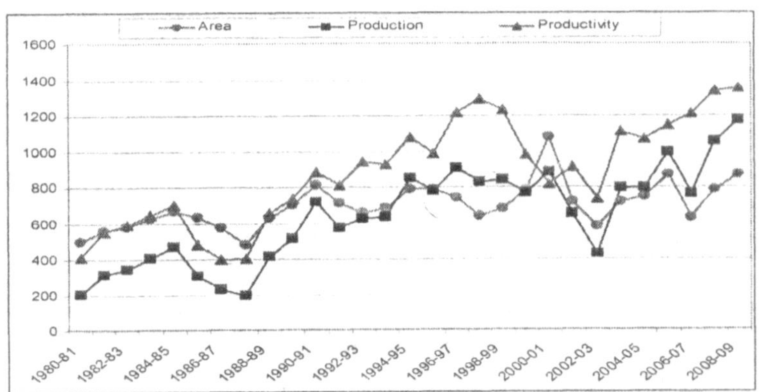

Methodology

In order to measure the RUM behaviour of castor growers in the country, it was decided to construct a summated-rating scale. An exhaustive list of RUM indicators was selected based on the review of literature, package of castor production technologies recommended and with thorough discussion with the researchers who have fairly sufficient research experience in castor. The list (18 indicators) was subjected to relevancy rating by castor researchers (41) working all over the country (in Directorate of Oilseeds Research (DOR), Hyderabad and in All India Coordinated Research Project [(AICRP) on Castor)] on a five-point continuum (Most relevant, relevant, some what relevant, irrelevant and most irrelevant). The rating was analyzed using mean and standard normal variant (Z) measures. All the indicators (9) with positive 'Z' values and with mean values higher than overall mean were selected for constructing the final scale. The final scale was constructed by writing items to the selected indicators based on the recommended castor production technologies. A total of 51 such items were constructed. The content validity of the scale after construction was assessed through the castor researchers who have fairly sufficient experience. The validity of the scale was assessed through split-half method (odd-even method) using Spearman-Brown prophecy formula reported by Singh, (2002) and the construct validity of the scale was assessed through correlation method (correlation between RUM behaviour of castor growers and their adoption behaviour towards recommended castor production technologies) in non-sampling area. Thus, the scale constructed was standardized to measure the RUM behaviour of castor growers all over the country. Surveys were conducted in Andhra Pradesh and Gujarat to assess the RUM behaviour of castor growers with the standardized scale following the methodology as furnished in Table 1. Post-survey stratification was done to categorize the respondents into small (farm size - less than or equal to 2 ha), medium (farm size - between 2 and 5 ha) and large (farm size - more than or equal to 5 ha) and analyze their distribution with respect to resource-use management towards castor cultivation.

Table 1 : Methodology Followed during Data Collection and Analysis

States*	**Andhra Pradesh (Rainfed)**	**Gujarat (Irrigated)**
Districts*	Mahabubngar (major castor growing district of Andhra Pradesh)	Banas Kantha, Sabar Kantha and Mehsena (Major castor growing districts of Gujarat)
Sampling procedure	Simple random sampling	Proportionate random sampling
Number of farmers	120	180
Data collection	January-March 2009	January 2010
Instrument	Standardized scale to measure RUM behaviour of castor growers	
Statistical tools used	Difference in RUM behaviour of farmers' categories	Kruskal —Wallis test, Chi-square test
	Level of RUM among farmers' categories	Mean, SD, percentage analysis
	Indicators of RUM contributing to castor productivity	Multiple regression
	Factor affecting RUM behaviour	Simple correlation

* Damodaram and Hegde, 2007

Findings and Discussion

Distribution of respondents according to their RUM behaviour

Majority of the castor growers had small to medium farm size. The difference between RUM of castor growers with small, medium and large farm size was highly significant at 1% level of probability both in Andhra Pradesh and Gujarat (Table 2). Most of the castor growers in Andhra Pradesh with small and medium farm size were having low to medium level, whereas most of the farmers with large farm size had medium to high level of RUM behaviour (Table 3). As far as Gujarat is concerned, most of the castor growers with small farm size were having low to medium level, whereas most of the farmers with medium and large farm size had medium to high level of RUM behaviour. The difference in distribution of castor growers with small, medium and large farm size with respect to their RUM behaviour was highly significant in both Andhra Pradesh and Gujarat. These results imply that intensive efforts are needed to improve the RUM behaviour of farmers with small and medium farm size in Andhra Pradesh and farmers with small farm size in Gujarat. Overall, the castor growers in Andhra Pradesh had low to medium level, whereas the castor growers in Gujarat had medium to high level of RUM behaviour. The indicator-wise distribution of respondents in Andhra Pradesh and Gujarat has been given in Table 4.

Table 2 : RUM Behaviour of Castor Growers

Farmers' categories	Andhra Pradesh (N=120)		Gujarat (N=180)	
	% of farmers	K-W value	% of farmers	K-W value
Small	35.0	14.80**	46.0	29.53**
Medium	40.0		37.0	
Large	25.0		17.0	
Total	100.0		100.0	

Table 3 : Distribution of Castor Growers according to RUM Behaviour

Level of RUM	Farmers' categories (%)							
	Andhra Pradesh				Gujarat			
	Small	Medium	Large	Mean	Small	Medium	Large	Mean
Low	42.9	35.4	6.7	30.8	22.9	6.0	6.7	13.9
Medium	50.0	52.1	60.0	53.3	71.1	67.2	83.3	71.7
High	7.1	14.5	33.3	15.9	6.0	26.8	10.0	14.4
Total	100.0	100.0	100.0	100.0	100.0	100.0	100.0	100.0
Mean	132				277			
SD	53				32			
X^2	**16.50****				**21.40****			

Table 4 : Indicator-wise Distribution of Castor Growers with respect to RUM

Indicators of RUM	Levels of RUM	
	Andhra Pradesh	**Gujarat**
Land management	Low-medium	Low-medium
Cropping systems management	Low-medium	Medium-high
Seeds management	Low-medium	Low-medium
Soil-moisture management	Low-medium	Medium-high
Nutrients management	Low-medium	Medium-high
Weed management	Medium-high	Low-medium
Insect pests management	Medium-high	Medium-high
Diseases management	Medium-high	Medium-high
Irrigation management	Medium-high	Medium-high
Overall	**Low-medium**	**Medium-high**

Indicators contributing to castor productivity and profitability

Management of cropping systems, seeds, soil-moisture, insect pests and diseases had highly significant contribution towards castor productivity in Andhra Pradesh, whereas nutrient management, disease management and irrigation management had highly significant contribution towards castor productivity in Gujarat (Table 5). Soil-moisture management had significant contribution towards castor productivity in Gujarat. The results implicate that the castor growers have to effectively manage the resources *viz,* cropping systems, seeds, soil-moisture, insect pests and diseases to increase castor productivity in Andhra Pradesh, whereas the resources *viz,* nutrient, disease, soil-moisture and irrigation in Gujarat. Intensive transfer of technology efforts like frontline demonstrations (FLDs), capacity building programmes and campaigns to popularize this information are to be implemented on priority basis.

Table 5 : RUM Indicators Contributing to Castor Productivity

Indicators of RUM	Unstandardised coefficients			
	Castor productivity-AP		Castor productivity-Gujarat	
	Beta value	SE	Beta value	SE
Constant	-77.80	53.9	-0.55	1081.65
Land management	0.17 NS	0.38	15.59 NS	40.92
Cropping systems management	5.69**	2.05	19.32 NS	35.91
Seeds management	3.72**	0.85	3.74 NS	5.63
Soil-moisture management	17.50**	3.20	53.40*	32.35
Nutrients management	5.87**	2.75	10.44**	3.43
Weeds management	1.14 NS	7.32	8.81 NS	54.45
Insect pests management	8.72**	3.07	12.09 NS	15.97
Diseases management	3.50**	1.45	135.23**	23.45
Irrigation management	2.76 NS	7.14	117.38**	32.33

SE=Standard error; AP=Andhra Pradesh

Relationship between independent variables and RUM behaviour of the respondents

The socio-economic variables of castor growers that influence the RUM of castor growers in Andhra Pradesh and Gujarat are given in Table 6. These results imply that young farmers with higher income and education in Andhra Pradesh, whereas experienced farmers with large farm size, higher educational status, annual income and those who directly participate in agriculture in Gujarat were effective in managing the resources with respect to castor cultivation. Hence, such farmers can be selected for demonstration of significantly contributing RUM indicators and improved production technologies. They may also be utilized as resource persons in capacity building programmes of development departments and research institutes to convince other castor growers towards adoption of effective RUM in castor cultivation.

Table 6 : Factors Affecting RUM behaviour of Castor Growers in India

Factors	RUM of Andhra Pradesh farmers	RUM of Gujarat farmers
Age	-	0.296**
Education	0.33**	-
Farming experience	-0.20*	-
Farm size	-	0.170*
Annual income	0.23*	0.269**
Nature of agricultural participation	-	0.191*

Conclusion

The study conducted to assess the RUM behaviour of castor growers in India concluded that the castor growers in Andhra Pradesh had low to medium level, whereas the castor growers in Gujarat had medium to high level of RUM behaviour. Castor growers ought to effectively manage the resources *viz,* cropping systems, seeds, soil-moisture, insect pests and diseases to increase castor productivity in Andhra Pradesh, whereas the resources viz, nutrient, disease, soil moisture and irrigation in Gujarat. Intensive extension efforts are needed to improve the RUM behaviour of farmers with small and medium farm size in Andhra Pradesh and farmers with small farm size in Gujarat. Young farmers with higher income and education in Andhra Pradesh, whereas experienced farmers with large farm size, higher educational status, annual income and those who directly participate in agriculture in Gujarat were effective in management of resources with respect to castor cultivation and hence, such factors may be selected for demonstration of significantly contributing RUM indicators and improved production technologies and as resource persons in capacity building programmes of development departments and research institutes to convince the castor growers towards adoption of effective RUM in castor cultivation.

References

Darnodaram T and Hegde D.M. 2007. Oilseeds situation: A statistical compendium. DOR. Hyderabad.

Padmaiah, M, 2007. Performance of castor hybrids in Andhra Pradesh. In: ISOR. 2007. Extended Summaries: National Seminar on "Changing Global Vegetable Oil Scenario: Issues and Challenges Before India". January 29-31, 2007. Indian Society of Oilseeds Research, Hyderabad. P (401).

Prasad, M.S. 2002. Adoption of recommended package of practices of dryland crops in relation to socio-economic and personal characteristics of farmers. Indian Journal of Dryland Agricultural Research and Development. 17 (1): 39-43.

Raghavaiah, C. V., Suresh, G and Hegde, DM. 2006. Strategies for enhancing castor production in India. Directorate of Oilseeds Research. Hyderabad. P (2-3).

Ramanjaneyulu, C.V and Padmaiah, M. 2003. Technology adoption and its impact. In: Hegde, D.M, Sujatha, M and Singh, N.B. (Eds.). Castor in India. Directorate of Oilseeds Research. Hyderabad.

Ray G.L. and Sagar Mondal. 2006. Research methods in social sciences and extension education. Kalyani Publishers. Ludhiana. P (321).

Singh, A.K. 2002. Tests, Measurements and Research Methods in Behavioural Sciences. Bharati Bhawan Publishers and Distributors. Patna. P (510).

●●●

Zeitfracht Medien GmbH
Ferdinand-Jühlke-Straße 7
99095 Erfurt, Deutschland
produktsicherheit@kolibri360.de